MONOGRAPHS ON STATISTICS AND APPLIED PROBABILITY

General Editors

1 Stochastic Population Models in Ecology and Epidemiology *M.S. Bartlett* (1960)

2 Queues *D.R. Cox and W.L. Smith* (1961)

3 Monte Carlo Methods *J.M. Hammersley and D.C. Handscomb* (1964)

4 The Statistical Analysis of Series of Events *D.R. Cox and P.A.W. Lewis* (1966)

5 Population Genetics *W.J. Ewens* (1969)

6 Probability, Statistics and Time *M.S. Bartlett* (1975)

7 Statistical Inference *S.D. Silvey* (1975)

8 The Analysis of Contingency Tables *B.S Everitt* (1977)

9 Multivariate Analysis in Behavioural Research *A.E. Maxwell* (1977)

10 Stochastic Abundance Models *S. Engen* (1978)

11 Some Basic Theory for Statistical Inference *E.J.G Pitman* (1979)

12 Point Processes *D.R. Cox and V. Isham* (1980)

13 Identification of Outliers *D.M. Hawkins* (1980)

14 Optimal Design *S.D. Silvey* (1980)

15 Finite Mixture Distributions *B.S. Everitt and D.J. Hand* (1981)

16 Classification *A.D. Gordon* (1981)

17 Distribution-free Statistical Methods, 2nd edition *J.S. Maritz* (1995)

18 Residuals and Influence in Regression *R.D. Cook and S. Weisberg* (1982)

19 Applications of Queueing Theory, 2nd edition *G.F. Newell* (1982)

20 Risk Theory, 3rd edition *R.E. Beard, T. Pentikainen and E. Pesonen* (1984)

21 Analysis of Survival Data *D.R. Cox and D. Oakes* (1984)

22 An Introduction to Latent Variable Models *B.S. Everitt* (1984)

23 Bandit Problems *D.A. Berry and B. Fristedt* (1985)

24 Stochastic Modelling and Control *M.H.A. Davis and R. Vinter* (1985)

25 The Statistical Analysis of Compositional Data *J. Aitchison* (1986)

26 Density Estimation for Statistics and Data Analysis *B.W. Silverman*

27 Regression Analysis with Applications *G.B. Wetherill* (1986)

28 Sequential Methods in Statistics, 3rd edition *G.B. Wetherill and K.D. Glazebrook* (1986)

29 Tensor Methods in Statistics *P. McCullagh* (1987)

30 Transformation and Weighting in Regression *R.J. Carroll and D. Ruppert* (1988)

31 Asymptotic Techniques for Use in Statistics *O.E. Barndorff-Nielsen and D.R. Cox* (1989)

32 Analysis of Binary Data, 2nd edition *D.R. Cox and E.J. Snell* (1989)

33 Analysis of Infectious Disease Data *N.G. Becker* (1989)

34 Design and Analysis of Cross-Over Trials *B. Jones and M.G. Kenward* (1989)

35 Empirical Bayes Methods, 2nd edition *J.S. Maritz and T. Lwin* (1989)

36 Symmetric Multivariate and Related Distributions *K.-T. Fang, S. Kotz and K.W. Ng* (1990)

37 Generalized Linear Models, 2nd edition *P. McCullagh and J.A. Nelder* (1989)

38 Cyclic and Computer Generated Designs, 2nd edition *J.A. John and E.R. Williams* (1995)

39 Analog Estimation Methods in Econometrics *C.F. Manski* (1988)

40 Subset Selection in Regression *A.J. Miller* (1990)

41 Analysis of Repeated Measures *M.J. Crowder and D.J. Hand* (1990)

42 Statistical Reasoning with Imprecise Probabilities *P. Walley* (1991)

43 Generalized Additive Models *T.J Hastie and R.J. Tibshirani* (1990)

44 Inspection Errors for Attributes in Quality Control *N.L. Johnson, S. Kotz and X. Wu* (1991)

45 The Analysis of Contingency Tables, 2nd edition *B.S. Everitt* (1992)

46 The Analysis of Quantal Response Data *B.J.T. Morgan* (1993)

47 Longitudinal Data with Serial Correlation: A State-space Approach *R.H. Jones* (1993)

48 Differential Geometry and Statistics *M.K. Murray and J.W. Rice* (1993)

49 Markov Models and Optimization *M.H.A. Davis* (1993)

50 Networks and Chaos – Statistical and Probabilistic Aspects *O.E. Barndorff-Nielsen, J.L. Jensen and W.S. Kendall* (1993)

51 Number-theoretic Methods in Statistics *K.-T. Fang and Y. Wang* (1994)

52 Inference and Asymptotics *O.E. Barndorff-Nielsen and D.R. Cox* (1994)

53 Practical Risk Theory for Actuaries *C.D. Daykin, T. Pentikäinen and M. Pesonen* (1994)

54 Biplots *J.C. Gower and D.J. Hand* (1996)

55 Predictive Inference: An Introduction *S. Geisser* (1993)

56 Model-Free Curve Estimation *M.E. Tarter and M.D. Lock* (1993)

57 An Introduction to the Bootstrap *B. Efron and R.J. Tibshirani* (1993)

58 Nonparametric Regression and Generalized Linear Models *P.J. Green and B.W. Silverman* (1994)

59 Multidimensional Scaling *T.F. Cox and M.A.A. Cox* (1994)

60 Kernel Smoothing *M.P. Wand and M.C. Jones* (1995)

61 Statistics for Long Memory Processes *J. Beran* (1995)

62 Nonlinear Models for Repeated Measurement Data *M. Davidian and D.M. Giltinan* (1995)

63 Measurement Error in Nonlinear Models *R.J. Carroll, D. Ruppert and L.A. Stefanski* (1995)

(Full details concerning this series are available from the publishers)

Biplots

J.C. GOWER

Former Head of Statistics Department and Biomathematics Division, Rothamsted Experimental Station, Harpenden, UK

and

D.J. HAND

Professor of Statistics, Open University, Milton Keynes, UK

CHAPMAN & HALL

London · Glasgow · Weinheim · New York · Tokyo · Melbourne · Madras

Published by Chapman & Hall, 2–6 Boundary Row, London SE1 8HN, UK

Chapman & Hall, 2–6 Boundary Row, London SE1 8HN, UK

Blackie Academic & Professional, Wester Cleddens Road, Bishopbriggs, Glasgow G64 2NZ, UK

Chapman & Hall GmbH, Pappelallee 3, 69469 Weinheim, Germany

Chapman & Hall USA, 115 Fifth Avenue, New York, NY 10003, USA

Chapman & Hall Japan, ITP-Japan, Kyowa Building, 3F, 2-2-1 Hirakawacho, Chiyoda-ku, Tokyo 102, Japan

Chapman & Hall Australia, 102 Dodds Street, South Melbourne, Victoria 3205, Australia

Chapman & Hall India, R. Seshadri, 32 Second Main Road, CIT East, Madras 600 035, India

First edition 1996

Typeset in 10/12pt Times by Thomson Press (India) Ltd, Madras
Printed in Great Britain by Hartnolls Ltd, Bodmin, Cornwall

ISBN 0 412 71630 5

A catalogue record for this book is available from the British Library

∞ Printed on permanent acid-free text paper, manufactured in accordance with ANSI/NISO Z39.48-1992 and ANSI/NISO Z39. 48-1984 (Permanence of Paper).

To Janet and Shelley

Contents

Preface

Biplots are the multivariate analogue of scatter plots. They approximate the multivariate distribution of a sample in a few dimensions, typically two, and they superimpose on this display representations of the variables on which the samples are measured. In this way, the relationships between the individual samples points can be easily seen and, as we shall see, they can also be related to values of the measurements. Thus, like scatter plots, biplots are useful for giving a graphical description of the data, for detecting patterns, and for displaying results found by more formal methods of analysis. Many uses fall under the headings of descriptive statistics and initial data analysis but it has to be recognized that often 'initial' is 'final'; just to be able to see the relationships between multivariate samples is a major step forward. The 'bi' in biplots arises from the fact that both samples and measured variables are represented, not that biplots are necessarily two-dimensional, though they usually are.

In recent years, the theory of biplots has been considerably extended, and we feel that the time is now right for a book such as this. We believe that the area is best looked at from what might be called a 'non-traditional' perspective of its own which allows well-known methods, such as principal components analysis, correspondence analysis and canonical variate analysis, as well as some newer and less well-known methods to be integrated into a unified presentation. Although these classical methods all have chapters of their own, other parts of the book are devoted to recently developed methods that have previously been described only in the journal literature and there is much that has not previously been published anywhere. Moreover, there is plenty of room for further development. The reader will find methods whose mathematical development still awaits computer implementation and practical experience, and will perceive many unsolved problems, not all of which are mentioned explicitly. We hope that others will be stimulated by this book to make their own contributions to the understanding and methodology of biplots.

Although the theory does not use any advanced mathematics, a good grasp of linear algebra and its geometrical underpinning is essential. Insofar as it was possible, without disrupting the flow of the ideas, we have abstracted the detailed algebra and the algebraic basis and put it into the appendix.

We thank Gillian Arnold and Simon Harding for programming many of the examples in Genstat 5.

J.C. Gower and D.J. Hand
The Open University, 1995

CHAPTER 1

Introduction

1.1 Overview

This book is concerned with representing information on samples and variables in a single diagram, and with how to interpret such diagrams.

The most elementary and well-known such display is the ordinary scatterplot of two variables. These are not only easy to produce, but they also have the merit of being straightforward to interpret, requiring very little, if any, formal training. Points in such scatterplots represent the samples, and the two axes represent the two variables concerned. When more than two variables are involved, however, things become more difficult: in general, multivariate data distributions are difficult to visualize. This book describes a strategy for ameliorating that difficulty: the distribution of multivariate samples is approximated by a distribution in a (usually) two-dimensional space, and the variables defining the multivariate space are approximated by graduated curves in the two-dimensional space. The advantage of such a representation is that two-dimensional representations can, of course, be easily viewed.

Approximations to the relationships between n samples can be achieved by the various methods of multidimensional scaling (MDS) but, except in the special cases of principal components analysis (PCA), multiple correspondence analysis (MCA), and canonical variate analysis (CVA), until recently methods for including information on the variables have been little developed. This book describes methodology to fill that gap.

The concept of inter-sample distance is central to all methods of multidimensional scaling, and this is the unifying concept that underpins all that follows. We use Pythagorean distance for PCA, χ^2 distance for MCA, Mahalanobis distance for CVA, any Euclidean embeddable distance for principal coordinates analysis (PCO), and many others. What we term Pythagorean distance is calculated by using Pythagoras' theorem on a pair of rows of **X** and is often termed Euclidean distance but, to avoid confusion, we reserve the latter term for all Euclidean embeddable distances which includes Pythagorean distance as a special case. We have confined our attention mostly to Euclidean distances and totally to Euclidean displays though, no doubt, this constraint could be relaxed as it is occasionally in MDS. Nevertheless, it is Euclidean displays with which most research workers are familiar and which form the overwhelming majority of published material, so we have few reservations about restricting our discussion in this way.

Mention of PCA and MCA indicates that we are concerned with both continuous and categorical variables. While a continuous variable will be represented in our plots by a biplot axis which is a continuous curve (not necessarily linear), a categorical variable will be represented by a biplot 'axis' which is a simplex of points labelled by category names. Both types of representation may occur in the same plot, so we refer to the set of such generalized axes as a reference system. None of our diagrams contain the usual rectangular coordinate axes (indeed, there are more 'axes' than displayed dimensions) but the reference system of biplot axes and category level points (CLPs) serves a similar purpose.

Much multivariate analysis is concerned with canonical axes of one kind or another. It could be said that the main thrust of this book is to return to coordinate axes representing the original variables, at the same time extending the notion to include representations of categorical variables.

We begin, in Chapter 2, with the best known, and oldest, method of multidimensional scaling: PCA. Classical PCA, of course, permits a staightforward representation of the variables by simple orthogonal projection. And, in fact, this very simplicity is misleading; it tends to conceal certain subtleties about the representation which do not become apparent until we generalize in various ways. In particular, it obscures the difference between interpolative and predictive biplot axes.

Given the values of the variables for a sample, **interpolation** describes the process of finding the position of the sample in the display. Conversely, given the position of a sample in the display, **prediction** is the process of inferring the values of its variables. In both cases, we do this by relating the given values to a set of axes. In exact representations, and with orthogonal axes, these two operations are consistent when using the same sets of axes. As we shall see, with PCA approximations, the two sets of axes remain linear and coincide but, for more general types of MDS, they are neither linear nor coincident. Moreover, even in the case of PCA, the axes are not identical; the graduations which are appropriate for interpolation are inversely related to those for prediction.

PCA is based on finding the two-dimensional subspace of the original measurement space which most accurately approximates the distribution of samples. 'Most accurately', in PCA, means that the sum of the differences between the inter-sample squared distances and the squared distances in the subspace is minimized. Other methods of MDS relax the constraint that the two-dimensional space should be a subspace of the original measurement space and use different measures of goodness of fit. In Chapter 3, we examine the effect that such generalizations have on the biplots. And this is where the obscuring simplicity of PCA begins to break down. We shall see that, while predictive biplot axes are linear projections of the original axes, optimal interpolative axes do not exist (though approximate linear ones can be found).

If PCA is the most popular biplot representation for continuous variables, then MCA is the most popular representation for categorical variables. MCA is normally presented as a generalization of correspondence analysis (CA, Chapter 9) in which the latter is applied to the coded indicator matrix of the categories, as explained in Chapter 4. To us, however, it seems more natural to present MCA as a form of PCA applied to categorical variables. The categorical nature of the variables in MCA means that the linear biplot axes of PCA are replaced, in MCA, by simplices of points, one simplex for each variable, one point for each category level (called, appropriately enough, **category level points**). We shall see that such simplices have many of the properties of ordinary coordinate axes, and so form a natural generalization to categorical variables. With MCA, the distinction between interpolation and prediction begins to become particularly dramatic. Samples interpolate to points in the display space, but whole regions of the display space predict particular category levels. Hence the display space is partitioned into convex **prediction regions**.

The distance used in PCA is Pythagorean distance and that used in MCA is χ^2 distance. In Chapter 5, we explore the use of biplots in canonical variate analysis (CVA), where Mahalanobis distance is used. This is fundamentally different from the earlier two distances: it is not additive, in the sense that the variables do not contribute independently to the overall measure of distance. What is especially interesting is that, in this situation, although both interpolative and predictive biplot axes exist and both are linear, their directions differ. The use of the different definition of distance has shown another property of biplots which was concealed in PCA.

In Chapter 6, we move on to more general distance measures for use with continuous variables. In particular, we show that if they are (a) Euclidean embeddable, (b) satisfy the additivity property defined above, and (c) the method of MDS used is principal coordinates analysis (PCO), then both interpolative and predictive biplot axes exist – but they are now curvilinear rather than linear. The ideas remain useful if we relax assumptions (a), (b), and (c), so providing a fairly general methodology for producing biplots with any form of metric scaling.

Chapter 6 provides a generalization of most of the methods described in previous chapters, but MCA and the categorical variables of Chapter 4 are an exception. Chapter 7 extends the theory of Chapter 6 to include such variables and general reference systems.

There is confusion between biadditive biplots (Chapter 8) and PCA, and there has been a tendency to regard the former as a variant of the latter. We have already remarked that there are links between MCA and CA and, to a great extent, these encourage a similar confusion. One source of the confusion is that all of these methods are based on the singular value decomposition of a matrix. Another source of confusion is that, for PCA and MCA, the matrix is a multivariate data-matrix (subjects × variables) whereas CA and biadditive models refer to a two-way table. Both data-matrices and two-way tables may be treated as matrices, but the exchangeability of the rows and columns of a table does not normally extend to exchangeability between sample units and variables in a multivariate sample. Thus, although there is an algebraic duality between rows and columns, which applies in both cases, the interpretations differ. We therefore treat biplots for two-way tables as distinct from multivariate biplots and discuss them towards the end of the book, in Chapter 8 (Quantitative variables) and 9

(Categorical variables). Biplots for two-way tables are more overtly model-based than are multivariate biplots; as a consequence, one of their more intriguing uses is for diagnostics, suggesting what additive and/or multiplicative terms to include in a model.

There is often an ambivalence as to whether one is approximating a data-matrix **X** or its inner product **X'X**. For example, this occurs in PCA when one might wish to approximate a covariance or correlation matrix rather than the data-matrix itself. The equivalent problem with categorical data is exacerbated because an entire two-way contingency table corresponds to a correlation coefficient. This leads to extra confusion between CA and MCA, the disentangling of which is the theme of Chapter 10.

Finally, in Chapter 11, we discuss some more generalizations and extensions which do not merit dedicated chapters in a book of this length. In particular, we describe biplots in non-metric scaling, biplots in generalized linear models, biplots in three-way analyses, biplots for special classes of matrices, and what can be done when there are missing values.

Gabriel (1971) coined the term 'biplot' and is the originator of PCA biplots (Chapter 2), canonical biplots (Chapter 5) and (with Bradu) biadditive model diagnostic biplots (Chapter 8); his considerable contribution to the subject is clear from his publications listed in the bibliography. For categorical variables, the key reference is to Benzécri *et al.* (1973) whose two-volume book on correspondence analysis (CA) has been very influential, especially in Francophone countries. Multidimensional scaling (MDS) in its metric and non-metric forms has a long history, beginning with Karl Pearson (1901) on PCA, extending to classical scaling of Torgerson (1958) and non-metric scaling of Shepard (1962a;b). Much of this work was developed in the field of psychometrics and only gradually became known to other statisticians. The graphical display of samples was central to this work, but lacked a display of variables, though from time to time the regression method (Chapter 3) seems to have been used. The regression method provides a good example of statistical folklore and a knowledge of it seems to have passed by word of mouth; we have no good reference for it and suspect that it was rediscovered independently on several occasions. Earlier work of Hirschfeld (1935) on CA and Fisher and Mackenzie (1923) on biadditive models gives some of the basic algebra but lacks the crucial graphical interpretations; in some respects, both of these early papers were before their time and were not taken up, presumably because ready computation had to await the arrival of the electronic computer.

1.2 Computation

Many of the methods described in this book are straightforward to compute, needing only eigenvalue/singular value decomposition procedures. Special stand-alone packages are available for some of the methods, especially correspondence analysis, and most packages have special facilities for the better known methods: principal components, correspondence and multiple correspondence analysis, and canonical variate analysis. Multidimensional scaling, especially in its non-metric forms, tends to come in stand-alone programs (e.g. KYST, ALSCAL, SMACOFF) but some methods are available in some packages (e.g. Genstat 5, SPSS, SAS). We do not give a comprehensive list here, but refer the reader to Cox and Cox (1994). The main computational problems are in integrating different bits of available software and in finding good portable graphic facilities. Some of the diagrams given in the following have been produced, at least initially, by Genstat 5; some have been drawn by Macdraw and others have been enhanced by Deltagraph 2. We regret that statistical methods used should depend on software availability. The conceptual unification that we have attempted has yet to be translated into software unification. Most of the examples have been computed in Genstat 5.

1.3 Notation

All the technical term that we need are defined in the text. However, we feel it necessary to mention here that the term **squared distance** occurs so often that we have abbreviated it to **ddistance**; this is not a misprint! Similarly for sstress.

Matrices are used extensively throughout this book. We use bold upper case letters for matrices and bold lower case letters for vectors. Thus, $\mathbf{A}$ is a matrix and $\mathbf{a}$ is a vector. Dimensions are defined at the outset but thereafter by implication. Vectors pertaining to samples are usually row vectors and those pertaining to variables are usually column vectors. Suffices and indices are used normally, except for (i) the use of ρ as a suffix or index (see Table 1.1.) and (ii) the special suffix notation ${}_{\mathrm{p}}\mathbf{A}_{\mathrm{q}}$, which is convenient for expressing conformity in matrix products such as ${}_{\mathrm{p}}\mathbf{A}_{\mathrm{q}}\mathbf{B}_{\mathrm{r}}$. Vectors and matrices may be defined in terms of their elements: $\mathbf{a} = \{a_i\}$ and $\mathbf{A} = \{a_{ij}\}$. The several vector spaces that are referred to in the text are denoted by curly upper case letters.

Table 1.1. Symbols, matrix dimensions and abbreviations

Scalars	
n	Number of sample-units
p	Number of variables (some, or all, can be categorical)
q	Number of dimensions for the exact representation of all the inter-sample distances
i, j	Indices denoting samples
h, k	Indices denoting variables
L_k	Number of values (levels, categories) of the kth categorical variable
L	$L = L_1 + \cdots L_p$
x_{ik}	Value of kth variable on ith sample-unit
d_{ij}	Distance between ith and jth samples
d_{ij}^2	$d_{ij}^2 = \sum_{k=1}^{p} d(x_{ik}, x_{jk})$, an important special class of distances
δ_{ij}	Distance between ith and jth samples in the ordination approximation
μ	A typical marker on an axis; often associated with a pseudosample
ρ	Number of dimensions in an MDS giving approximations to all d_{ij}. When ρ occurs in a suffix position of a matrix, it denotes the first ρ columns of a matrix and when as an index (superfix) it denotes the first ρ rows of the inverse of the matrix.
Vectors (dimensions)	
$\mathbf{e}_k(p \times 1)$	$\mathbf{e}_k = (0, 0, \ldots, 0, 1, 0, \ldots, 0)'$ zero except for kth element
$\mathbf{1}$	$\mathbf{1} = (1, \ldots, 1)'$, dimensions by implication but often $(n \times 1)$
$\mathbf{d}$	$\mathbf{d} = \{d_i^2\} = (\mathrm{diag}\,\mathbf{B})\mathbf{1}$, the ddistances from the centroid (origin)
$\mathbf{d}_{n+1}$	$\mathbf{d}_{n+1} = \{d_{n+1,i}^2\}$
Matrices (dimensions)	
$\mathbf{N}(n \times n)$	$= \mathbf{1}\mathbf{1}'/n$, the centring matrix
$\mathbf{X}(n \times p)$	$= \{x_{ij}\}$, the data-matrix
$\mathbf{G}(n \times L)$	$= \{\mathbf{G}_1, \ldots, \mathbf{G}_p\}$, the indicator matrix (categorical variables only)
$\mathbf{G}_k(n \times L_k)$	The indicator matrix for the kth categorical variable
$\mathbf{I}$	A unit matrix, usually $(n \times n)$
$\mathbf{J}_\rho(p \times \rho)$	The first ρ columns of a p-dimensional unit matrix
$\mathbf{L}(L \times L)$	$= \mathrm{diag}\,(\mathbf{G}'\mathbf{G})$, gives frequencies of category-levels
$\mathbf{L}_k(L_k \times L_k)$	$= \mathrm{diag}\,(\mathbf{G}_k'\,\mathbf{G}_k)$, frequencies of levels of kth categorical variable
$\mathbf{D}(n \times n)$	$\{-\frac{1}{2}\mathrm{d}_{ij}^2\}$, the distance matrix
$\mathbf{D}_k(n \times n)$	$\{-\frac{1}{2}\mathrm{d}\,(x_{ik}, x_{jk})\}$, the distance matrix for the kth variable

$\mathbf{B}(n \times n)$	$= (\mathbf{I} - \mathbf{N})\mathbf{D}(\mathbf{I} - \mathbf{N})$
$\mathbf{B}_k(n \times n)$	$= (\mathbf{I} - \mathbf{N})\mathbf{D}_k(\mathbf{I} - \mathbf{N})$
$\mathbf{B}_k^*(n \times L_k)$	The distinct columns of $\mathbf{B}_k$ (categorical variables only)
$\mathbf{\Delta}(n \times n)$	$\{-\frac{1}{2}\delta_{ij}^2\}$, the fitted distance matrix
$\mathbf{Y}(n \times q)$	The coordinates in $\mathscr{R}$ that generate the ddistances d_{ij}^2; if principal coordinates then: $\mathbf{B} = \mathbf{YY}', \mathbf{Y}'\mathbf{Y} = \Lambda$ (eigenvalues)
$\mathbf{Z}(n \times \rho)$	The multidimensional coordinates in $\mathscr{L}$ that generate $\mathbf{\Delta}$; for PCA, $\mathbf{Z} = \mathbf{Y}_\rho$
$\mathbf{Z}_k(n \times \rho)$	$= \mathbf{B}_k\mathbf{Y}\mathbf{\Lambda}^{-1}$, gives basic coordinates on the kth trajectory
$\mathbf{Z}_k^*(L_k \times \rho)$	$= \mathbf{B}_k^*\mathbf{Y}\mathbf{\Lambda}^{-1}$, gives coordinates of category-levels for the kth categorical variable
$\mathbf{Z}^*(L \times \rho)$	$= \begin{pmatrix} \mathbf{Z}_1^* \\ \vdots \\ \mathbf{Z}_p^* \end{pmatrix}$, gives coordinates of category-levels
$\mathbf{\Pi}(n \times n(p+1))$	$= \{\mathbf{B}, \mathbf{B}_1, \ldots, \mathbf{B}_p\}$, matrices derived for simultaneous ordination
$\mathbf{\Pi}^*(n \times (n+L))$	$= \{\mathbf{B}^*, \mathbf{B}_1^*, \ldots, \mathbf{B}_p^*\}$, as for $\mathbf{\Pi}$ but without redundancies
$\mathbf{J}(L \times L)$	$= \text{diag}\,(\mathbf{1}_{L_1}\mathbf{1}'_{L_1}, \ldots, \mathbf{1}_{L_p}\mathbf{1}'_{L_p})$

Spaces

$\mathscr{B}$	The ρ-dimensional approximation of χ in $\mathscr{L}$. Two forms, $\mathscr{B}_\text{I}$ (for interpolation) and $\mathscr{B}_\text{p}$ (for prediction), are required
$\mathscr{F}_k$	The L_k-dimensional prediction-region in $\mathscr{R}$
$\mathscr{L}$	The ρ-dimensional space containing the MDS approxi mation
$\mathscr{N}_\mu$	The normal space to ξ_k at a marker μ
$\mathscr{R}$	The q-dimensional space containing $\mathbf{Y}$ that generates $\mathbf{D}$ exactly
$\mathscr{R}^+$	The m-dimensional space containing $\mathbf{Y}$ and the reference system χ
χ	The reference system in $\mathscr{R}^+$
ξ_k	The kth coordinate axis in $\mathscr{R}$ (not necessarily linear and including the kth simplex of CLPs)
β_k	The kth biplot axis in $\mathscr{L}$ (including CLPs) – approximates ξ_k

Abbreviations

CA	Correspondence analysis
CLP	Category-level-point
CVA	Canonical variate analysis
JCA	Joint correspondence analysis
MCA	Multiple correspondence analysis
MDS	Multidimensional scaling
PCA	Principal components analysis
PCO	Principal coordinates analysis
p.d.	positive definite
p.s.d.	positive semi-definite
SVD	Singular value decomposition

CHAPTER 2

Principal components analysis (PCA)

2.1 Introduction

Given the p-variate measurement vectors describing n samples, how might we best display n points representing those samples? We want the display to portray the distances between the n samples accurately, in some sense, so that similar samples are represented by nearby points. Multidimensional scaling (MDS) methods are statistical methods which do precisely this: a measure of discrepancy between the observed and fitted inter-sample distances is minimized.

The most common approach is the principal components method; Chapter 3 discusses others. Principal components analysis (PCA) is a dimension reduction technique originally described by Pearson (1901) and Hotelling (1933). Pearson's geometric approach is more in the spirit of the following.

We are seeking to display the points, so we will normally be concerned with a two-dimensional representation, but higher dimensional representations are possible. For generality, we shall continue the discussion assuming that we want to produce a 'display' in ρ-dimensions. We begin with the standard geometric representation of the n points in the p-dimensional space $\mathscr{R}_p$ defined by Cartesian axes representing the measurements. Thus, the rows of the $n \times p$ data matrix $\mathbf{X}$ give the coordinates of the n samples in $\mathscr{R}_p$. The distance d_{ij} between a pair of points is given by Pythagoras' theorem and therefore we shall refer to **Pythagorean distance** (section 1.1). The method of principal components then chooses as the ρ-dimensional display space that subspace $\mathscr{L}$ of the p-space which is best fitting in the least squares sense. Given any ρ-dimensional subspace, we can orthogonally project the n points into it, and that particular ρ-dimensional subspace is chosen which has the smallest sum of squared residuals between the original points and their projections. When $\rho=2$, as will usually be the case, $\mathscr{L}$ will be a **plane of best fit.**

It is easy to show that the best fitting subspace defined above necessarily passes through the centroid of the data. The sum of squared distances from the sample points to some arbitrary point $\mathbf{c}=(c_1, c_2, c_3, \ldots, c_p)$ is $\|\mathbf{X}-\mathbf{1c}\|$ and we have the easily verified identity (Huygens' Principle)

$$\|\mathbf{X}-\mathbf{1c}\| = \|\mathbf{X}-\frac{1}{n}\mathbf{11}'\mathbf{X}\| + n\|\frac{1}{n}\mathbf{1}'\mathbf{X}-\mathbf{c}\|$$

which is minimum when $\mathbf{c}$ is at the centroid, i.e.

$$\mathbf{c}=\frac{1}{n}\mathbf{1}'\mathbf{X}$$

Thus, the sum of squares of the residuals from $\mathscr{L}$ is minimized when $\mathscr{L}$ passes through the centroid of the n points. Subsequent discussion will be simplified if we take the centroid of the data matrix $\mathbf{X}$ to be the origin, so that $\mathbf{1}'\mathbf{X}=0$.

Figure 2.1 illustrates the geometry, showing the centroid of all n points, G, lying in the best fitting plane $\mathscr{L}$, two sample points P_i and P_j, their projections Q_i, Q_j onto $\mathscr{L}$ and the residuals r_i, r_j. The distance between P_i and P_j is d_{ij} and is approximated by the distance δ_{ij} between Q_i and Q_j. We have

$$\sum_{i=1}^{n} \mathrm{GP}_i^2 = \sum_{i=1}^{n} \mathrm{GQ}_i^2 + \sum_{i=1}^{n} r_i^2$$

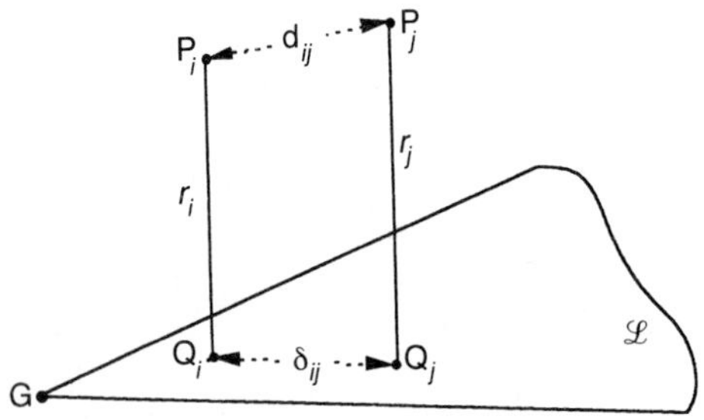

Fig. 2.1. *G is the centroid of all n points and lies in* $\mathscr{L}$. r_i *and* r_j *are the residuals from* $\mathscr{L}$ *whose total sum-of-squares is minimized by PCA.*

which, from the result derived after (A-52), may be written

$$\sum_{i<j}^{n} d_{ij}^2 = \sum_{i<j}^{n} \delta_{ij}^2 + n \sum_{i=1}^{n} r_i^2 \tag{2.1}$$

Thus, because the sum-of-squares of residuals is minimized, it follows that the difference between the sums-of-squares of observed and fitted distances is minimized. Note that the fitted distances are constrained to be derived from orthogonal projections onto $\mathscr{L}$, otherwise it would be trivial to get exact agreement between the sums of squares of observed and fitted distances. The equation (2.1) may be compared with the similar results (3.9) and (3.10) derived in Chapter 3 for other measures of fit.

Once the best fitting subspace $\mathscr{L}$ has been found, the representations of the sample points in it are found by simple projections. To produce a *biplot*, we also need to display the p Cartesian axes. Finding the biplot axes – the representations of the p Cartesian variables describing the samples – is slightly more difficult than finding points to represent the samples. It turns out that although the two types of biplot axes – interpolative and predictive – are both linear and lie in the same direction, they need to have different graduations. In later chapters, we shall see that different types of biplot lead to axes no longer in the same direction and not necessarily even linear.

2.2 Finding the subspace, interpolation and prediction

The eigenvectors (section A.1) of $\mathbf{X}'\mathbf{X}$ form the columns of an orthogonal matrix $\mathbf{V}$ satisfying $\mathbf{X}'\mathbf{X} = \mathbf{V}\mathbf{\Lambda}\mathbf{V}'$ and $\mathbf{V}'\mathbf{V} = \mathbf{I}$. Because $\mathbf{X}'\mathbf{X}$ is positive semi-definite (p.s.d.), the eigenvalues in the diagonal of $\mathbf{\Lambda}$ are necessarily non-negative. The eigenvectors form an alternative basis for the p-dimensional space in which the samples are described – they can be regarded simply as a rotation, given by the orthogonal matrix $\mathbf{V}$, of the p Cartesian axes. (And one can then seek to describe particular eigenvectors in terms of the properties of the samples which are characterized by those eigenvectors – a process known as **reification**. Reification receives very little mention in this book, partly because we regard it as being open to over-enthusiastic misinterpretation but more because our main objective is to interpret multidimensional displays in the terms of the original variables. We regard principal axes and similar constructs as providing essential scaffolding on which to base our plots, but the scaffolding is removed for display purposes.) We show in section A.4 that the best fitting ρ-dimensional subspace, $\mathscr{L}$, in the sense defined above, is spanned by the first ρ principal eigenvectors of $\mathbf{X}'\mathbf{X}$ (forming the columns of a matrix which we shall write $\mathbf{V}_\rho$). These ρ eigenvectors (or components) define a natural set of orthogonal coordinate axes for the ρ-dimensional subspace. However, as axes for the subspace, they are the scaffolding: what concerns us are representaions of the original axes – the variables – in the ρ-dimensional subspace. These representations are the biplot axes and we shall return to them below. Relative to the ρ principal axes, the coordinates of the projections (A.74) of the samples onto the subspace $\mathscr{L}$, are given by $\mathbf{Z} = \mathbf{X}\mathbf{V}_\rho$. That is, the best display of the n points, in the sense above, is given by the n rows of the matrix $\mathbf{Z}$.

To interpolate a new sample, with measurements $\mathbf{x}$, into the display, we need to project it into the ρ-dimensional subspace, and give its coordinates in terms of a basis of this subspace. The projection $\mathbf{z}$ of $\mathbf{x}$ onto the subspace spanned by $\mathbf{V}_\rho$ is given by $\mathbf{z} = \mathbf{x}\mathbf{V}_\rho$ (A.74).

We can predict the p-space coordinates of a point $\mathbf{z} = (z_1, \ldots, z_\rho)$ described in terms of the coordinate system, of $\mathscr{L}$ as follows. First, we note that because $\mathscr{L}$ is a subspace of $\mathscr{R}$, the point representing $\mathbf{z}$ in $\mathscr{L}$ also has a coordinate representation $\mathbf{x}$ relative to the axes of $\mathscr{R}$; it is these, which refer to the actual values of the variables, that we wish to predict. A point $\mathbf{x} = (x_1, \ldots, x_p)$ described in terms of the p-space coordinates but which lies in the subspace, is projected into itself, i.e. $\mathbf{x} = \mathbf{x}\mathbf{V}_\rho(\mathbf{V}'_\rho\mathbf{V}_\rho)^{-1}\mathbf{V}'_\rho$ (A.71). But from the above we know that $\mathbf{z} = \mathbf{x}\mathbf{V}_\rho$ is its description in terms of the subspace coordinates. Hence $\mathbf{x} = \mathbf{z}(\mathbf{V}'_\rho\mathbf{V}_\rho)^{-1}\mathbf{V}'_\rho$. This is a general argument, but since we are taking the columns of $\mathbf{V}_\rho$ to be orthonormal we have that $\mathbf{V}'_\rho\mathbf{V}_\rho = \mathbf{I}$, so that $\mathbf{x} = \mathbf{z}\mathbf{V}'_\rho$.

To summarize, interpolation is achieved by $\mathbf{z} = \mathbf{x}\mathbf{V}_\rho$ and prediction by $\mathbf{x} = \mathbf{z}\mathbf{V}'_\rho$. It follows from this that the coordinates, in the p-dimensional space, of the interpolate of $\mathbf{x}$ are $\mathbf{x}\mathbf{V}_\rho\mathbf{V}'_\rho$ and those of the interpolated original samples are given by $\mathbf{X}\mathbf{V}_\rho\mathbf{V}'_\rho$. The graphical implications of these formulae are discussed in the next section.

2.3 The biplot axes

Writing $\mathbf{e}_k$ for the unit vector along the kth coordinate axis in $\mathscr{R}$, the point $\mathbf{x}$ with coordinates $(x_1, x_2, \ldots, x_p)$ may be written

$$\mathbf{x} = \sum_{k=1}^{p} x_k \mathbf{e}_k$$

which will be interpolated to

$$\mathbf{x}\mathbf{V}_\rho = \sum_{k=1}^{p} x_k(\mathbf{e}_k \mathbf{V}_\rho)$$

where $\mathbf{e}_k\mathbf{V}_\rho$ is the interpolant of the unit point on the kth axis.

This shows that the interpolant of a general point $\mathbf{x}$ is given simply by the vector-sum of the unit points weighted by $x_1, x_2, \ldots, x_p$. Graphically, if the points $\mu\mathbf{e}_k\mathbf{V}_\rho$ for $\mu = 0, \pm 1, \pm 2, \pm 3, \cdots$ are plotted, then the point $x_k(\mathbf{e}_k\mathbf{V}_\rho)$ may be found by inspection (if necessary by visual interpolation between successive values of μ that bracket x_k). This may be done for all p variables and the vector-sum obtained as is described shortly. The directions $\mathbf{e}_k\mathbf{V}_\rho(k = 1, 2, \ldots, p)$, which are the projections of the original axes onto $\mathscr{L}$, define the biplot axes and the points defined by the different values of μ are termed **markers** or **graduations**. Markers are precisely the same thing as scale markers on ordinary graphs and similarly will have associated measurement scales and need not be confined to integer values. Section 2.6.1 discusses some practical considerations for plotting markers.

The usual graphical method for evaluating sums of vectors is to 'complete parallelograms' but this is very cumbersome when there are several vectors to sum. Figure 2.2 illustrates a simple method for interpolating a point by summing the vectors for its three markers on the biplot axes β_1, β_2, and β_3. We show the markers and illustrate the interpolation of $(2, -3, 4)$. G is the centroid of the three markers and the interpolated point is at three (the number of axes) times the vector $\overrightarrow{\mathrm{OG}}$ We often use this method in following chapters and refer to it as the **vector sum method**.

The above tells us the directions of the biplot axes and the graduations along them when we are interested in interpolating points from the p-dimensional measurement space. Can something similar be done for prediction?

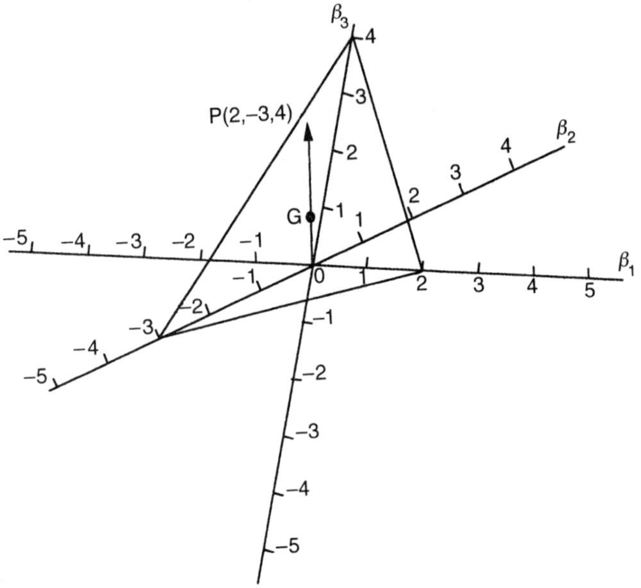

Fig. 2.2. *Interpolation of the point (2, −3, 4).*

Consider a hyperplane $\mathcal{N}$ (Fig. 2.3) perpendicular to the kth Cartesian axis ξ_k going through the unit marker on that axis. This will intersect the display subspace in a $(\rho-1)$-dimensional subspace (a line when the display space is two-dimensional) $\mathcal{L} \cap \mathcal{N}$. All points in this subspace should therefore predict the unit value. Now, we can define a prediction axis as some line (through the origin) in the display subspace which intersects $\mathcal{L} \cap \mathcal{N}$ at a point which may be marked with the unit marker. It will also intersect all other $(\rho-1)$-dimensional subspaces parallel to $\mathcal{L} \cap \mathcal{N}$ and other markers may be added at equal intervals to produce a conventional scale. To predict the value, in the p-space, of a display space point $\mathbf{z}$, simply identify that subspace (line) parallel to $\mathcal{L} \cap \mathcal{N}$ which goes through $\mathbf{z}$. The marker attached to the point at which it intersects the kth prediction axis will be the predicted value corresponding to $\mathbf{z}$. So far, however, this method is not unique. We chose some line in the display subspace which intersected $\mathcal{L} \cap \mathcal{N}$. To avoid non-orthogonal projections and to ensure uniqueness, we choose the

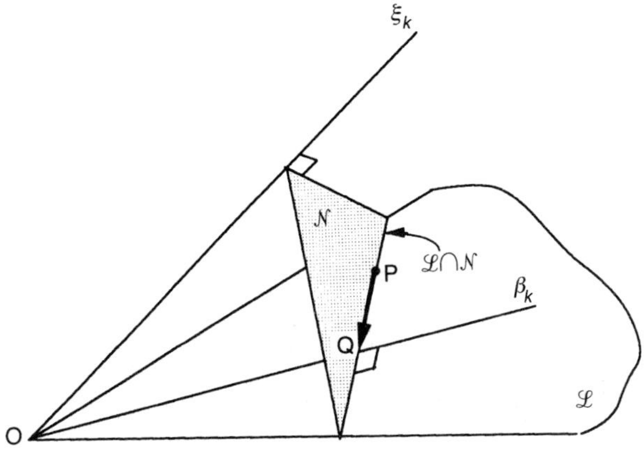

Fig. 2.3. *The plane $\mathcal{N}$ contains all points with coordinate x_k on the axis ξ_k. The intersection $\mathcal{L} \cap \mathcal{N}$ contains all points in $\mathcal{L}$ that predict the value x_k. The projection of the point* **P** *onto biplot axis β_k gives the correct prediction, provided a suitable scale is marked on β_k.*

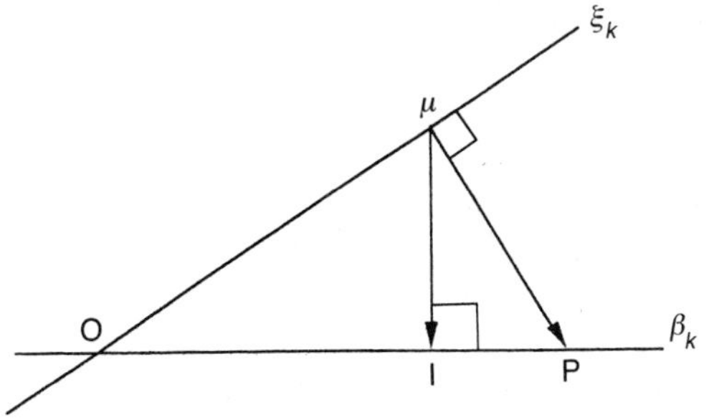

Fig. 2.4. *Relationship between the markers for prediction and interpolation.*

line which is orthogonal to $\mathcal{L} \cap \mathcal{N}$. This line is the projection of ξ_k onto $\mathcal{L}$ and so coincides in direction with the interpolation biplot axis β_k discussed above. However, the markers induced by the intersection with $\mathcal{N}$ are not projections onto $\mathcal{L}$ and so differ from the interpolation markers. Figure 2.4 illustrates the relationship between the two sets of markers.

ξ_k is one of the axes in $\mathcal{R}$ and β_k is the corresponding biplot axis in $\mathcal{L}$. The marker μ projects to I for interpolation and back-projects (A.9) to P for prediction. Figure 2.4 shows that $\mathrm{OI} \times \mathrm{OP} = \mu^2$ and we know that I is given by $\mu\mathbf{e}_k\mathbf{V}_\rho$. It follows that P is given by

$$\left[\frac{\mu\mathbf{e}_k\mathbf{V}_\rho}{\mathbf{e}_k\mathbf{V}_\rho\mathbf{V}'_\rho\mathbf{e}'_k}\right]$$

As μ varies, this gives the markers for prediction.

2.4 Equivalent definitions

An alternative way of expressing the above description is via the singular value decomposition $\mathbf{X} = \mathbf{U\Sigma V}'$ (section A.3). The rows of $\mathbf{U\Sigma}$ given n p-dimensional coordinates for the n samples. The columns of $\mathbf{V}'$ – and hence the rows of $\mathbf{V}$ – give the directions of the biplot axes, as above. $\mathbf{\Lambda}$ above is equal to $\mathbf{\Sigma}'\mathbf{\Sigma}$.

Prediction corresponds to the Eckart–Young theorem (section A.4): the rank ρ matrix $\hat{\mathbf{X}}$ which minimizes $\|\mathbf{X} - \hat{\mathbf{X}}\|$ is given by

$$\hat{\mathbf{X}} = \mathbf{U\Sigma}_\rho\mathbf{V}'_\rho = \mathbf{XV}_\rho\mathbf{V}'_\rho$$

where $\mathbf{\Sigma}_\rho$ is the matrix consisting of the first ρ columns of $\mathbf{\Sigma}$.

Other equivalent characterizations of the display subspace (Jolliffe, 1986) are that subspace:

(i) which maximizes the sum of the variances of the projections of the n points in ρ orthogonal directions;
(ii) such that the sum of ${}^nC_2$ ddistances between the projections of the n points in the subspace is maximized (A.56); and
(iii) which minimizes $\|\mathbf{XX}' - \mathbf{ZZ}'\|$.

2.5 How good is the representation?

So far we have found the best ρ-dimensional subspace in which to project the points, and have found the coordinates of the points when projected into that subspace (both relative to a set of axes spanning that space – the first ρ principal axes – and relative to the original axes). However, we have not, as yet, considered how good a representation is provided by the projections of the points. This is important since we will want to know how adequate our representation of the data is – how well the display represents the true relationships in the data.

From the above, the residuals are given by

$$\mathbf{X} - \hat{\mathbf{X}} = \mathbf{X}(\mathbf{I} - \mathbf{V}_\rho \mathbf{V}'_\rho)$$

Hence the sum of squared residuals has value

$$\text{trace}\,(\mathbf{X} - \hat{\mathbf{X}})'(\mathbf{X} - \hat{\mathbf{X}}) = \text{trace}\,\mathbf{X}(\mathbf{I} - \mathbf{V}_\rho \mathbf{V}'_\rho)\mathbf{X}' = \text{trace}\,(\mathbf{\Lambda} - \mathbf{\Lambda}_\rho)$$

That is, the sum of squared residuals is equal to the sum of the eigenvalues corresponding to the eigenvectors which do *not* span the best fitting subspace. The sum of all the eigenvalues is equal to the total variation (sum of the variances of all the p variables) and is the same thing as the sum of the squared distances of the n points from their centroid. Hence the sum of all the eigenvalues can be viewed as a measure of 'total variation within the data'. This being the case, the ratio of the sum of the first ρ eigenvalues to the total sum gives a measure of quality of fit ranging between 0 and 1, and can be viewed as the proportion of variation explained by the subspace.

There is also interest in the quality of display of the variables. To an extent, this is visually given by the relative lengths of the range between the minimum and maximum markers for observed values on each axis. Poorly represented variables will be relatively short and therefore will contribute little to the approximation – this is because they do not contribute much to inter-sample distances in $\mathscr{R}$. A numerical measure of the quality of the variables is given by the sums-of-squares of the rows of $\mathbf{V}_\rho$, i.e. $\mathbf{V}_\rho \mathbf{V}'_\rho \mathbf{1}$. This gives the ddistances of each unit point from the origin, which in exact representations should be unity, as in $\mathscr{R}$ where $\mathbf{V}$ is orthogonal. For this reason, unit circles are sometimes shown on PCA biplots and related methods.

All of the above aims at displaying the n points in a ρ-dimensional subspace such that the relationships between the points are well approximated. However, unless the p variables are measured in the same units there is inevitably a certain arbitrariness in the results. Because of this, most authors recommend standardizing the raw variables before undertaking the PCA. One method of standardization is to take the logarithms of the raw scores; of course, this can be done only when the variables can take only positive values. Differences in units of measurement then simply become additive constants which do not contribute to Pythagorean distances. An alternative is to standardize by dividing each variable by its standard deviation so that the resulting matrix $\mathbf{X}'\mathbf{X}$ is a correlation matrix. We have more to say about correlation matrices below. Before that, however, some words about interpreting the components of $\mathbf{V}$ are appropriate.

The sample scores in the direction of the kth component of $\mathbf{V}$ are given by $\mathbf{y}_k = \mathbf{X}\mathbf{v}_k$, where $\mathbf{X}'\mathbf{X}\mathbf{v}_k = \lambda_k \mathbf{v}_k$. Denote the vector of sample scores on the hth variable by $\mathbf{x}_h$. Then the correlation between $\mathbf{x}_h$ and $\mathbf{y}_k$ is

$$r^2 = \frac{(\mathbf{x}_h'\mathbf{X}\mathbf{v}_k)^2}{(\mathbf{x}_h'\mathbf{x}_h)(\mathbf{v}_k'\mathbf{X}'\mathbf{X}\mathbf{v}_k)} = \frac{\lambda_k^2 v_{kh}^2}{(s_{hh}\lambda_k)}$$

where v_{kh} is the hth component of $\mathbf{v}_k$ and where s_{hh} is the hth element of the diagonal of $\mathbf{X}'\mathbf{X}$.

When the data has been standardized by the standard deviation before analysis, then $s_{hh} = 1$ as $\mathbf{X}'\mathbf{X}$ is the correlation matrix, as noted above. In this case, $r^2 = \lambda_k v_{kh}^2$, so that $\lambda_k^{\frac{1}{2}} v_{kh}$ is the correlation of the hth variable with the kth component. Sometimes, the $\mathbf{v}_k$ are given multiplied by $\lambda_k^{\frac{1}{2}}$ to facilitate this interpretation.

If we project the n samples onto $\mathscr{L}$, we obtain $\mathbf{X}\mathbf{V}_\rho$. Projecting these further onto some direction β in $\mathscr{L}$ we obtain values $\mathbf{X}\mathbf{V}_\rho\beta$. The direction $\beta = \hat{\beta}$ which leads to the maximum correlation between $\mathbf{X}\mathbf{V}_\rho\beta$ and the original n scores, $\mathbf{x}_h$, on the hth variable, is given by the vector of regression coefficients when $\mathbf{x}_h$ is regressed on $\mathbf{X}\mathbf{V}_\rho$, i.e. $\mathbf{x}_h = \mathbf{X}\mathbf{V}_\rho\hat{\beta}$, yielding

$$\hat{\beta} = (\mathbf{V}_\rho'\mathbf{X}'\mathbf{X}\mathbf{V}_\rho)^{-1}\mathbf{V}_\rho'\mathbf{X}'\mathbf{x}_h = \mathbf{\Lambda}_\rho^{-1}\mathbf{\Lambda}_\rho\mathbf{v}_h = \mathbf{v}_h$$

as asserted above. Thus, $\hat{\beta}$ is the hth row of $\mathbf{V}_\rho$ and coincides with the direction of the kth biplot axis. This shows that among all directions in $\mathscr{L}$, projections onto the kth biplot axis have maximal correlation with the observed values for the kth variable; this property is exploited further in section 3.3.2.

PCA is sometimes considered an analysis of correlations. This view, of course, depends on the columns of $\mathbf{X}$ being normalized so that $\mathbf{X}'\mathbf{X}=\mathbf{R}$ is a correlation matrix. Then it is certainly true that

(i) $\mathbf{V}\mathbf{\Lambda}^{\frac{1}{2}}$ is such that $\mathbf{R}=(\mathbf{V}\mathbf{\Lambda}^{\frac{1}{2}})(\mathbf{V}\mathbf{\Lambda}^{\frac{1}{2}})'$ so that the inner product reproduces $\mathbf{R}$ and
(ii) that ρ-dimensional approximations are obtained when $\mathbf{V}$ and $\mathbf{\Lambda}$ are replaced by $\mathbf{V}_\rho$ and $\mathbf{\Lambda}_\rho$.

Indeed, this approximation is that given by the Eckart–Young theorem but note that it implies a scaling of the eigenvectors that differs from that of a conventional PCA – this representation is discussed further in section 11.5.1. The SVD of a correlation matrix suffers from the disadvantage that the unit diagonal values of $\mathbf{R}$ are approximated as well as the correlations which are of substantive interest. The problem can be avoided if, instead, we use principal factor analysis, which replaces the diagonal by estimated communalities (Chapter 10) and approximates only the correlations. Alternative approaches to representing approximations to the correlations are discussed in section 11.5.3. It is no surprise to us that the representation of correlations in PCA is not fully satisfactory, because the fundamental assumption of Pythagorean distance assumes that variables contribute independently. Furthermore, product-moment correlations strictly apply to multinormal distributions and no such assumption is necessary for the geometrical approach that we have adopted. Nevertheless, correlations do have an effect on PCA displays. Complete correlation between two variables induces a loss in rank, and so automatic dimension reduction, which manifests itself in coincidence between the corresponding projected biplot axes. High correlations induce similar effects so that a small angle between two biplot axes suggests a high correlation between the corresponding variables. Note that, in the full space $\mathscr{R}$, all axes are orthogonal, whatever the correlations.

2.6 Some practical considerations concerned with plotting variables and scales

The remarks made in this section apply at several places in this book but it seems best to make them here, just before discussing our first example. It has already been made clear that our approach to biplots is not the conventional one. Indeed, in the following chapters there are several examples of novel forms of biplot and other examples which we regard as biplots but which would not normally be regarded as such. Here, our comments are related to PCA but the reader should have no difficulty in making the necessary adaptations required in other contexts. Even with classical biplots, the topic of this chapter, our representation differs in several ways from what is normally plotted. The differences are of two kinds:

(i) we think that many things that are plotted should not be plotted and
(ii) we think that there are several things that are not plotted that should be.

It would be invidious to quote published examples of what we regard as poor biplots, so we have concocted an artificial example that demonstrates many deficiencies. Thus, our example has unusually many faults, but probably not all that may be found in the published literature; things are not usually quite as bad as we depict. We are conscious that we too are far from perfect and sometimes even break our own rules. The example of what we consider to be bad biplot practice is shown in Fig. 2.5. In this figure, there are supposed to be ten samples, labelled S1, S2, ..., S10 and five variables, labelled V1, V2, ..., V5.

In Fig. 2.5, both the scaling of the axes and the plotting of the variables are unhelpful. A unit change in the horizontal direction is not equal to a unit in the vertical direction so that a true square would be shown as a rectangle. Such a scaling is useless for appreciating distances and angles, including orthogonal projections, which are vital for interpreting biplots. We recognize that those who have made similar elementary errors may have been more sinned against than sinners, because they are at the mercy of graphical software that is designed to make plots that fit comfortably into a page, thus doing violence to the true aspect ratio. The scales of Fig. 2.5 are in the appalling E-format and do not give sensible values; it is, of course, a far from trivial problem to develop a general algorithm for plotting ‘nice’ whole number or decimal scales. Further, there are two sets of scales; those at the bottom and to the left give coordinates of samples and those at the top and to the right refer to variables – this seems to us to be very hard to justify in joint representations of the relationships between samples and variables. These remarks are beside the point if we take the view, as we do, that there is little interest in the values of the

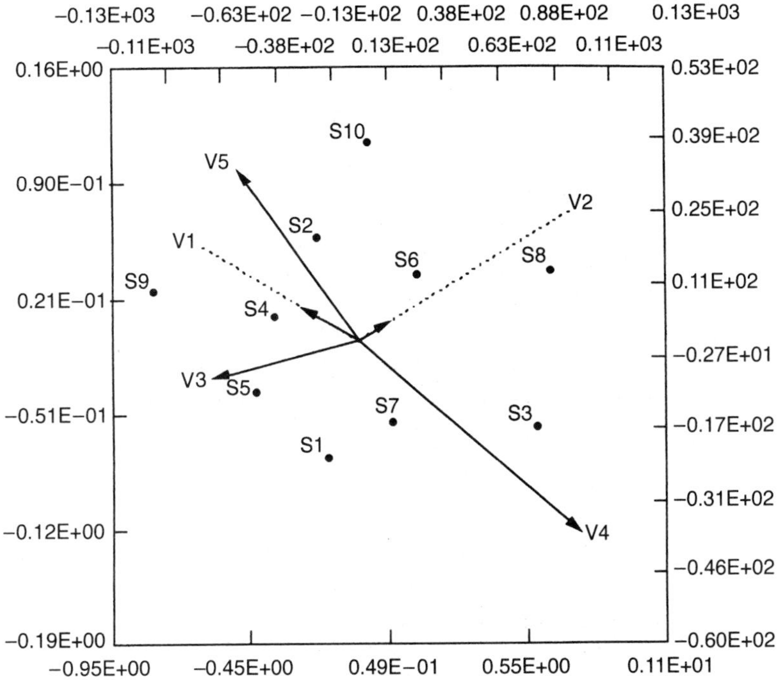

Fig. 2.5. *An artificial example of a biplot that exhibits many faults.*

principal components themselves (section 2.2) apart from their convenience for plotting purposes. We return below (section 2.6.1) to the plotting of scales for the original variables. Even though we do not show scales for the components, it is vital that biplot displays should always have the true aspect ratio.

The plotting of the samples in Fig. 2.5 is acceptable but not that of the variables. The latter are shown as arrowed vectors, the length from the origin to the head of the arrow indicating some interesting statistic, typically one standard deviation of the labelled variable. This is acceptable, especially when a primary reason for the representation is to approximate variances, covariances and correlation (section 11.5.1), but even then it is not totally satisfactory. When a plotted vector is short, it has to be extended to facilitate an appreciation of orthogonal projections onto it and this explains the dotted lines in the figure. In this book, we regard biplot axes as variants of conventional coordinate axes so they should be marked with scales and should, normally, pass through the origin in both directions. If one wishes, the position of one standard deviation may be marked on such an axis. Thus, the plotting of variables as shown in Fig. 2.5 needs modifications of the kind discussed in the next section.

2.6.1 *Scales for the variables*

Because PCA works in terms of the centred matrix **X**, computer output is normally expressed in terms of deviations from the means. Yet it is our aim to relate interpretation to the original values of the variables. We have seen that the rows of the matrix $\mathbf{V}_\rho$ gives the coordinates of unit markers on each of the interpolative biplot axes. What this means is that the markers indicate one unit in the scale of the original variables. However, if we print the value 1 opposite a marker (and $0, \pm 1, \pm 2, \pm 3, \ldots$ at equal steps along the axes) then this does not reproduce the original measurements but only deviations from their means. In PCA, and other methods, the origin G is associated with the zero marker and represents the mean values of all the variables, as well as the centroid of the points representing the n samples. What is required, is a way of replacing the deviations by true values. This is easily done as follows.

For the kth variable, G represents its mean $\bar{x}_k$ and the unit marker U_k on the kth axis represents $\bar{x}_k + 1$ with coordinates $\mathbf{v}_k$, the kth row of $\mathbf{V}_\rho$. Somewhere between G and U_k there will be a point representing a standard value of the original variable. By a standard value, we mean a convenient unit for plotting purposes and this will depend on the scale and range of values of the kth variable; for discussion purposes, we assume that integer values are convenient but, for example, fractions or multiples of 10 will often be preferable. Assuming that an integer scale is used and that $\bar{x}_k + \lambda$ (for $0 < \lambda < 1$) is an integer I, say, then the coordinates for the marker I, are at $\lambda \mathbf{v}_k$ and the other markers on this axis are at multiples of $\mathbf{v}_k$ on either side of I. The automation of this process is far from trivial and we have not attempted it here but note that many graphical computer programs incorporate facilities of this kind, though not necessarily in accessible form.

When standard markers have been provided for all p biplot axes, we note that the point of concurrency G normally corresponds neither to standard values nor to the same value on all axes. With conventional orthogonal axes, it is often convenient to assign all the zero markers to the origin. There seems to be no way of doing this for predictive biplots, but it can be done for interpolative biplots when the centroid method of interpolation is valid, as it is with PCA, CVA (Chapter 5) and PCO (Chapter 6). Let $\mu_1, \mu_2, \ldots, \mu_p$ be selected as standard markers on the biplot axes, with their centroid at

$$\mu = \frac{1}{p}(\mu_1 + \mu_2 + \cdots + \mu_p)$$

Then for any other point $\xi_1, \xi_2, \ldots, \xi_p$ we have that

$$\xi_1 + \xi_2 + \cdots + \xi_p = \sum_{k=1}^{p} [\xi_k + (\mu_k - \mu)]$$

The implication is that if we translate every point of the kth axis by an amount $(\mu_k - \mu)$ $(k = 1, 2, \ldots, p)$, then the vector sum is unchanged. Thus the origin (i.e. point of concurrency of the axes) is translated to μ which is associated with the standard markers for $\mu_1, \mu_2, \ldots, \mu_p$. We may choose these standard markers as we please, including all zero, if that is appropriate. Figure 2.7 in the following section, shows the effect on Fig. 2.6 of this type of translation, where the standard values are all chosen to be 3 units. Note that the centroid G remains unaffected and remains the origin for extending vector sums in interpolation.

2.7 Example

The data for this example are taken from Cook and Weisberg (1982, Table 2.3.1), who extracted them from a 1979 RAND Corporation report. They show the values of four variables on each of 21 fighter aircraft. The variables are:

SPR = specific power, proportional to power per unit weight
RGF = flight range factor
PLF = payload as a fraction of gross weight of aircraft
SLF = sustained load factor

The raw data are given in Table 2.1.

Table 2.1 The values of four variables for 21 types of aircraft

	Aircraft	*SPR*	*RGF*	*PLF*	*SLF*
a	FH-1	1.468	3.30	0.166	0.10
b	FJ-1	1.605	3.64	0.154	0.10
c	F-86A	2.168	4.87	0.177	2.90
d	F9F-2	2.054	4.72	0.275	1.10
e	F-94A	2.467	4.11	0.298	1.00
f	F3D-1	1.294	3.75	0.150	0.90
g	F-89A	2.183	3.97	0.000	2.40
h	XF10F-1	2.426	4.65	0.117	1.80
i	F9F-6	2.607	3.84	0.155	2.30
j	F100-A	4.567	4.92	0.138	3.20
k	F4D-1	4.588	3.82	0.249	3.50
m	F11F-1	3.618	4.32	0.143	2.80
n	F-101A	5.855	4.53	0.172	2.50
p	F3H-2	2.898	4.48	0.178	3.00
q	F102-A	3.880	5.39	0.101	3.00
r	F-8A	0.455	4.99	0.008	2.64
s	F-104A	8.088	4.50	0.251	2.70
t	F-105B	6.502	5.20	0.366	2.90
u	YF-107A	6.081	5.65	0.106	2.90
v	F-106A	7.105	5.40	0.089	3.20
w	F-4B	8.548	4.20	0.222	2.90

The alphabetic code (omitting the letters l and o) given in the first column of Table 2.1 is used in the diagrams rather than the full name of each aircraft. Inspection of the table shows that the variable PLF is not commensurable with the other three variables and that, therefore, some form of transformation would be desirable before embarking on a PCA. We do not transform here because we wish to demonstrate how incommensurability manifests itself in a biplot. The means of the variables are

SPR	RGF	PLF	SLF
3.831	4.488	0.1671	2.278

The eigenvalues of $\mathbf{X'X}$ (corrected for the means) are

121.176	5.475	4.447	0.119

The eigenvectors of $\mathbf{X'X}$ (corrected for the means) are

SPR	0.9545	0.2969	−0.0176	−0.0219
RGF	0.1200	−0.4371	−0.8912	0.0178

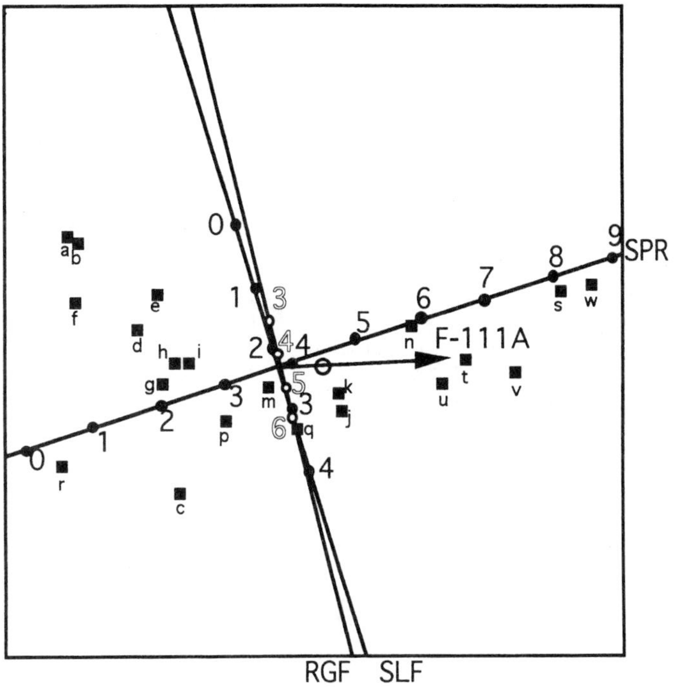

Fig. 2.6. *An interpolative biplot for the data of Table 2.1.*

PLF	0.0104	0.0406	0.0015	0.9991
SLF	0.2729	−0.8480	0.4532	0.0309

The first dimension gives an excellent approximation but, in the following, we shall discuss the two-dimensional solution. The quality of representation in ρ dimensions of each variable may be assessed by noting that the first ρ columns of $\mathbf{V}$ give the projections of the unit values for each variable and the remaining columns give the residuals of the unit points. Because $\mathbf{V}$ is an orthogonal matrix, its rows have unit sums-of-squares and it follows that the sums-of-squares of the first ρ columns gives a measure of the adequacy of fit for each variable. With these data, the values for $\rho = 2$ are

SPR	RGF	PLF	SLF
0.9992	0.2054	0.0018	0.7936

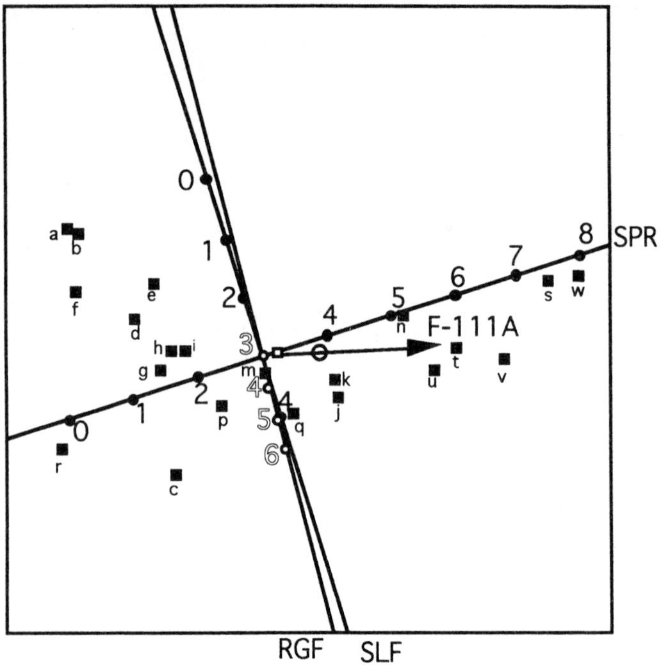

Fig. 2.7. *The same as Fig. 2.6 but with axes translated to give more easily read scales.*

Thus SPR and SLF are very well represented, RGF is not unreasonably represented but PLF is very poorly represented, as its incommensurability would lead us to expect. Indeed, the residual fit for PLF is 0.9982 indicating that the true axis for this variable is nearly orthogonal to the plane of projection–inspection of the vectors shows that it is largely in dimension number four. The numerical values of this variable are so trivial that PCA makes little attempt to approximate it in the two-dimensional display.

The plot, relative to the first two components, of the projections of the points representing the 21 aircraft is shown in Fig. 2.6. Although the first component is in the horizontal direction and the second in the vertical direction, we do not show these axes explicitly but confine attention to the biplot axes which carry information on the original variables. However, we note that, in this example, the first principal axes may be reified convincingly because the alphabetic labelling progresses fairly systematically from left to right, and it is known that this represents year of manufacture – similar remarks could be made for projections onto the axis representing the variable SPR which indicates an increase of power over time. The variable PLF does not appear in the figure because it has insufficient dispersion in the two dimensions shown to permit a meaningful display, reflecting that because of its scale (Table 2.1) it contributes little to inter-sample distance and, as we have seen, is represented by an axis that is nearly orthogonal to the display. Thus the following remarks are concerned with the other three variables. The scales for interpolation of these three variables, covering their ranges in the data, are shown and it is manifest that their biplot axes do not intersect at integer values of any of the variables (section 2.6.1). Although the three biplot axes are extended to the boundaries of the figure, the scale values shown, which are relevant to Table 2.1, use only part of the lengths of the axes. Beyond the range of the exhibited scale, values are extrapolated or, if negative, impossible. One of the benefits of showing scales is, as we have already seen, that a variable that is contributing little to the approximation will not occupy a significant range of its biplot axis and, on inspection, may be identified and disregarded in subsequent interpretations. Another benefit is that attention is drawn to extrapolated or impossible values. It is not normal practice to include scales in biplots and this deficiency encourages misinterpretation or overinterpretation of features discerned in the display. We believe that scales are just as important in the multivariate context as they are with conventional x–y plots. The notion of inspecting, and perhaps discarding, biplot axes, implies the desirability of developing interactive software with appropriate facilities.

Suppose we wish to interpolate a new sample – the 22nd in the table given in Cook and Weisberg:

	SPR	RGF	PLF	SLF
F-111A	6.321	6.45	0.187	2.00

The interpolated point is $\mathbf{z} = \mathbf{x}\mathbf{V}_\rho = (2.536, 0.118)$. The graphical interpolate is shown in Fig. 2.6 with the centroid of the markers for values of the variables of F-111A shown as a small open circle and extended from the origin $p = 4$ times to obtain its final location.

This interpolation demonstrates the vector-sum method (section 2.3); in this case, we could equally have chosen $p = 3$ and excluded the origin, which represents to sufficient accuracy all values for PLF, when finding the centroid – a zero contribution does not affect the vector sum. The F-111A is the most recently developed of the aircraft given in the table and its interpolated position is to the right of the diagram among the other recent developments.

Figure 2.7 shows the same information but with the axes translated as described in section 2.6.1. It can be seen that the labelling of the scales is now clearer and that interpolation is unaffected. The value 3 for all three exhibited variables has been chosen as the new origin while the centroid of samples, represented by a small open square, at G remains unaffected. In this example, the displacement of the origin from G is very slight.

To predict the values of the four variabes for a point $\mathbf{z} = (3.485, -0.211)$, we need to evaluate $\mathbf{z}\mathbf{V}'_\rho$, with $\rho = 2$, yielding $(7.194, 4.999, 0.203, 3.408)$ after remembering to add in the means. These are the values predicted by the rank two Eckart–Young approximation. The values of $\mathbf{z}$ happen to be the coordinates of the F-106A (labelled v) and the predicted values will be seen to agree well with those given in Table 2.1.

Figure 2.8 shows the predictive biplot that allows the calculations for prediction to be done visually. The common directions of the interpolative and predictive biplot axes are given by the lines through $\mathbf{e}_k\mathbf{V}_\rho$, i.e. by the lines from the origin through the points represented by the rows of $\mathbf{V}_\rho$ but the scales bear an inverse relationship, as described in section 2.3. Figure 2.8 shows that the predictive scales are coarser than those for interpolation, as is evident from the relationships shown in Fig. 2.4. In contrast to interpolation, the standard scale markers cannot be translated to coincide with an origin.

The directions of the variables RGF and SLF are nearly coincident and hence projections from any point onto these two directions will predict values which are nearly linearly related. It follows that, if the approximation is a good one, the variables must be highly correlated. However, we saw above that although SLF is well represented, RGF is not; in fact, the correlation between these variables is 0.6004. The fourth variable, PLF, has a respectable scale for prediction purposes and is now included. However, because the variable PLF is orthogonal to the exhibited plane of

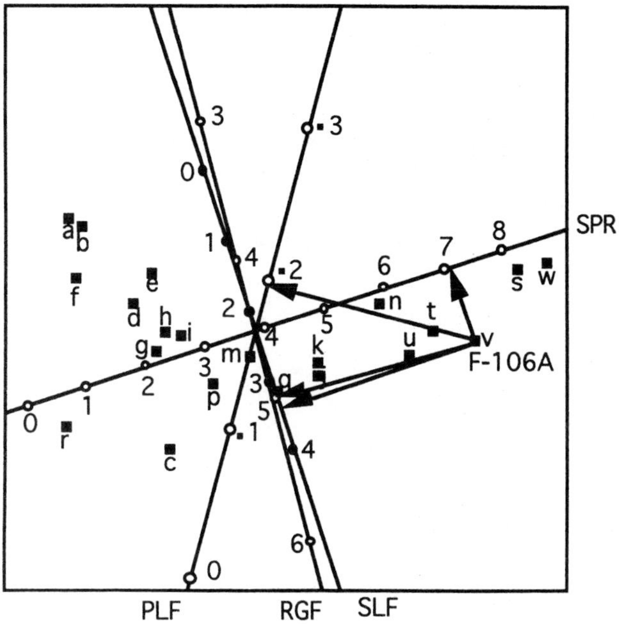

Fig. 2.8. *Predictive scales on the biplot axes. The projections giving the predicted values of F-106 A are shown.*

approximation, predictions for PLF are likely to be unreliable. The details of predicting F-106A are shown in the figure. It will be seen that the predicted values are as close as can be expected from a graphical technique to those calculated numerically, above.

Even without the scales, the ordering of the projections onto a particular axis will rank correctly the predicted values for the corresponding variable according to the Eckart–Young theorem. However, note that equal intervals on different axes do not correspond to equal contributions.

The prediction of the characteristics of the F-106A gave the Eckart–Young approximation for the relevant row of Table 2.1. Yet this could be regarded as a rather uninteresting exercise as the true values may easily be obtained directly from the table. Of more interest are predictions associated with a point not representing one of the original samples. Thus, in market research, for example, the samples might refer to existing products and inspection might reveal regions of the display with no sample. The researcher might then regard these regions as representing possible gaps in the market and wish to predict the characteristics of possible new products that could exploit the gaps. The prediction process proceeds exactly as above but is no longer associated with the Eckart–Young approximation.

CHAPTER 3

Other linear biplots

3.1 Introduction

In Chapter 2, we saw how PCA provides a display in $\rho < p$ dimensions of the points whose inter-point distances approximate the distances d_{ij} between the samples given by Pythagorean ddistance

$$d_{ij}^2 = \sum_{k=1}^{p} (x_{ik} - x_{jk})^2$$

We saw how, with PCA, linear biplot axes could be included in the display. These biplot axes could be used to interpolate new samples – i.e. to locate the display positions corresponding to new samples – and also to predict – i.e. to find the values of the original variables which should be associated with any point in the display.

However, PCA is but one method of multidimensional scaling (MDS), albeit the best known, and there are many others. In this chapter, we take a first look at more general forms of MDS and examine interpolation, prediction, and how to include linear biplot axes in these more general forms. We shall continue with the PCA assumption that **X** is referred to Cartesian axes and generates Pythagorean distances (2.1). This assumption is not necessary for most forms of MDS but is relevant for the linearity of biplots; the perceptive reader will notice several places in this chapter where the Pythagorean assumption is superfluous and, in Chapter 6, it is fully relaxed. We explore least-squares-scaling and least-squares-squared-scaling in some detail, since these methods are the most popular methods of metric MDS other than PCA, but the methodology developed has wider applicability. First, we must discuss the fundamentals of MDS.

3.2 Multidimensional scaling (MDS)

MDS, the term which we shall use throughout this book, arises in psychometrics; in the biometric literature it is often referred to as **ordination**. The problem of MDS is to find a set of n points in ρ dimensions, with coordinates given as the rows of a matrix $\mathbf{Z}$, which generate an $n \times n$ matrix of interpoint distances $\{\delta_{ij}\}$ that approximate the distances $\{d_{ij}\}$. To achieve this, some criterion measuring the difference between $\{d_{ij}\}$ and $\{\delta_{ij}\}$ must be optimized, which we write

$$S_M = S(\tau(d_{ij}), \delta_{ij}) \tag{3.1}$$

where the suffix M refers to the particular method of MDS chosen. The function τ (.) represents a transformation of the given distances that is sometimes permitted, especially in non-metric MDS, but whose further consideration is deferred until later chapters (Section 11.1); thus, in this chapter, we are concerned with metric scaling and with τ the identity transformation. We have already seen that S_{PCA} takes the form of a residual sum-of-squares that has to be minimized, subject to projection constraints on the permitted points $\mathbf{Z}$ that generate the fitted distances $\{\delta_{ij}\}$. The ρ-dimensional space spanned by the columns of $\mathbf{Z}$ will be denoted by $\mathscr{L}$ and the n-dimensional space spanned by the columns of $\mathbf{X}$ will be denoted by $\mathscr{R}$. In PCA, $\mathscr{L}$ is a subspace of $\mathscr{R}$ but this is not so for other methods of MDS; this absence of a direct link between the two spaces is at the root of many of the problems that we now discuss.

Apart from PCA, the best-known methods of metric scaling are **least-squares-scaling** in which

$$S_{\text{LSS}} = stress = \sum_{i<j=1}^{n} (d_{ij} - \delta_{ij})^2 \tag{3.2}$$

and **least-squares-squared-scaling** in which:

$$S_{\text{LSSS}} = sstress = \sum_{i<j=1}^{n} (d_{ij}^2 - \delta_{ij}^2)^2 \tag{3.3}$$

(Often these criteria are divided by

$$\sum_{i<j=1}^{n} d_{ij}^2 \quad \text{and} \quad \sum_{i<j=1}^{n} d_{ij}^4$$

respectively, to provide normalized forms that help when comparing the fits of different MDSs.) Thus, the *stress* and *sstress* criteria provide coordinates **Z** that generate the distances $\{\delta_{ij}\}$ that approximate $\{d_{ij}\}$. We have

$$\delta_{ij}^2 = \sum_{k=1}^{r} (z_{ik} - z_{jk})^2$$

so differentiating *stress* (3.2) with respect to z_{ik} involves only those terms in δ_{ij} $(j = 1, 2, \ldots, n)$ and therefore leads to normal equations

$$\sum_{j=1}^{n} \frac{(d_{ij} - \delta_{ij})}{\delta_{ij}} (z_{jk} - z_{ik}) = 0$$

All such normal equations may be gathered together to give

$$\mathbf{FZ} = \mathbf{0} \tag{3.4}$$

where

$$\left.\begin{aligned} f_{ij} &= \frac{(d_{ij} - \delta_{ij})}{\delta_{ij}} && \text{when } i \neq j \\ &= -\sum_{i \neq j}^{n} f_{ij} && \text{otherwise} \end{aligned}\right\} \tag{3.5}$$

Similarly, the normal equations for *sstress* are

$$\mathbf{GZ} = \mathbf{0} \tag{3.6}$$

where

$$\left.\begin{aligned} g_{ij} &= (d_{ij}^2 - \delta_{ij}^2) && \text{when } i \neq j \\ &= -\sum_{i \neq j}^{n} g_{ij} && \text{otherwise} \end{aligned}\right\} \tag{3.7}$$

It is not our purpose to discuss the algorithmic solutions to these normal equations because this is a major subject in its own right with a substantial literature (Cox and Cox, 1994). Here, it suffices to note the existence of good publicly available software (Cox and Cox, 1994) which provides the coordinates **Z** which generate the distances $\{\delta_{ij}\}$ for different settings of M in (3.1). Thus, existing software for these and other methods of metric MDS provide the required display of the samples. An example of sample display based on the stress criterion is shown in Fig. 3.3 and, as can be seen, it is of precisely the same nature as that of PCA. Thus the main outstanding problem is to discuss how biplot methodology can be extended to supply information on the variables.

First, we establish a few algebraic results that follow from (3.4) and (3.6). Note that for any orthogonal matrix $\mathbf{Q}$, $\mathbf{ZQ}$ may replace $\mathbf{Z}$ as solutions, because $\{\delta_{ij}\}$ is invariant to orthogonal transformations. Also, from the definitions of $\mathbf{F}$ and $\mathbf{G}$ we have that $\mathbf{F1} = \mathbf{G1} = \mathbf{0}$; it follows that any translation of $\mathbf{Z}$ to $\mathbf{Z} + \mathbf{1m}'$ is also a solution of (3.4). Thus solutions are invariant to translations and therefore we may assume that the origin of $\mathbf{Z}$ is at its centroid and so satisfies $\mathbf{1}'\mathbf{Z} = \mathbf{0}$. Similar remarks apply to $\mathbf{G}$ and solutions of (3.6).

With the generating coordinates centred at their centroid we can express $\mathbf{ZZ}'$ as $\mathbf{ZZ}' = (\mathbf{I} - \mathbf{N})\boldsymbol{\Delta}(\mathbf{I} - \mathbf{N})$, where $\boldsymbol{\Delta} = \{-\frac{1}{2}\delta_{ij}^2\}$ (A.53, A.54). From (3.4), $\mathbf{FZZ}' = \mathbf{0}$ so

$$\mathbf{F}(\mathbf{I} - \mathbf{N})\boldsymbol{\Delta}(\mathbf{I} - \mathbf{N}) = \mathbf{0}$$

which simplifies to

$$\mathbf{F}\boldsymbol{\Delta}(\mathbf{I} - \mathbf{N}) = \mathbf{0} \tag{3.8}$$

Equation (3.8) is expressed entirely in terms of $\boldsymbol{\Delta}$, which is satisfactory because it does not require $\mathbf{Z}$ with its arbitrary rotations and translations. Once $\boldsymbol{\Delta}$ is known, $\mathbf{Z}$ may be evaluated as any decomposition satisfying $\mathbf{ZZ}' = (\mathbf{I} - \mathbf{N})\boldsymbol{\Delta}(\mathbf{I} - \mathbf{N})$. (Even though one is fitting the distances in $\boldsymbol{\Delta}$, existing algorithms for minimizing *stress* work by iterating on the ρ-dimensional coordinates $\mathbf{Z}$ that generate $\boldsymbol{\Delta}$; unfortunately, despite the suggestive form of (3.8), it has not yet led to an algorithm which works by iterating directly on $\boldsymbol{\Delta}$.) Taking the trace of (3.8) gives

$$\text{trace } \mathbf{F}\boldsymbol{\Delta}(\mathbf{I} - \mathbf{N}) = \text{trace } \mathbf{F}\boldsymbol{\Delta} = \mathbf{0}$$

or

$$\sum_{i<j}^{n} (d_{ij}\delta_{ij} - \delta_{ij}^2) = 0$$

from which we obtain the analysis of variance

$$\sum_{i<j}^{n} (d_{ij}^2) = \sum_{i<j}^{n} (\delta_{ij}^2) + \sum_{i<j}^{n} (d_{ij} - \delta_{ij})^2 \tag{3.9}$$

expressing the total sum-of-squares of distances as equal to the fitted sum-of-squares of distances plus the sum-of-squares of residuals. Similar arguments give

$$\text{trace } \mathbf{G}\boldsymbol{\Delta} = \mathbf{0}$$

or

$$\sum_{i<j}^{n} (d_{ij}^2 \delta_{ij}^2 - \delta_{ij}^4) = 0$$

and the analysis of variance

$$\sum_{i<j}^{n} (d_{ij}^4) = \sum_{i<j}^{n} (\delta_{ij}^4) + \sum_{i<j}^{n} (d_{ij}^2 - \delta_{ij}^2)^2 \tag{3.10}$$

Compare (3.9) and (3.10) with (2.1), which gives the equivalent analysis of variance for PCA where $\{\delta_{ij}\}$ is obtained from projections; the resemblance is striking, although we must remember that δ_{ij} obtained by the different methods are not the same.

3.3 Prediction

Suppose $\mathbf{z}_{n+1}$ is a point in the space $\mathscr{L}$, spanned by the columns of $\mathbf{Z}$, with distances from the n points of $\mathscr{L}$ given by $\delta_{n+1} = \{\delta_{i,n+1}\}$. We wish to predict $\mathbf{x}_{n+1}$, which generates distances $\mathbf{d}_{n+1} = \{d_{i,n+1}\}$, the corresponding distances in $\mathscr{R}$. Thus, the interpolant is the value of $\mathbf{z}$ which minimizes

$$S_M = S(\tau(d_{i,n+1}), \delta_{i,n+1}) \tag{3.11}$$

giving rise to similar normal equations to those discussed in section 3.2. When the function in the MDS criterion (3.1) is the same as the function in (3.11), the prediction is said to be **coherent**. In PCA, $\mathscr{L}$ is a subspace of $\mathscr{R}$ and then S_M may be made zero by choosing $\mathbf{x}_{n+1}$ to be the same as $\mathbf{z}_{n+1}$ expressed in the coordinate system of $\mathscr{R}$. In general, the problem of minimizing (3.11) with respect to $\mathbf{x}_{n+1}$ is the converse of the MDS problem and, so far as we are aware, it has not been studied. Hence, we have to fall back on approximate solutions which are not coherent, i.e. the criteria differ for the MDS step and for the prediction step (section 3.4). Two incoherent methods are discussed in the following.

3.3.1 *Procrustean prediction*

In the above, no assumption of Pythagorean distance was needed. These distances can be calculated in any way or may even be observed directly; indeed distances need not even be Euclidean.

However, recall that, in this chapter, we are assuming that distances are Pythagorean. Although there has been no overt mention of the data-matrix **X**, it has appeared implicitly in assuming that the distances $(d_{1,n+1}, d_{2,n+1}, \ldots, d_{n,n+1})$ can be calculated. To predict what values of variables should be associated with a point in the MDS, it is essential that there be some link between the space of **Z** and that of **X**. With PCA we saw, in Chapter 2, that this link was simple, because $\mathbf{Z}_{\text{PCA}}$ is defined to be in a sub-space of **X**. Note that $\mathbf{Z}_{\text{PCA}}$ has two forms: $\mathbf{XV}_\rho\mathbf{V}'_\rho$, given ρ-dimensional coordinates relative to the original p axes; and $\mathbf{XV}_\rho$, giving coordinates relative to the ρ principal axes (section A.8). The Cartesian coordinate axes to which **X** is referred are the basis of prediction; every point of $\mathbf{Z}_{\text{PCA}}$ predicts the values of its coordinates $\mathbf{XV}_\rho\mathbf{V}'_\rho$ relative to these Cartesian axes. Further, the Eckart–Young theorem shows that the $\mathbf{Z}_{\text{PCA}}$ is oriented such that $\|\mathbf{X}-\mathbf{Z}\|$ is minimized over all matrices **Z** of rank ρ. This suggests that the **Z** found by other methods of metric scaling should also be oriented to fit **X** optimally. Now, **Z** has only ρ columns but we are interested in the location in the p-dimensional space so that the columns of **Z** should be augmented by p-ρ zero columns or any other values that have no effect on the distances generated by the rows of **Z**; indeed, any variant of **Z** may be used that represents different initial orientations/translations of **Z** without affecting the final optimal orientation. With such a p-dimensional form of **Z**, we seek an orthogonal matrix **Q** representing the optimal orientation, such that $\|\mathbf{X}-\mathbf{ZQ}\|$ is minimized. The reorientation **ZQ** does not affect the fitted distances $\boldsymbol{\Delta}$ of the MDS. The minimization is that of **Orthogonal Procrustes Analysis** with the solution $\mathbf{Q}=\mathbf{VU}'$ where $\mathbf{U\Sigma V}'$ is the SVD of $\mathbf{X}'\mathbf{Z}$ (section A.10).

Thus we have minimized the residual sum-of-squares conditional on the given $\boldsymbol{\Delta}$, while PCA minimizes the criterion globally. Hence, here we are concerned with a form of the Eckart–Young theorem which gives a best least-squares rank ρ approximation **ZQ** to **X** conditional on the approximation generating the distances $\boldsymbol{\Delta}$ in ρ dimensions given by a chosen form of MDS. At the end of this process, we have embedded the approximation space $\mathscr{L}$ in the space $\mathscr{R}$ of **X**. It is shown below that this means that precisely the same form of linear predictive biplot can be used as for PCA. The only difference is that the back-projections are now onto a different subspace. In particular, because we are assuming that **X** is referred to orthogonal Cartesian axes which induce Pythagorean distances, then the biplot axes are the projections of the Cartesian axes and hence are linear.

Geometry of linear biplot prediction

In this section, we justify the statement made earlier that the mechanism for predictive biplots in PCA is applicable to any linear subspace and not just to the space spanned by the ρ principal eigenvectors of $\mathbf{X}'\mathbf{X}$. Recall that we are assuming Cartesian axes and suppose that these span a Euclidean space $\mathscr{R}$ and that one of these axes, ξ_k say, is associated with one of the variables of $\mathbf{X}$. The key observation is that if x is a scale-marker on ξ_k, then all points in $\mathscr{R}$ with the value x on that variable lie in the plane $\mathscr{N}$ that is normal to ξ_k at x. This is no profound statement; it is merely the basic component in the conventional way of positioning a point $(x_1, x_2, \ldots, x_p)$ as the intersection of the p normal planes $\mathscr{N}_1, \mathscr{N}_2, \ldots, \mathscr{N}_p$. Suppose now that $\mathscr{L}$ is any linear ρ-dimensional subspace of $\mathscr{R}$, then all points in $\mathscr{L}$ with the value x on ξ_k lie in the intersection space $\mathscr{L} \cap \mathscr{N}$; these are the points in $\mathscr{L}$ that predict x. Figure 3.1 shows this geometry.

Different values of x on ξ_k will define different parallel intersection spaces. Prediction for any point in $\mathscr{L}$ is a matter of deciding in which intersection space the point lies and then predicting the associated value of x. We may draw any line in $\mathscr{L}$ and mark on it the different values of x at the point where this line cuts the successive intersection spaces and, provided we know the angle of intersection, the complete intersection spaces may be reconstructed just from this information. The most simple line to choose is the one that is orthogonal to the intersection spaces; this line is indicated in Fig. 3.1 with the label β_k. The marker is at the shortest distance from O of any point in $\mathscr{L} \cap \mathscr{N}$; it is also at the shortest distance from the corresponding marker on ξ_k and hence the orthogonalities indicated in the figure. Prediction is achieved by orthogonally projecting onto β_k and reading off the marker as shown in Fig. 3.1. β_k is the predictive biplot axis corresponding to the kth variable associated with the Cartesian axis ξ_k. In this way, each Cartesian axis defines an associated biplot axis in $\mathscr{L}$. For prediction, they are used just like conventional axes, i.e. by projection onto them and reading off the marker. The markers give scale values just as with ordinary graphs and as usual, to avoid clutter, judgement has to be used in deciding how densely they should be plotted. When $\mathscr{L}$ is obtained from an MDS, all information necessary for prediction lies in $\mathscr{L}$ and $\mathscr{R}$ is no longer needed. Note that when P in Fig. 3.1 has arisen from a sample approximated in an MDS, then although 5 is predicted for the value of its kth variable, this will rarely be correct. For example, in PCA, P arises from the projection of some point R, say, in $\mathscr{R}$ which is normally outside $\mathscr{N}$ and cannot have value 5; the only exception is when R itself lies in $\mathscr{L} \cap \mathscr{N}$.

The coordinates of the markers

The coordinates of the markers on β_k can be found as follows. The equation of $\mathscr{N}$ corresponding to a marker μ on ξ_k is

$$\mathbf{x}\mathbf{e}_k = \mu \tag{3.12}$$

The point $\mathbf{z}$ that we require lies in both $\mathscr{N}$ and $\mathscr{L}$, so we have that

$$\mathbf{z}\mathbf{e}_k = \mu \quad \text{and} \quad \mathbf{z} = \mathbf{b}\mathbf{V}' \tag{3.13}$$

where $\mathbf{V}$ is a $p \times \rho$ matrix whose columns, assumed orthonormal without loss of generality, span $\mathscr{L}$, and $\mathbf{b}$ is a vector of coefficients that are to be chosen so that $\mathbf{z}\mathbf{z}'$ is minimized. Thus we must minimize $\mathbf{z}\mathbf{z}' = \mathbf{b}\mathbf{b}'$ subject to $\mathbf{b}\mathbf{V}'\mathbf{e}_k = \mu$, giving

$$\mathbf{b} = \lambda \mathbf{e}'_k \mathbf{V}$$

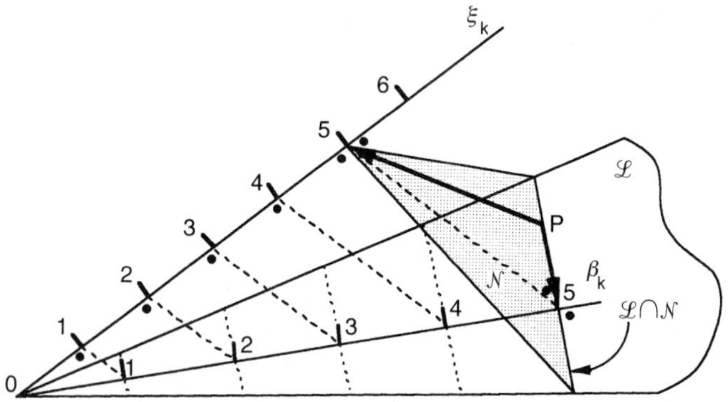

Fig. 3.1. *ξ_k is the kth Cartesian coordinate axis with scale-markers, where $x = 5$ is selected for illustration. Parallel intersection spaces marked by dotted lines induce related markers on β_k as shown. (Right angles are indicated by black circular dots).*

where λ is a Lagrange multiplier determined from $\mu = \lambda \mathbf{e}_k' \mathbf{V}\mathbf{V}' \mathbf{e}_k$. Thus, finally

$$\mathbf{z} = \frac{\mu}{\mathbf{e}'_k \mathbf{V}\mathbf{V}' \mathbf{e}_k} \mathbf{e}'_k \mathbf{V}\mathbf{V}' \tag{3.14}$$

This confirms that equally spaced markers μ on ξ_k induce equally spaced markers on β_k. Also, the ordinary orthogonal projection $\mathbf{p}$ of $\mu\mathbf{e}_k$ onto $\mathscr{L}$ is given by $\mathbf{p} = \mu\mathbf{e}_k'\mathbf{V}\mathbf{V}'$ so (3.14) defines a point on the same vector from the origin. This confirms that β_k is the orthogonal projection of ξ_k onto $\mathscr{L}$ but that (3.14) differs from the ordinary projection of $\mu\mathbf{e}_k$; in fact, (3.14) is the simplest example of a back-projection (section A.9) of a point from one subspace, ξ_k, onto another, $\mathscr{L}$, a concept which is generalized in Chapter 7. Note that the unit markers ($\mu = 1$) for projection and back-projection are inversely related, for $\mathbf{p}\mathbf{p}' = \mathbf{e}'_k \mathbf{V}\mathbf{V}' \mathbf{e}_k$ while

$$\mathbf{z}\mathbf{z}' = \frac{1}{\mathbf{e}'_k \mathbf{V}\mathbf{V}' \mathbf{e}_k}$$

These results are not quite in a form suitable for practical use. This is because they refer to the p-dimensional coordinate system of $\mathscr{R}$ and not to the ρ dimensions $\mathscr{L}$ which contains the results of the MDS and the biplot axes β_k $(k = 1, 2, \ldots, p)$. Indeed, the precise coordinate system of $\mathscr{L}$ has been left vague. With PCA, it is given by the column-vectors of $\mathbf{V}_\rho$, the ρ principal axes of $\mathbf{X}$ (or more precisely of $\mathbf{X}'\mathbf{X}$) but, in the above, $\mathbf{V}$ is any orthonormal set of vectors that span $\mathscr{L}$; in the notation of section 3.3.1 $\mathbf{V}$, for prediction, is given by the first ρ columns of the matrix $\mathbf{Q}'$ derived from the Procrustean fit, where the transpose occurs because we are now rotating $\mathbf{X}$ to fit $\mathbf{Z}$ rather than $\mathbf{Z}$ to fit $\mathbf{X}$, as previously. The orientation of $\mathbf{Z}$ itself is arbitrary, but commonly $\mathbf{Z}$ is referred to its principal axes (or more precisely those of $\mathbf{Z}'\mathbf{Z}$) not for any essential reason but rather to give a unique representation; different orientations of $\mathbf{Z}$ are all taken care of by the Procrustean fit and make no essential difference to the final graphical plot, affecting only its orientation in $\mathscr{L}$. Relative to these axes, the projection of $\mu\mathbf{e}_k$ and its back projection onto $\mathscr{L}$ are respectively given by

$$\mu\mathbf{V}'\mathbf{e}_k \quad \text{and} \quad \frac{\mu}{\mathbf{e}'_k \mathbf{V}\mathbf{V}' \mathbf{e}_k} \mathbf{V}'\mathbf{e}_k \tag{3.15}$$

These are the values needed for plotting markers on the kth axis in practical ρ-dimensional graphical interpolative and predictive biplots, respectively. These results have been developed for prediction and, in section 3.4, it is suggested that, while retaining the same formulae, a better setting of **V** can be derived for interpolation.

Because $\mathscr{L}$ can be any linear subspace of $\mathscr{R}$, this geometry applies equally to the subspace given by PCA and to the Procrustean embedded subspaces described in section 3.3.1 for general methods of MDS that are based on Pythagorean distance. The argument given above is the most simple example of a generalization of this geometry described in Chapters 6 and 7 which underpins more general forms of biplot.

3.3.2 Regression biplots – prediction

An alternative approach arises from the observation that one is trying to predict **X** given **Z** and it is then natural to do this by using the multiple regression $\mathbf{X} = \mathbf{Z}\boldsymbol{\beta}'$, with the usual estimate $\hat{\boldsymbol{\beta}}' = (\mathbf{Z}'\mathbf{Z})^{-1}\mathbf{Z}'\mathbf{X}$. With PCA, we have that $\mathbf{Z} = \mathbf{X}\mathbf{V}_\rho$ so that $\hat{\boldsymbol{\beta}} = \mathbf{V}_\rho$, correctly predicting that $\hat{\mathbf{X}} = \mathbf{Z}\mathbf{V}_\rho' = \mathbf{X}\mathbf{V}_\rho\mathbf{V}'_\rho$. Indeed, plotting the p rows of $\boldsymbol{\beta}$ onto the space of **Z** gives the unit points for interpolation on linear biplot axes. It follows from (3.15), and also the discussion of prediction in Chapter 2, that the unit points for prediction can be obtained by inversion. Thus, in the context of PCA, this regression approach correctly gives the biplot axes and their scale-markers and interpretation is as already described.

Precisely the same process can be used even when **Z** is given by other methods of MDS and it can be hoped that the method of interpretation remains, at least approximately, valid. Indeed, the prediction for a position **z** in $\mathscr{L}$ is given by $\mathbf{x} = \mathbf{z}\hat{\boldsymbol{\beta}}'$ and, for the kth variable, by $x_k = \mathbf{z}\hat{\boldsymbol{\beta}}'_k$, where $\hat{\boldsymbol{\beta}}'_k$ is the transpose of the kth row of $\boldsymbol{\beta}$. Further, all points **z** predicting the same value x_k lie in the plane $x_k = \mathbf{z}\hat{\boldsymbol{\beta}}'_k$ whose normal $\hat{\boldsymbol{\beta}}_k$ is in the direction of the biplot axis. It follows that graphical prediction remains by orthogonal projection onto the axis and reading off an appropriate scale marker. To establish the appropriate scale, we note that the setting $\mathbf{z} = \hat{\boldsymbol{\beta}}_k/(\hat{\boldsymbol{\beta}}_k\hat{\boldsymbol{\beta}}'_k)$ predicts the unit value $x_k = 1$ for the kth variable and may be used to establish a scale for the kth biplot axis as described in section 2.6; this is essentially the result on predictive scales found in section 2.3. Thus predictive biplots obtained by regression are used in the same way as those previously described. The regression biplot method is widely used and is mentioned by Kruskal and Wish (1978), but its origins are obscure.

Multiple regression minimizes $\|\mathbf{X}-\mathbf{Z}\boldsymbol{\beta}\|$ which can be compared with section 3.3.1 which minimizes $\|\mathbf{X}-\mathbf{Z}\mathbf{Q}\|$. Whereas $\boldsymbol{\beta}$ is a general matrix, $\mathbf{Q}$ is restricted to being orthogonal. It follows that the regression method must give the smaller residual sum-of-squares but it does not follow that it gives a smaller value of (3.11). The relative merits of the two approaches need further evaluation (section 3.4.3 gives comments on regression biplots and interpolation); in the example discussed in section 3.5, there is little to choose between the two methods.

3.3.3 Multiple axes

We close this section with some remarks which we think are interesting, and may help understanding, but which give graphical plots which we regard as curiosities rather than being of serious practical importance. It has been pointed out that biplot axes need not meet the intersection spaces orthogonally and this fact can be exploited to allow two, or indeed more, variables to be referred to the same biplot axis. Figure 3.2 gives an example for two variables h and k in a two-dimensional biplot with a single double-biplot axis β_{hk}. The dotted lines show the parallel intersection spaces for the two variables, two of which have been labelled, one for the value 2 of h and the other for the value 3 of k. (For practical use, only β_{hk} with its scale and the directions of projection are needed.) The interpretation of this biplot requires some form of icon giving the non-orthogonal directions of projection for each variable; it is sufficient

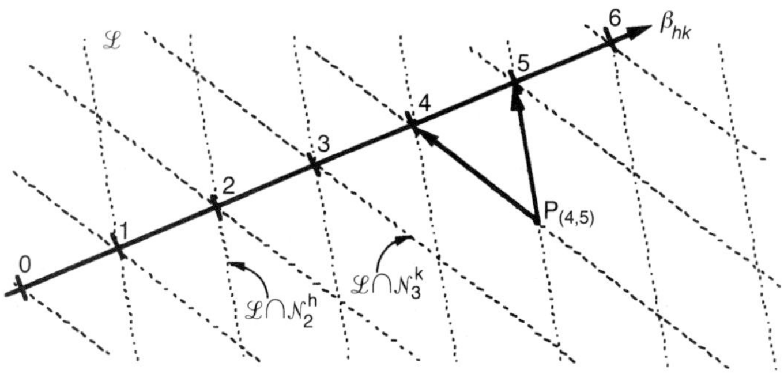

Fig. 3.2. *A single biplot axis β_{hk} for two variables, h and k in the approximation space $\mathscr{L}$.*

to provide this for one point, such as P(4, 5), in the figure. The points on β_{hk} itself predict $x_h = x_k$. In general, a ρ-dimensional biplot can be presented in a similar manner with ρ variables using the same scale markers on a single axis; p variables may be represented in ρ dimensions with a single axis, provided, $p-1$ different scales are marked. Another possibility is to use several multiple axes; for example, we could represent four variables with two axes β_{12} and β_{34}. These ideas are not confined to biplot axes; similar devices could replace conventional coordinate systems.

3.4 Interpolation

With the same notation as in section 3.3, suppose we have an $(n+1)$th sample $\mathbf{x}_{n+1}$ and know its distances $\mathbf{d}_{n+1} = \{d_{i,n+1}\}$ from the original n samples, then its coordinates $\mathbf{z}_{n+1}$ can be placed in the space of $\mathbf{Z}$ by solving equations that are special case of (3.1). The interpolant is the value of $\mathbf{z}_{n+1}$ which minimizes (3.11). Note that, for prediction, we solve (3.11) for $\mathbf{x}_{n+1}$ given $\mathbf{z}_{n+1}$ while interpolation requires the solution to the converse problem. Thus, the process of interpolating is similar to MDS and gives rise to similar normal equations to those discussed in section 3.1. When the function in the MDS method is the same as the function in (3.11), the interpolation is said to be **coherent.** Equation (3.11) usually has only algorithmic solutions and therefore graphical biplots for interpolation are not coherent. Non-coherent graphical interpolation methods are discussed in sections 3.4.2 and 3.4.3.

3.4.1 *Interpolation with* stress *and* sstress

For the *stress* criterion (3.2) we require the minimization of

$$\sum_{i=1}^{n}(d_{n+1,i} - \delta_{n+1,i})^2$$

with respect to $\mathbf{z} = (z_{n+1,1}, z_{n+1,2}, \ldots, z_{n+1,\rho})$ and similarly for *sstress*. This minimization ensures that when $\mathbf{x}_{n+1}$ coincides with one of rows of $\mathbf{X}$, then $\mathbf{z}$ is given by the corresponding row of $\mathbf{Z}$.

The normal equations, essentially an extra row to (3.4) and (3.6), are found to be

$$\mathbf{z} = \frac{\mathbf{f}'\mathbf{Z}}{\mathbf{f}'\mathbf{1}} \quad \text{and} \quad \mathbf{z} = \frac{\mathbf{g}'\mathbf{Z}}{\mathbf{g}'\mathbf{1}} \tag{3.16}$$

where $f_i = (d_{n+1,i} - \delta_{n+1,i})/\delta_{n+1,i}$ and $g_i = (d^2_{n+1,i} - \delta^2_{n+1,i})$. Note that $\mathbf{f}'\mathbf{1}$ and $\mathbf{g}'\mathbf{1}$ arise from the equivalent of the diagonal terms defined in (3.5) and (3.7) and that $\mathbf{Z}$ is given. The interpolates $\mathbf{z}$ are weighted means of the coordinates $\mathbf{Z}$, the weights being the elements of the vector

$$\mathbf{s} = \frac{\mathbf{f}}{\mathbf{f}'\mathbf{1}} \quad \text{and} \quad \mathbf{s} = \frac{\mathbf{g}}{\mathbf{g}'\mathbf{1}}$$

respectively, where $\mathbf{s}'\mathbf{1} = 1$. Because $\mathbf{f}$ and $\mathbf{g}$ involve the unknowns $\delta_{n+1,j}$ which depend on $\mathbf{z}$, this precludes the use of simple vector-sums for graphical interpolation. Indeed, it seems unlikely that these interpolation formulae admit any direct graphical interpretation but the core of any algorithm for minimizing *stress* or *sstress*, as appropriate, may be used to do the computations and support interactive computer graphics.

With current knowledge, the core part of algorithms for solving (3.8), and the corresponding equations for minimizing *sstress*, must be used; standard metric scaling software usually provides facilities for this kind of interpolation. Another approach that is worth consideration arises from rewriting (3.16) as $\mathbf{z} = \mathbf{s}'\mathbf{Z}$. We may temporarily move the origin to $\mathbf{z}$ by taking any decomposition of $(\mathbf{I} - \mathbf{1}\mathbf{s}')\mathbf{\Delta}(\mathbf{I} - \mathbf{s}\mathbf{1}')$. The diagonal of this matrix (A.52) gives the ddistances $\delta = \{\delta^2_{n+1,i}\}$ of the fitted distances of $\mathbf{z}$ from the n original fitted values in $\mathscr{L}$. Thus

$$(\mathbf{s}'\mathbf{\Delta}\mathbf{s})\mathbf{1} - 2\mathbf{\Delta}\mathbf{s} = \boldsymbol{\delta}$$

Pre-multiplication by $\mathbf{s}'$ shows that

$$(\mathbf{s}'\mathbf{\Delta}\mathbf{s}) = -\mathbf{s}'\boldsymbol{\delta}$$

and hence

$$(\mathbf{s}'\boldsymbol{\delta})\mathbf{1} + 2\mathbf{\Delta}\mathbf{s} = -\boldsymbol{\delta} \tag{3.17}$$

Equation (3.17) applies both to *stress* and *sstress*, with the obviously different definitions of $\mathbf{s}$. It is interesting to speculate that (3.17) might provide an iterative algorithm for computing **s**, and hence **z**, but this needs investigation.

3.4.2 Interpolation biplots by projection

Having seen how **Z** can be embedded in the space of **X** for prediction, raises the question whether something similar may be done for interpolation. Indeed, the Procrustean embedding of section 3.3.1 developed for prediction has also led to an interpolative biplot whose unit markers are given in (3.15). Except with PCA, any **Z** obtained by a metric scaling method will not normally be a projection of **X**. However, we can seek the orientation **ZQ** that is nearest to being a projection of **X**. This is the **minimal error projection Procrustes problem** (see Gower, 1994 for a discussion of and brief history of this problem) which requires the minimization of $\|\mathbf{Z} - \mathbf{XP}\|$ where **P** is a $p \times \rho$ projection matrix with orthonormal columns; an algorithmic solution is given in section A.10. What this means is that **X** in the p-dimensional space $\mathscr{R}$ is projected onto the smaller ρ-dimensional space $\mathscr{L}$ that contains **Z**. Then we may write the orthogonal matrix

$$\mathbf{Q}' = (\mathbf{P}, \bar{\mathbf{P}})$$

where $\bar{\mathbf{P}}$ is a matrix of any orthonormal columns which are orthogonal to the columns of **P**; the columns of $\bar{\mathbf{P}}$ refer to dimensions of **X** that are orthogonal to the space of **Z** and so are irrelevant for projection onto the space of **Z**. Just as with PCA, we can project the Cartesian axes onto **Z**, so oriented, to give linear biplot axes and use the vector-sum method of interpolation. The rows of **P** give the coordinates of the unit points in **Z** for the p interpolation biplot axes. Except in PCA, **Z** does not match with any projection of **X**; even the original samples will not interpolate into their correct positions as given by the formulae of section 3.4.1. Thus, the method is not so accurate as the algebraic solution discussed in section 3.4.1 but it is more simple and allows paper-and-pencil graphical interpolation, using vector-sums. When MDS is by PCA, **P** is given by the first ρ eigenvectors of $\mathbf{X}'\mathbf{X}$ and the methods coincide. As yet, we have little information on the accuracy of the method when used with other methods of metric scaling, but see the example of section 3.5.

3.4.3 *Regression biplots – interpolation*

It was shown in section 3.3.2 that the regression method can be used to give predictive biplot axes, which are correct for PCA. The regression method gives the same directions for both interpolation and prediction and therefore, except for PCA, at least one of these must be suboptimal. The degree of approximation needs investigation but we know that, in general, neither the predictive nor interpolative forms of (3.11) are satisfied by regression biplots. In Chapter 6, which deals with nonlinear biplot axes, the regression method may still be used as an approximation which gives linear biplots but its status is then very uncertain and the convenience of linearity could be a high price to pay.

3.5 Examples

This example uses data originally collated by Lyons (1980) and subsequently published by Krzanowski (1988) and Hand *et al.* (1994). It gives the yields of winter wheat in the four years 1970–1973 at 12 sites. The yields are shown in Table 3.1. The abbreviated names given in the first column are used to label the diagrams.

Table 3.1 Yields of winter wheat (kgm per unit area) at 12 sites for the years 1970–3

	Site	*1970*	*1971*	*1972*	*1973*
CA	Cambridge	46.81	39.40	55.64	32.61
CP	Cockle Park	46.49	34.07	45.06	41.02
HA	Harper Adams	44.03	42.03	40.32	50.23
HH	Headley Hall	52.24	36.19	47.03	34.56
MO	Morely	36.55	43.06	38.07	43.17
MY	Myerscough	34.88	49.72	40.86	50.08
RO	Rosemaund	56.14	47.67	43.48	38.99
SH	Seale-Hayne	45.67	27.30	45.48	50.32
SP	Sparsholt	42.97	46.87	38.78	47.49
SB	Sutton Bonnington	54.44	49.34	24.48	46.94
TE	Terrington	54.95	52.05	50.91	39.13
WY	Wye	48.94	48.63	31.69	59.72
	Mean Yield	47.009	43.028	41.817	44.522

A two-dimensional least-squares scaling had a normalized stress value of 0.1194 and gave the coordinates listed in Table 3.2. Thus Table 3.1 gives the matrix **X** (in uncentred form) and Table 3.2 gives **Z**. When **X** is rotated to fit **Z** by orthogonal Procrustes analysis, the rotation matrix **Q** is found to be

1970	−0.1141	−0.7357	−0.6677	0.0000
1971	0.4338	−0.5703	0.5543	0.4234
1972	−0.6903	0.0596	0.0522	0.7192
1973	0.5677	0.3605	−0.4942	0.5509

The first two columns $\mathbf{Q}_2$ of **Q** give the directions of the predictive biplot axes in the two-dimensional least squares scaling. The scales to be attached are found in precisely the same way as for PCA (sections 2.3 and 2.6) giving the predictive biplot shown in Fig. 3.3.

We pointed out in section 3.4 that coherent interpolation cannot be done graphically but needs the computational method of section 3.4.1. In section 3.4.2, we gave an approximate method based on minimal error projection Procrustes analysis.

The projection matrix **P** giving the best projection of the four-dimensional data onto the two-dimensional least-squares scaling fit was found to be

1970	−0.1761	−0.7675
1971	0.4899	−0.5434
1971	−0.6777	0.0626
1972	0.5194	0.3341

Table 3.2 The coordinates given by a least-squares scaling of the data of Table 3.1

CA	−18.987	−3.023
CP	−8.731	4.569
HA	3.670	3.329
HH	−12.987	−2.614
MO	3.743	9.500
MY	10.598	10.372
RO	−2.954	−10.778
SH	−9.138	15.091
SP	6.320	1.453
SB	15.830	−12.435
TE	−6.872	−14.641
WY	19.508	−0.823

These give the directions of interpolative biplot axes to which scales can be attached as described in sections 2.3 and 2.6. The resulting biplot is shown in Fig. 3.4. The black squares give the positions of the sites as in Fig. 3.3, and the black circles are the positions of the sites as given by a minimal error projection Procrustes fit.

The vector-sum method of interpolation for the sites recovers the minimal error projection positions and as these are satisfactorily close to the least-squares scaling positions, the approximation to coherent interpolation is very good in this case and can be expected to be adequate when interpolating new sites by the vector-sum method.

Finally, we demonstrate the regression method. For each variable, we require the regression coefficients of the raw data **X** (Table 3.1) on the least-squares scaling coordinates **Z** (Table 3.2). This gives the matrix **B** containing the following regression coefficients

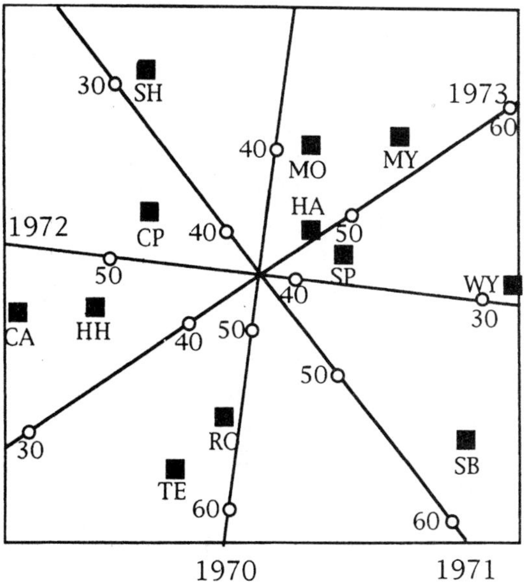

Fig. 3.3. *Yields of winter wheat at 12 sites in four years: linear predictive biplot for least squares scaling obtained by the orthogonal Procrustes method.*

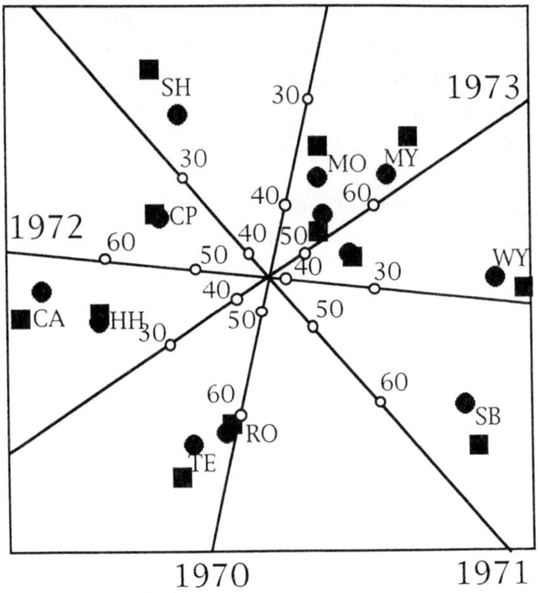

Fig. 3.4. *Approximate interpolative linear biplot for least squares scaling.*

1970	−0.1143	−0.6087
1971	0.3887	−0.4604
1972	−0.6301	0.0339
1973	0.5238	0.3096

from which, we obtain the unit points for prediction by the method of Section 3.3.2

1970	−0.2980	−1.5869
1971	1.0706	−1.2681
1972	−1.5825	0.0851
1973	1.4148	0.8363

The scales are derived as before, using the methods of sections 2.6 and 2.3 and the resulting biplot is shown in Fig. 3.5. It can be seen that the directions of the axes are very similar to those of the Procrustes biplot of Fig. 3.3 but with slightly lengthened scales.

In this book, we are primarily concerned with multivariate graphical representations but it is interesting to see the numerical

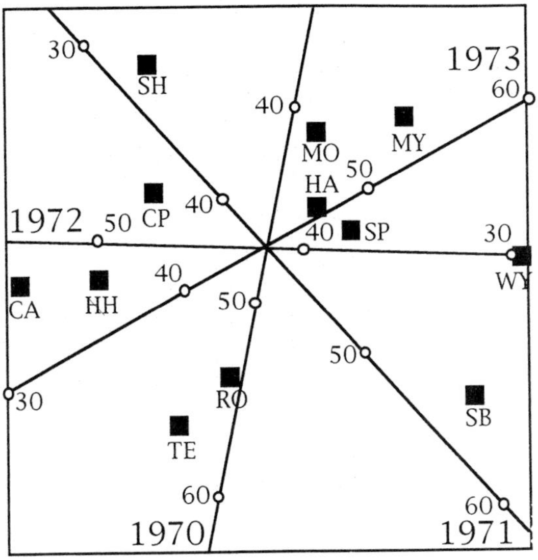

Fig. 3.5. *Predictive linear biplot for least-squares scaling obtained by the regression method.*

values of the predictions. Algebraically, the orthogonal Procrustes predictions are given by $\mathbf{ZQ}_2'$ and the predictions based on regression by $\mathbf{ZB}'$. The first two columns of $\mathbf{Q}$ are not very different from the columns of $\mathbf{P}$, as is evident from examining the direction of the axes in Figs. 3.3 and 3.4. It follows that the directions given by $\mathbf{P}$, with a suitable prediction scale attached, might be used for prediction projection Procrustes by $\mathbf{ZP}'$. The predictions for all these methods are shown in Table 3.3 and may be compared with the original data of Table 3.1. These predicted values give the values more accurately than can be determined graphically. As one would expect from the general agreement of the biplots, the differences in the predictions are very minor. The residual sums-of-squares are

$$\|\mathbf{X} - \mathbf{ZB}\| = 600.435$$
$$\|\mathbf{X} - \mathbf{ZQ}\| = 642.467$$
$$\|\mathbf{X} - \mathbf{ZP}\| = 658.080$$

Table 3.3 Predicted yields: **Q** by the orthogonal Procrustes method; **B** by the regression method and **P** by the minimum error projection Procrustes method

Site	*1970*			*1971*			*1972*			*1973*		
	Q	*B*	*P*	*Q*	*B*	*P*	*Q*	*B*	*P*	*Q*	*B*	*P*
CA	51.40	51.02	52.67	36.51	37.04	35.37	54.74	53.68	54.49	32.65	33.64	33.65
CP	44.64	45.23	45.04	36.63	37.53	36.27	48.12	47.47	48.02	41.21	41.36	41.51
HA	44.14	44.56	43.81	42.72	42.92	43.02	39.48	39.62	39.54	47.81	47.47	47.54
HH	50.41	50.08	51.30	38.88	39.18	38.09	50.63	49.91	50.45	36.21	36.91	36.90
MO	39.59	40.80	39.06	38.23	40.11	39.70	39.80	39.78	39.87	50.07	49.42	49.64
MY	38.17	39.48	37.18	41.71	42.37	42.58	35.12	35.49	35.28	54.28	53.28	53.49
RO	55.28	53.91	55.80	47.89	46.84	47.44	43.21	43.31	43.14	38.96	39.64	39.39
SH	36.95	38.87	37.04	30.46	32.53	30.35	49.02	48.09	48.95	44.77	44.41	44.82
SP	45.22	45.40	44.78	44.94	44.82	45.33	37.54	37.88	37.62	48.63	48.28	48.29
SB	54.35	52.77	53.77	56.99	54.91	57.54	30.15	31.42	30.31	49.03	48.96	48.59
TE	58.56	56.71	59.46	48.40	47.10	47.62	45.69	45.65	45.56	35.34	36.39	36.06
WY	45.39	45.28	44.21	51.96	50.99	53.03	28.30	29.50	28.54	55.30	54.48	54.38

CHAPTER 4

Multiple correspondence analysis

4.1 Introduction

It may seem perverse to discuss multiple correspondence analysis (MCA) before classical correspondence analysis (CA) (Chapter 9) but there is a good reason. MCA is concerned with observations on p categorical variables for each of n samples and may be viewed as a form of principal components analysis (PCA) applicable to categorical rather than quantitative variables. Indeed, we shall introduce the method by stressing its relationship with PCA so that its discussion follows naturally those of Chapters 2 and 3. However, this is but one way of proceeding and certainly not the usual one, which is to regard MCA as the CA of a two-way contingency table formally applied to the coded indicator matrix (section 4.2). Conversely, classical CA may be regarded as a variant of MCA when $p = 2$. This interplay between CA and MCA raises several difficult issues which cannot be resolved until Chapter 10. Our primary intention here is to explore the biplot properties of MCA but other derivations which emphasize different properties will be presented later in this chapter.

Rectangular axes for quantitative variables are very familiar and we have seen that they can be approximated in low-dimensional representations by linear non-orthogonal biplot axes. Are there corresponding representations for categorical variables? If we consider a categorical variable *colour* with category-levels 'red', 'blonde', 'brown', and 'black' (say) then these do not form a continum and so cannot be represented on an axis. An ordered categorical variable like *wealth*, with levels 'rich', 'well-off', 'poor' and 'bankrupt', might be represented as a set of four points on a linear axis but there would be no special reason to space them evenly. We shall not here be concerned with ordered categories but shall represent unordered categories by non-collinear points, one for each category-level; these points will therefore be termed **category-level-points** (CLPs). It will be shown that the simplex formed by the set of CLPs associated with a categorical variable has many of the properties of a coordinate axis. One way of choosing the positions of these points will be discussed in section 4.3.2 and another in section 4.5.

4.2 The indicator matrix

It is convenient to represent categorical variables in a coded numerical form. The ordinary arithmetic operations are not valid, so care has to be taken to operate on the coded values in ways that are appropriate for categories. Table 4.1 shows a small data-matrix for $n = 6$ samples and $p = 3$ categorical variables.

Thus, the categorical variable *sex* has two levels (male and female), the variable *colour* has four levels (red, blonde, brown and black) and the variable *wealth* also has four levels (rich, well-off, poor and bankrupt). We may now define dummy variables, two for *sex*, four for *colour* and four for *wealth* and assign binary values as shown in Table 4.2.

Thus each dummy variable corresponds to a single category-level and is scored one in all samples where that category occurs and otherwise is scored zero. The labels associated with the columns of Table 4.2 are: male, female; red, blonde, brown, black; rich, well-off, poor, bankrupt. Because every sample must have one sex, one colour and one level of wealth, every row of Table 4.2 must sum to three and in general when there are p categorical variables the row sums must be p. A table such as Table 4.2 is termed an **indicator**

Table 4.1 The levels of three categorical variables associated with six samples

variable	*sex*	*colour*	*wealth*
	male	brown	rich
	male	red	poor
	female	red	well-off
	male	black	bankrupt
	female	blonde	poor
	male	blonde	poor

matrix and is denoted by the symbol $\mathbf{G}$. The indicator matrix for the kth categorical variable is written $\mathbf{G}_k$ so that $\mathbf{G} = (\mathbf{G}_1, \mathbf{G}_2, \ldots, \mathbf{G}_p)$. Let the kth variable have L_k levels so that $\mathbf{G}_k$ is of order n by L_k and writing

$$L = L_1 + L_2 + L_3 + \cdots + L_p$$

for the total number of levels, then $\mathbf{G}$ is of order n by L. Thus in the table, $n = 6, L_1 = 2, L_2 = 4, L_3 = 4$ and $L = 10$. In general, we have that

$$\mathbf{G}_k\mathbf{1} = \mathbf{1} \quad \text{and} \quad \mathbf{G1} = p\mathbf{1} \tag{4.1}$$

We shall also write diag $(\mathbf{1}'\mathbf{G}_k) = \mathbf{L}_k$ and diag $(\mathbf{1}'\mathbf{G}) = \mathbf{L}$ giving the frequencies of the levels of the kth variable, and all variables, respectively. Thus, in the table, $\mathbf{L}_1 = \text{diag}\ (4, 2)$, $\mathbf{L}_2 = \text{diag}\ (2, 2, 1, 1)$, $\mathbf{L}_3 = (1, 1, 3, 1)$ and, in general

$$\mathbf{1}'\mathbf{G}_k = \mathbf{1}'\mathbf{L}_k \quad \mathbf{1}'\mathbf{G} = \mathbf{1}'\mathbf{L}$$

and

$$\mathbf{1}'\mathbf{G}_k\mathbf{1} = \mathbf{1}'\mathbf{L}_k\mathbf{1} = n, \quad \mathbf{1}'\mathbf{G1} = \mathbf{1}'\mathbf{L1} = np \tag{4.2}$$

4.3 Multiple correspondence analysis (MCA)

With these preliminaries, and having established some notation, we can define MCA as the PCA of the derived data-matrix $\mathbf{X} = p^{-1/2}\mathbf{GL}^{-1/2}$. The factor $p^{-1/2}$ is not essential but is retained in this chapter for comparison with equivalent formulae for classical CA

Table 4.2 The information of Table 4.1 in coded form using two dummy variables for *sex*, and four dummy variables each for *colour* and *wealth*

variable	*sex*	*colour*	*wealth*
	1 0	0 0 1 0	1 0 0 0
	1 0	1 0 0 0	0 0 1 0
	0 1	1 0 0 0	0 1 0 0
	1 0	0 0 0 1	0 0 0 1
	0 1	0 1 0 0	0 0 1 0
	1 0	0 1 0 0	0 0 1 0

(Chapter 9). The Pythagorean d distance between the ith and jth samples is given by

$$d_{ij}^2 = \frac{1}{p}(\mathbf{g}_i - \mathbf{g}_j)\mathbf{L}^{-1}(\mathbf{g}_i - \mathbf{g}_j)' \tag{4.3}$$

where $\mathbf{g}_i$ is the ith row of $\mathbf{G}$. Thus, when the levels of the kth variable differ between the ith and jth samples and have frequencies l_i and l_j, then its contribution to ddistance is

$$\frac{1}{p}\left(\frac{1}{l_i} + \frac{1}{l_j}\right) \tag{4.4}$$

Total ddistance is the sum over all variables of the contributions (4.4). This is the special form of chi-squared distance (section 9.1) that occurs in MCA. Note that weighting is inversely proportional to frequency, so that relative to common characters, rare categories have high weight when calculating chi-square distance. It is far from being the only distance available for categorical variables (see section 4.5) and its suitability has been criticized (Greenacre, 1989). Nevertheless, this is the distance associated with the method and we shall retain it for the following discussion.

Having identified the problem as a variant of PCA, the methodology of Chapter 2 is immediately applicable, or almost so. However, there are two issues that first need attention:

(i) the matter of centring; and
(ii) the variables are binary so that linear, or any form of continuous, biplot axes are inappropriate.

These issues are now discussed.

4.3.1 *Centring*

Principal components pass through the centroid of the points that generate the Pythagorean distances (Huygens' Principle, section 2.1) and therefore it was assumed in Chapter 2 that the data-matrix had been centred so that $\mathbf{1}'\mathbf{X} = 0$. The $\mathbf{X}$ of MCA may be centred in the usual way by the operation

$$(\mathbf{I} - \mathbf{N})\mathbf{X} = \mathbf{X} - \frac{1}{n}\mathbf{1}\mathbf{1}'\mathbf{X}$$

and that would settle this matter, were it not for the following interesting algebraic result. Using the identities (4.1) and (4.2), we may write

$$\mathbf{X}(\mathbf{L}^{1/2}\mathbf{1}) = p^{-1/2}\mathbf{G}\mathbf{1} = p^{1/2}\mathbf{1}$$
$$\mathbf{1}'\mathbf{X} = p^{-1/2}\mathbf{1}'\mathbf{G}\mathbf{L}^{-1/2}\mathbf{1}'\mathbf{L}^{1/2} \tag{4.5}$$

which may be rewritten

$$\mathbf{X}\frac{\mathbf{L}^{1/2}\mathbf{1}}{(\sqrt{np})} = \frac{\mathbf{1}}{\sqrt{n}}$$
$$\frac{\mathbf{1}'}{\sqrt{n}}\mathbf{X} = \frac{\mathbf{1}'\mathbf{L}^{1/2}}{\sqrt{np}} \tag{4.6}$$

where

$$\frac{1}{\sqrt{np}}\mathbf{L}^{1/2}\mathbf{1} \quad \text{and} \quad \frac{\mathbf{1}}{\sqrt{n}}$$

are unit vectors. It follows that these are singular vectors of $\mathbf{X}$ corresponding to a singular value $\sigma = 1$. Further, because these are non-negative vectors and $\mathbf{X}$ is a non-negative matrix, it follows from the Frobenius theorem on non-negative matrices (section A.11) that $\sigma = 1$ is the largest singular value of $\mathbf{X}$. Thus, we may write the SVD of $\mathbf{X}$ as

$$\mathbf{X} = \frac{1}{n\sqrt{p}}\mathbf{1}\mathbf{1}'\mathbf{L}^{1/2} + \sum_{k=2}^{L} \sigma_k \mathbf{u}_k \mathbf{v}_k' \tag{4.7}$$

Now

$$\frac{1}{n\sqrt{p}}\mathbf{1}\mathbf{1}'\mathbf{L}^{1/2} = \frac{1}{n}\mathbf{1}\mathbf{1}'\mathbf{X} = \mathbf{N}\mathbf{X}$$

so that the first term of the decomposition (4.7) represents the centring, or correction for the mean. Thus, the PCA can be done without centring $\mathbf{X}$, provided that the first term of the SVD, or equivalently the first eigenvector of $\mathbf{X}'\mathbf{X}$ is disregarded. Let us rewrite (4.7) as

$$\mathbf{X} = \frac{1}{n\sqrt{p}}\mathbf{1}\mathbf{1}'\mathbf{L}^{1/2} + \mathbf{U}\boldsymbol{\Sigma}\mathbf{V}'$$

then the orthogonality properties of SVD yield

$$\mathbf{1}'\mathbf{U} = 0 \quad \text{and} \quad \mathbf{1}'\mathbf{L}^{1/2}\mathbf{V} = 0 \tag{4.8}$$

4.3.2 Category-level points (CLPs)

Every dummy variable of the indicator matrix takes only two values: zero and unity. The CLPs, whose basic coordinates are given by the rows of the $L \times L$ identity matrix, when weighted by

$$\frac{1}{\sqrt{p}}\mathbf{L}^{-1/2}$$

as is required by the definition of $\mathbf{X}$, have coordinates relative to the principal axes given by the rows of

$$\mathbf{Z} = \frac{1}{\sqrt{p}}\mathbf{L}^{-1/2}\mathbf{V} \tag{4.9}$$

The first ρ columns of $\mathbf{Z}$ give the projections of the L CLPs onto the ρ-dimensional PCA approximation. These are the CLPs for MCA. The orthogonality property (4.8) shows that $\mathbf{1}'\mathbf{L}\mathbf{Z} = 0$ so that the coordinates of the CLPs weighted by the frequencies ($\mathbf{1}'\mathbf{L}$) are at the origin. Indeed, we show below (section 4.3.3) that this is true for the category-level-points of the individual variables and not just globally, i.e. $\mathbf{1}'\mathbf{L}_k\mathbf{Z}_k = 0$, where $\mathbf{Z}_k$ is the part of $\mathbf{Z}$ pertaining to the category-level coordinates

$$\frac{1}{\sqrt{p}}\mathbf{L}_k^{-1/2}\mathbf{V}_k$$

of the kth variable.

Being a PCA, the points plotted for the projections of the samples are given as before. Denoting the sample coordinates by $\mathbf{Z}_0$

$$\mathbf{Z}_0 = \mathbf{X}\mathbf{V} = \frac{1}{\sqrt{p}}\mathbf{G}\mathbf{L}^{-1/2}\mathbf{V} = \mathbf{G}\mathbf{Z} \tag{4.10}$$

so that the samples are plotted precisely at the point that is the vector-sum of the CLPs for the categories that actually occur in the sample. This shows that interpolation is occurring correctly. To interpolate a new point given by a row-vector **g** of category-levels in the same form as in a row of the indicator matrix, merely requires the vector-sum

$$\mathbf{z} = \mathbf{gZ} \tag{4.11}$$

The variables are now represented by CLPs and not axes, let alone linear axes, but this representation still shows the units **GZ** and the variables **Z** simultaneously and is to be considered as a biplot. Interpolation remains as a vector-sum as in Chapters 2 and 3 (section 3.4).

Transition formulae

Formulae relating the coordinates of the samples with those of the CLPs are a special feature of all forms of CA. We have

$$\mathbf{Z}_0 = \mathbf{XV} = \mathbf{U}\sum = \mathbf{GZ}$$

and

$$\mathbf{Z} = p^{-1/2}\mathbf{L}^{-1/2}\mathbf{V} = p^{-1}\mathbf{L}^{-1}\mathbf{G}'\mathbf{Z}_0\mathbf{\Sigma}^{-2} \tag{4.12}$$

The first of the pair of relationships (4.12) is the most interesting. It shows that the position of the samples is centred among the CLPs pertaining to each sample; the only problem is the lack of the divisor p which would give the centroids. This can be remedied quite simply by redefining

$$\mathbf{Z} = p^{1/2}\mathbf{L}^{-1/2}\mathbf{V}$$

giving

$$\mathbf{Z}_0 = \mathbf{XV} = p^{-1/2}\mathbf{GL}^{-1/2}\mathbf{V} = p^{-1}\mathbf{GZ}$$

as required. The rescaling of **Z** is trivial and does not affect distances between samples in the biplot but it does affect the expression of the samples as a vector sum of the CLPs. Classical CA (Chapter 9) uses other more symmetric forms of scaling, appropriate to the equal status of rows and columns of a two-way contingency table, but not to the data-matrix of MCA (section 9.3).

4.3.3 *The Burt matrix*

It is normal to perform a PCA on the sums-of-squares-and-products matrix of **X** rather than on its SVD; these are merely alternative modes of computation that allow one to operate on the smaller matrix (usually $p < n$) and do not affect the numerical results. Because of the special form of the first term in the singular value decomposition of the $\mathbf{X}$ of MCA, we may proceed by finding the spectral decomposition of the uncentred sums-of-squares-and-products matrix $\mathbf{X}$ which may be expressed as

$$\mathbf{X'X} = \frac{1}{p}\mathbf{L}^{-1/2}\mathbf{G'GL}^{-1/2} = \mathbf{V\Lambda V'} \tag{4.13}$$

where the eigenvalues$\mathbf{\Lambda} = (\lambda_1, \lambda_2, \lambda_3, \ldots, \lambda_L)$ are related to the singular values by $\mathbf{\Lambda} = \mathbf{\Sigma}^2$. (Note that $\mathbf{V}$ is now redefined to include the initial vector

$$\frac{1}{\sqrt{np}}\mathbf{L}^{1/2}\mathbf{1}$$

and $\lambda_1 = 1$). Rearranging (4.13) gives:

$$\frac{1}{p}\mathbf{G'G}(\mathbf{L}^{-1/2}\mathbf{V}) = \mathbf{L}(\mathbf{L}^{-1/2}\mathbf{V})\mathbf{\Lambda} \tag{4.14}$$

Equation (4.14) is in the form of the two-sided eigenvalue problem (section A.2) for the matrices

$$\frac{1}{p}\mathbf{G'G} \quad \text{and} \quad \mathbf{L}$$

with eigenvalues $\mathbf{\Lambda}$ and eigenvectors $\mathbf{Z} = p^{-1/2}\mathbf{L}^{-1/2}\mathbf{V}$, the coordinates of the CLPs. Because $\mathbf{V}$ is an orthogonal matrix, the eigenvectors are scaled to give $\mathbf{ZZ'} = p^{-1}\mathbf{L}^{-1}$, or, equivalently $\mathbf{Z'LZ} = p^{-1}\mathbf{I}$. With this scaling, $\mathbf{GZZ'G'} = p^{-1}\mathbf{GL}^{-1}\mathbf{G'}$ and hence by (A.50) and (A.51) the rows of $\mathbf{GZ}$ generate the χ^2 distances (4.4). Thus, finally we may write (4.14) as

$$\frac{1}{p}\mathbf{BZ} = \mathbf{LZ\Lambda} \tag{4.15}$$

where $\mathbf{Z'LZ} = p^{-1}\mathbf{I}$ and $\mathbf{B} = \mathbf{G'G}$. The matrix $\mathbf{B}$ is known as the **Burt matrix**. The eigenvalue equation (4.15) has at most min $(L - p, n - 1)$ non-trivial eigenvalues derived from the L columns of $\mathbf{G}$, less the p linear constraints $\mathbf{G}_k\mathbf{1} = \mathbf{1}$, $p - 1$ of which are linearly independent, giving rise to $p - 1$ zero eigenvalues, plus the other trivial eigenvalue $\lambda = 1$. Other zero eigenvalues may arise by

Table 4.3 The Burt matrix derived from the indicator matrix of Table 4.2

variable	*sex*	*colour*	*wealth*
	4, 0	1, 1, 1, 1	1, 0, 2, 1,
	0, 2	1, 1, 0, 0	0, 1, 1, 0
	1, 1	2, 0, 0, 0	0, 1, 1, 0
	1, 1	0, 2, 0, 0	0, 0, 2, 0
	1, 0	0, 0, 1, 0	1, 0, 0, 0
	1, 0	0, 0, 0, 1	0, 0, 0, 1
	1, 0	0, 0, 1, 0	1, 0, 0, 0
	0, 1	1, 0, 0, 0	0, 1, 0, 0
	2, 1	1, 2, 0, 0	0, 0, 3, 0
	1, 0	0, 0, 0, 1	0, 0, 0, 1

chance. The Burt matrix derived from the indicator matrix of Table 4.2 is shown in Table 4.3.

From Table 4.3, we see that the Burt matrix is a symmetric block-matrix $\{\mathbf{B}_{hk}\} = \{\mathbf{G}_h'\mathbf{G}_k\}$ with diagonal blocks which are themselves diagonal; indeed, $\mathbf{B}_{hh} = \mathbf{L}_h$. The off-diagonal block $\mathbf{B}_{hk}$ is the contingency table for the hth and kth categorical variables. As a consequence of (A.16), it is often convenient (see especially Chapter 9) to work in terms of the *normalized Burt Matrix* $\mathbf{B}^*$ with (h, k)th block $\mathbf{L}_h^{-1/2}\mathbf{B}_{hk}\mathbf{L}_k^{-1/2}$ and hence, unit diagonal blocks. Classical CA is concerned with the analysis of a single two-way contingency table (Chapter 9) so we see from the Burt matrix that the multivariate nature of MCA manifests itself in a simultaneous analysis of all two-way contingency tables and therefore cannot directly accommodate three- or more-way interactions. This difficulty may be circumvented by combining two or more factors into a new factor which includes interaction effects (van der Heijden, *et al.* 1989).

When viewed as a matrix approximation technique, one may have little interest in approximating the frequencies of the diagonal blocks of $\mathbf{B}$. The situation is close to that of the factor-analytical approximation of a correlation matrix where the unit diagonal values need no representation. At this stage, we note that just as PCA does not approximate a correlation matrix optimally (section 2.5) and that better approximations are available (section 11.6.3), so similarly, MCA does not approximate $\mathbf{B}$ optimally and, if that is what is wanted, better methods are available (section 10.3).

Goodness of fit

From the Eckart–Young theorem and PCA, the goodness of fit of a ρ-dimensional MCA is given by

$$\frac{(\lambda_2 + \lambda_3 + \cdots + \lambda_\rho)}{(\lambda_2 + \lambda_3 + \cdots + \lambda_L)}$$

$$= \frac{(\sigma_2^2 + \sigma_3^2 + \cdots + \sigma_\rho^2)}{(\sigma_2^2 + \sigma_3^2 + \cdots + \sigma_L^2)}$$

where the unit singular value and eigenvalues relating to the mean have been removed. In MCA, these measures usually give what are universally regarded as very pessimistic assessments of the fit. An understanding of the nature of this fact depends on the interrelationships between MCA and CA, so we defer further discussion until after the introduction of CA in Chapter 9. The point is taken up again in section 10.3 where alternative measures are proposed.

Centring of CLP

It was shown in section 4.3.2 that $\mathbf{1}'\mathbf{LZ} = 0$; now we show that this is true for the individual variables, i.e. $\mathbf{1}'\mathbf{L}_k\mathbf{Z}_k = 0$ for $k = 1, 2, \ldots, p$. From (4.1) and (4.2) we have

$$\mathbf{1}'\mathbf{B}_{jk} = \mathbf{1}'\mathbf{G}_k = \mathbf{1}'\mathbf{L}_k \tag{4.16}$$

Definding $\mathbf{z}' = (\mathbf{z}_1, \mathbf{z}_2, \ldots, \mathbf{z}_p)'$ to be one of the columns of $\mathbf{Z}$, partitioned conformably with the blocking of $\mathbf{B}$ and expanding $\mathbf{Bz} = p\lambda\mathbf{Lz}$ gives for the kth row of blocks

$$\sum_{j=1}^{p} \mathbf{B}_{kj}\mathbf{z}_j = p\lambda\mathbf{L}_k\mathbf{z}_k \tag{4·17}$$

Pre-multiplying (4.17) by $\mathbf{1}'$ and using (4.16) gives

$$\sum_{j=1}^{p} \mathbf{1}'\mathbf{L}_j\mathbf{z}_j = p\lambda\mathbf{1}'\mathbf{L}_k\mathbf{z}_k \tag{4·18}$$

for $k = 1, 2, \ldots, p$. Except when $\lambda = 1$, the left-hand side of (4.18) is a column of $\mathbf{1}'\mathbf{LZ}$ and hence is zero so the result follows; when $\lambda = 1$, $\mathbf{z} = \mathbf{L}^{1/2}\mathbf{1}/n$, the first eigenvector, and $\mathbf{1}'\mathbf{L}_k\mathbf{z}_k \neq 0$ but this is the centring eigenvector which is ignored in MCA solutions (section

4.3.1). Thus, for every variable, the centroid of the CLP weighted by their frequencies is at the origin.

4.4 Other approaches to MCA

Just as PCA may be derived in several ways, so may be MCA. An important class of problems is concerned with assigning numerical scores to category levels. This is the approach of Guttman (1941) scaling, otherwise known as **homogeneity analysis** (Gifi, 1990). For links between this type of approach and Fisher's method of **optimal scores** (Fisher, 1938), see Gower (1990b).

4.4.1 Homogeneity analysis

Suppose scores given in a vector **z** are assigned to all the category-levels. Thus **G** is converted into a two-way table of numerical values by substituting these scores for the coded indicators. In psychometrics, this process is often termed **quantification**, referring to the quantitative values, or scores, given to the levels of the categorical variables. In homogeneity analysis/Guttman scaling, the scores are chosen to maximize the ratio of the variance between the scores of the sample means to the variance within samples. The row totals for the samples are given by **Gz**. We may set up the analysis of variance shown in Table 4.4.

The between/within ratio is maximized when the between/total ratio is maximized. The latter criterion is easier to handle and differentiation with respect to **z** leads to

$$\left(\frac{1}{p}\mathbf{G}'\mathbf{G} - \frac{1}{np}\mathbf{L}\mathbf{1}\mathbf{1}'\mathbf{L})\mathbf{z} = \lambda(\mathbf{L} - \frac{1}{np}\mathbf{L}\mathbf{1}\mathbf{1}'\mathbf{L}\right)\mathbf{z} \tag{4.19}$$

Table 4.4 Between and within group analysis of variance for an indicator matrix **G** associated with a scorces vector **z**

Source	*d.f.*	*Sums-of-squares*
Between sample	$n-1$	$\frac{1}{p}\mathbf{z}'\mathbf{G}'\mathbf{G}\mathbf{z} - \frac{1}{np}(\mathbf{1}'\mathbf{L}\mathbf{z})^2$
Within samples	$n(p-1)$	By subtraction
Total	$np-1$	$\mathbf{z}'\mathbf{L}\mathbf{z} - \frac{1}{np}(\mathbf{1}'\mathbf{L}\mathbf{z})^2$

We have seen in (4.14) that

$$\frac{1}{p}\mathbf{G}'\mathbf{G}\mathbf{z} = \lambda\mathbf{L}\mathbf{z}$$

has solutions for $\lambda \neq 1$ for which $\mathbf{1}'\mathbf{L}\mathbf{z} = 0$ and hence these are also solutions to (4.19). Equation (4.19) also has the solution $\mathbf{z} = \mathbf{1}$ but then the row-totals are constant and both sides of (4.19) vanish. This solution corresponds to the uninteresting solution $\mathbf{v} = \mathbf{L}^{1/2}\mathbf{1}$ in the SVD (4.7), the vector for centring the samples at their centroid (section 4.3.1). There is, of course, an arbitrariness in the scaling of the size of $\mathbf{z}$ but, because we are maximizing a ratio, whatever size constraint is adopted is irrelevant; MCA adopts $\mathbf{z}'\mathbf{L}\mathbf{z} = p^{-1}$ which is consistent with χ^2 distance (section 4.3.3). The condition $\mathbf{1}'\mathbf{L}\mathbf{z} = 0$ is sometimes regarded as a further constraint but we have seen that in MCA it arises naturally as a consequence of centring the samples which is itself a consequence of Huygens' principle in PCA (section 2.1). When the approach is as above via maximizing a variance ratio as in Guttman scaling then the constraint may be imposed, but with no effect. Alternatively, we might maximize the ratio $\mathbf{z}'\mathbf{G}'\mathbf{G}\mathbf{z}/\mathbf{z}'\mathbf{L}\mathbf{z}$, in which the constraint is essential if we are to eliminate the uninteresting solution which gives a maximal unit value to the ratio. An advantage of the analysis of variance formulation is that it provides a ready means for handling missing values in **G**. A between/within analysis of variance does not require equal 'replication' within groups and so (4.19) is easily modified to accommodate unequal row sums and then we may proceed as before.

Many problems, including MCA, that can be expressed as maximizing ratios of sums-of-squares, may be formulated alternatively in constrained least-squares form (Gifi, 1990). Thus, the least-squares formulation of homogeneity analysis is to minimize

$$\sum_{k=1}^{p} \| \mathbf{z}_0 - \mathbf{G}_k\mathbf{z}_k \|$$

where $\mathbf{z}_0$ represent ideal *homogeneous* scores for each row, which for given $\mathbf{z}_k$ are clearly given by (Huygens' principle again) the row means

$$\mathbf{z}_0 = \frac{1}{p}\sum_{k=1}^{p} \mathbf{G}_k\mathbf{z}_k$$

In terms of Table 4.4, the homogeneity analysis criterion is to choose $\mathbf{z}_k (k = 1, 2, \ldots, p)$ that minimize the within-sample sum-of-squares. Constraints have to be imposed, if the trivial solution

$$\mathbf{z}_k = 0 \text{ for } (k = 1, 2, \ldots, p)$$

is to be avoided. If the MCA solution is to be found, then it is essential that the imposed constraints are consistent with those of MCA. As noted by Healy and Goldstein (1976), if some other constraint, such as $\mathbf{1}'\mathbf{z} = 1$, is imposed to fix the otherwise arbitrary scaling of the eigenvectors of (4.18) then different scores will be found. These alternative constraints may be of interest to those motivated by the need to find scores but give solutions to problems that are immaterial for MCA based on χ^2 distance. An essential difference between the MCA and homogeneity approaches is that scores are basically a one-dimensional concept, whereas MCA is concerned with multidimensional representations. However, it is simple to replace $\mathbf{z}_0$ by $\mathbf{Z}_0$ and $\mathbf{z}$ by $\mathbf{Z}$ to get multidimensional MCA solutions – provided appropriate orthogonality constraints are imposed.

In general, a least-squares formulation has computational advantages, leading naturally to the use of alternating least-squares algorithms in which one or more sets of parameters are held fixed while the remaining parameters are estimated. Different sets of parameters are estimated in successive cycles, the residual sum-of-squares being reduced at each cycle until the process converges – at least to a local optimum. Gifi (1990) gives many examples of this approach, where indeed it is adopted as a unifying principle for many methods of multidimensional data analysis.

4.4.2 Generalized canonical correlation analysis (GCCA)

This method, due to Carroll (1968) and proposed as an extension of canonical correlation analysis from two to many sets of variables, is discussed by Tenenhaus and Young (1985) who show its equivalance to MCA. We define the scores $\mathbf{v}' = (\mathbf{v}_1', \mathbf{v}_2', \ldots, \mathbf{v}_p')$ for the category levels and scores $\mathbf{s}$ for the n samples. These symbols differ from those used previously, because we have yet to show that they are the same as for MCA; indeed, that is the objective of the remainder of this section. The squared correlation between the scores for variable k and those for all variables is

$$r_k^2(\mathbf{G}_k\mathbf{v}_k,\mathbf{s}) = \frac{(\mathbf{s}'\mathbf{G}_k\mathbf{v}_k)^2}{(\mathbf{s}'\mathbf{s})(\mathbf{v}_k'\mathbf{G}_k'\mathbf{G}_k\mathbf{v}_k)} \tag{4.20}$$

The GCCA criterion is to choose scores that maximize

$$\sum_{k=1}^{p} r_k^2(\mathbf{G}_k\mathbf{v}_k,\mathbf{s})$$

over $\mathbf{s}$ and all $\mathbf{v}_k$.

Correlations are invariant to the scaling of these vectors so if we are to relate to MCA, appropriate scalings will have to be supplied. The maximal correlation property of multiple regression shows that for given $\mathbf{s}$ the maximum is achieved by setting

$$\mathbf{v}_k = \rho_k(\mathbf{G}_k'\mathbf{G}_k)^{-1}\mathbf{G}_k'\mathbf{s} \quad \text{for} \quad k = 1, 2, \ldots, p \tag{4.21}$$

where ρ_k is an arbitrary scaling factor determined later. Substituting for $\mathbf{v}_k$ in (4.20) gives

$$\begin{aligned}\max r_k^2(\mathbf{G}_k\mathbf{v}_k,\mathbf{s}) &= \frac{(\mathbf{s}'\mathbf{G}_k(\mathbf{G}_k'\mathbf{G}_k)^{-1}\mathbf{G}_k'\mathbf{s})^2}{(\mathbf{s}'\mathbf{s})(\mathbf{s}'\mathbf{G}_k(\mathbf{G}_k'\mathbf{G}_k)^{-1}\mathbf{G}_k'\mathbf{s})} \\ &= \frac{\mathbf{s}'\mathbf{G}_k(\mathbf{G}_k'\mathbf{G}_k)^{-1}\mathbf{G}_k'\mathbf{s}}{\mathbf{s}'\mathbf{s}} \\ &= \frac{\mathbf{s}'\mathbf{G}_k\mathbf{L}_k^{-1}\mathbf{G}_k'\mathbf{s}}{\mathbf{s}'\mathbf{s}}\end{aligned}$$

Thus for given $\mathbf{s}$

$$\max \sum_{k=1}^{p} r_k^2(\mathbf{G}_k\mathbf{v}_k,\mathbf{s}) = \frac{\mathbf{s}'(\mathbf{G}\mathbf{L}^{-1}\mathbf{G})\mathbf{s}}{\mathbf{s}'\mathbf{s}}$$

which is a ratio of quadratic forms and hence is maximized over $\mathbf{s}$ by finding the maximal eigenvalue μ and associated vector $\mathbf{s}$ of

$$(\mathbf{G}\mathbf{L}^{-1}\mathbf{G}')\mathbf{s} = \mu\mathbf{s} \tag{4.22}$$

Premultiplying both sides of (4.15) by $\mathbf{G}\mathbf{L}^{-1}$ gives

$$\frac{1}{p}(\mathbf{G}\mathbf{L}^{-1}\mathbf{G}')\mathbf{G}\mathbf{z} = \lambda\mathbf{G}\mathbf{z} \tag{4.23}$$

so comparison with (4.22) shows that

$$\mathbf{s} = \kappa \mathbf{G z} \text{ and } \mu = p\lambda$$

where κ is a factor of proportionality which we shall take to be unity. Then

$$\mathbf{s} = \mathbf{G z} \tag{4.24}$$

Substituting (4.24) into (4.21) and using (4.15) premultipled by $\mathbf{L}^{-1}$ gives

$$\mathbf{v}_k = \rho_k (\mathbf{G}_k' \mathbf{G}_k)^{-1} \mathbf{G}_k' \mathbf{G z} = \lambda \rho_k \mathbf{z}_k \quad \text{for} \quad k = 1, 2, \ldots, p \tag{4.25}$$

Thus, by choosing the hitherto arbitrary scaling factor ρ_k to be λ^{-1} for all k, we have

$$\mathbf{v} = \mathbf{z} \tag{4.26}$$

The results (4.24) and (4.26) show that, with the scalings found above, the solution of the GCCA problem is the same as that of MCA.

The other eigenvalues and vectors of (4.22) also give solutions, which on arguments similar to those given in section A.4 for PCA, give ρ-dimensional solutions that maximize

$$\sum_{k=1}^{p} r_k^2 (\mathbf{G}_k \mathbf{V}_k, \mathbf{S})$$

where $\mathbf{V}_1, \mathbf{V}_2, \ldots, \mathbf{V}_p$ and $\mathbf{S}$ have ρ columns and $r_k^2(\mathbf{G}_k \mathbf{V}_k, \mathbf{S})$ is to be interpreted as the sum of the squares of the ρ correlations between corresponding columns of $\mathbf{G}_k \mathbf{V}_k$ and $\mathbf{S}$.

4.5 Alternatives to χ^2 distance

4.5.1 The extended matching coefficient

MCA and all the above discussion is based on the use of χ^2 distance. This ddistance was developed in the context of classical CA where it has its main justification. Chapter 9 discusses χ^2 distance as a measure of distance between two rows or two columns of a two-way contingency table. A consequence of transferring the CA methodology to MCA is that $\mathbf{G}$ is treated as if it were a contingency table, even though it takes only binary values and these are dummies. This gives the special case of χ^2 distance shown in (4.4). In the context of market research, this distance has been criticized by Greenacre (1990) and defended by Carroll, Green and Schaffer (1986; 1987). Much of this discussion is concerned with the justification of interpreting row–column distances. Be this as it may, it seems to us that the row-distance of MCA which weights categories inversely by their frequencies of occurrence is hard to justify for general use, although there may be instances when one wants to given specially high weight to rare events. We believe that it is undesirable to be obliged to apply weights as in (4.4) and that further distances should be sought that are applicable to categorical variables. The most simple, and one we favour, is the **extended matching coefficient** (EMC) which, like the Pythagorean distance of PCA, gives equal weight to all variables. For two sample-units, this counts the proportion of matches among the p variables, so is a simple generalization of the simple matching coefficient (Sneath and Sokal, 1973) which is defined only for $l_k = 2\ (k = 1, 2, \ldots, p)$. The matrix giving the extended matching coefficients between all pairs of samples may be written

$$\frac{1}{p}\mathbf{G}\mathbf{G}' \tag{4.27}$$

which should be compared with the inner-product

$$\frac{1}{p}\mathbf{G}\mathbf{L}^{-1}\mathbf{G}' \tag{4.28}$$

from which the χ^2 distances of MCA may be obtained. The factor $1/p$ is included in (4.27) to give the usual form of similarity matrix with unit diagonal and also to reinforce the comparison with (4.28) but for most practical work it is better ignored (see below).

Like the matrix of χ^2 distances, the dissimilarity matrix

$$\mathbf{1}\mathbf{1}' - \frac{1}{p}\mathbf{G}\mathbf{G}'$$

derived from the similarity matrix (4.27) should be treated as a matrix of ddistances, the corresponding distance being the square root of the extended matching dissimilarity.

A biplot using the extended matching coefficient may be obtained as a PCA of

$$\frac{1}{\sqrt{p}}\mathbf{G}$$

with CLPs

$$\frac{1}{\sqrt{p}}\mathbf{I}$$

The first singular vectors are not special, so the normal centring for PCA

$$\frac{1}{\sqrt{p}}(\mathbf{I}-\mathbf{N})\mathbf{G}=\mathbf{U\Sigma V}'$$

is required; the same translation

$$\frac{1}{\sqrt{p}}\frac{{}_L\mathbf{1}_1\mathbf{1}'_n}{n}\mathbf{G}$$

must be applied to the CLPs. The effect of centring is to replace the Burt matrix $\mathbf{B}$ by

$$\frac{1}{p}\left(\mathbf{B}-\frac{1}{n}\mathbf{L11'L}\right)$$

from which the eigenvectors $\mathbf{V}$ are derived; the sum of the eigenvalues is

$$\frac{1}{p}\left(np-\frac{1}{n}\mathbf{1}'\mathbf{L}^2\mathbf{1}\right)$$

Thus, as for PCA, we could plot the first ρ columns of

$$\mathbf{Z}_0=\mathbf{U\Sigma}=\frac{1}{\sqrt{p}}(\mathbf{I}-\mathbf{N})\mathbf{GV}$$

for the samples and

$$\frac{1}{\sqrt{p}}\mathbf{V}$$

for the 'variables'. The variables have only zero/one values that refer to categories, so there is little justification for representing them as a vector, as in PCA; only the end-points giving the CLPs need be plotted. For practical purposes, it is better to drop the factor $1/\sqrt{p}$ and the resulting formulae are shown in the PCA row of Table 4.5; the remaining rows of the table refer to MCA, for comparison, and variant forms of plotting associated with the EMC that are discussed below.

From Table 4.5, it can be seen that the PCA and MCA plots become equivalent when $\mathbf{L}=\mathbf{I}$. The orthogonality constraint (4.8) makes it unnecessary to centre before the PCA or SVD of $\mathbf{GL}^{1/2}$ provided the 'uninteresting' first vectors are discarded. A disadvantage, compared with MCA, of the unmodified PCA plot is that to interpolate a new sample $\mathbf{g}$ requires the graphical evaluation of

$$\left(\mathbf{g}-\frac{1}{n}\mathbf{1}'\mathbf{G}\right)\mathbf{V}$$

Even though $\mathbf{g}$ has only p non-zero unit values,

$$\mathbf{f}=\mathbf{g}-\frac{1}{n}\mathbf{1}'\mathbf{G}$$

has L fractional values, so interpolation is given by the vector-sum $\mathbf{fV}$ of all L CLPs weighted by the elements of the vector $\mathbf{f}$. In contrast, interpolation in MCA is given by the the simple vector-sum (4.11), which denotes the unweighted sum of p CLPs, as is clear from the MCA row of Table 4.5. The rows labelled 'sample-shift' and 'CLP-shift' of Table 4.5 show two-ways of overcoming this difficulty. The basic solution remains that of PCA, with its least-squares best fit in the ρ-dimensional space $\mathscr{L}$ that passes through the centroid of the sample-points (Huygens' principal, section 2.1). However, there is no reason why the projections of samples and CLPs onto $\mathscr{L}$ should not be translated, or even rotated, within $\mathscr{L}$, perhaps independently from each other. Such translations leave distances unaltered and hence the ρ-dimensional approximation to the extended matching dissimilarity distance remains unaltered. Other properties of the decomposition do change and may be exploited with advantage.

Sample-shift representation

In the samaple-shift representation, the sample-points are translated from their PCA positions by an amount

$$\frac{1}{n}\mathbf{1}'\mathbf{GV}$$

Table 4.5 Variant forms of plot to be used with the extended matching coefficient

Scaling	*Samples*	*CLPs*($\mathbf{Z}$)	$\mathbf{GZ}$	*Inner-product*
MCA	$(\mathbf{I}-\mathbf{N})\mathbf{GL}^{-1/2}\mathbf{V}=\mathbf{GL}^{-1/2}\mathbf{V}$	$\mathbf{L}^{-1/2}\mathbf{V}$	$\mathbf{GL}^{-1/2}\mathbf{V}$	$\mathbf{GL}^{-1}$
PCA	$(\mathbf{I}-\mathbf{N})\mathbf{GV}=\mathbf{Z}_0$	$\mathbf{V}$	$\mathbf{GV}=\mathbf{Z}_0+\mathbf{NGV}$	$(\mathbf{I}-\mathbf{N})\mathbf{G}$
Sample-shift	$\mathbf{Z}_0+\mathbf{NGV}$	$\mathbf{V}$	$\mathbf{GV}=\mathbf{Z}_0+\mathbf{NGV}$	$\mathbf{G}$
CLP-shift	$\mathbf{Z}_0$	$\left(\mathbf{I}-\frac{\mathbf{11'G}}{np}\right)\mathbf{V}$	$\mathbf{Z}_0$	$(\mathbf{I}-\mathbf{N})\mathbf{G}\left(\mathbf{I}-\frac{\mathbf{G'11'}}{np}\right)$

while the CLPs remain untranslated. It can be seen that the interpolation **GV** now reproduces the (new) positions of the samples so is as simple to use as in MCA. Further, the inner-product now simplifies to **G** which is more useful than the PCA inner-product of $(\mathbf{I}-\mathbf{N})\mathbf{G}$ because right-angles subtended at the origin denote zero elements of **G**. Collinearities with the origin suggest unit values but lengths and sense of vectors must also be taken into account. It must also be remembered that the approximation does not give a least-squares approximation to **G** in ρ dimensions, which would require the singular value decomposition of **G** itself.

CLP-shift representation

In the CLP-shift representation, the CLPs are translated from their PCA positions by an amount

$$-\frac{1}{np}\mathbf{1}'\mathbf{GV}$$

while the samples remain untranslated. Again, interpolation is obtained as the vector sum of the relevant CLPs as in MCA and now, because

$$\mathbf{1}'\mathbf{L}\left(\mathbf{I}-\frac{\mathbf{11}'\mathbf{G}}{np}\right)=0$$

we also have the MCA property that the weighted sum of the CLPs is at the origin. The inner-product is given by the complicated function shown in Table 4.5 and seems to be of little interest.

In all the representations, rather than the values given in Table 4.5, p times the given coordinates of the CLPs may be plotted; then interpolation is by finding the centroids of the CLPs. In such representations, the sample-points lie at the centroids of their active CLPs as in (section 4.3.2.) MCAs. Finding a centroid is an easier operation than finding a vector-sum; indeed, we have seen that vector-sums are most simply found by first constructing a centroid and extending it p times from the origin – section 2.3. Further, it follows that any sample-point must lie within the convex hull of the set of CLPs pertaining to the p different variables – which is helpful in interpretation. It may be noted that multiplying the CLPs of Table 4.5 by p shows that the translations of the CLP-shift representation is merely the negative of the translation for the sample-shift method. However, the different origins involved have a crucial effect on the form of the inner-product. Inner-products depend on the origin; distances do not. If we were to recommend a single representation, we would choose the sample-shift method and this is what is used in the example of section 4.6.

4.5.2 *Other matching coefficients*

In Chapter 7, it is shown that any dissimilarity coefficient applicable to categorical variables may be used as the basis for biplots. Gower (1992) showed that many such dissimilarity coefficients are monotonically related to the extended matching coefficient. This result may be seen as follows. The comparison between the *i*th and *j*th individuals in the indicator matrix **G**, is often based on the a_{ij} 1–1 matches, b_{ij} 1–0 matches, c_{ij} 0–1 matches and d_{ij} 0–0 matches. This notation is the usual one for 2×2 'contingency' tables and, as usual, for simplicity we shall drop the suffices. The indicator matrix is special because every row sums to p, giving

$$a + b = a + c = p$$

as well as

$$a + b + c + d = L$$

Thus

$$b = p - a \quad c = p - a \quad \text{and} \quad d = L - 2p + a$$

We also have that $L \geqslant 2p$, equality occurring only when all categorical variables are binary. Thus, unlike the usual situation, where there are three free parameters, there is only one free parameter, which we shall take to be a or, rather the **extended matching coefficient** $E = 1 - a/p$. It follows that all dissimilarity coefficients based on a, b, c and d may be expressed as functions of the single parameter a. Of special importance are the dissimilarity coefficients discussed by Gower and Legendre (1986):

$$1 - S_\theta = \frac{\theta(b + c)}{a + d + \theta(b + c)} \quad \text{and} \quad 1 - T_\theta = \frac{\theta(b + c)}{a + \theta(b + c)}$$

where $\theta > 0$. For different values of θ, most of the coefficients for presence/absence data are included in these families. Thus, when $\theta = 1$, S_θ is the **simple matching coefficient** and T_θ is the **Jaccard**

coefficient. For the special case of an indicator matrix these coefficients become

$$1 - S_\theta = \frac{2\theta E}{L/p - 2(1-\theta)E} \quad \text{and} \quad 1 - T_\theta = \frac{2\theta E}{1-(1-2\theta)E}$$

(note that, in the current context, L, rather than p, is the number of variables concerned.) When $\theta = \frac{1}{2}$, $1 - T_\theta$ becomes E, while when $\theta = 1$, $1 - S_\theta$ is proportional to E. It is easily verified that:

(i) for fixed values of E, $1 - S_\theta$ and $1 - T_\theta$ are both monotonic increasing functions of θ, and
(ii) for fixed values of $\theta, 1 - S_\theta$ and $1 - T_\theta$ are both monotonic increasing functions of E.

The implication is that, so far as non-metric MDS is concerned (section 11.2), all these coefficients are equivalent, leading to the same results. For metric methods of MDS, different coefficients of this class will give different results and Chapter 7 gives a general methodology for their use in constructing biplots. One might suggest that the best coefficient to use is the one given by the value of θ that minimizes the chosen MDS goodness-of-fit criterion (3.1). Figure 4.1 shows these two coefficients for a range of values of θ; to avoid problems of the dependency of S_θ on p and L, $1 - S_\theta$ is plotted against $F = pE/L$.

Thus there are many alternatives to the use of χ^2 distance and Chapter 7 discusses how these may be used in MDSs that incorporate biplot information. We have already seen that many of the properties that are often regarded as special to MCA are also valid for distances defined to be the square root of the extended dissimilarity distances and, in section 7.3.1, we shall see that this applies more generally. In particular, the centring properties of CLPs (sections 4.3.2. and 4.3.3) and a version of the transition formulae (section 4.3.2) all have their counterparts in more general biplot methods. Indeed, these centroid properties are less remarkable than they may seem. With any method of MDS based on categorical variables, CLPs may be defined by placing them at the centroids of those sample-points with the same category-levels. This provides a not unreasonable basis for sample interpolation but is less justifiable for prediction. These properties help in the interpretation of MCA, based roughly on the observation that something which is at

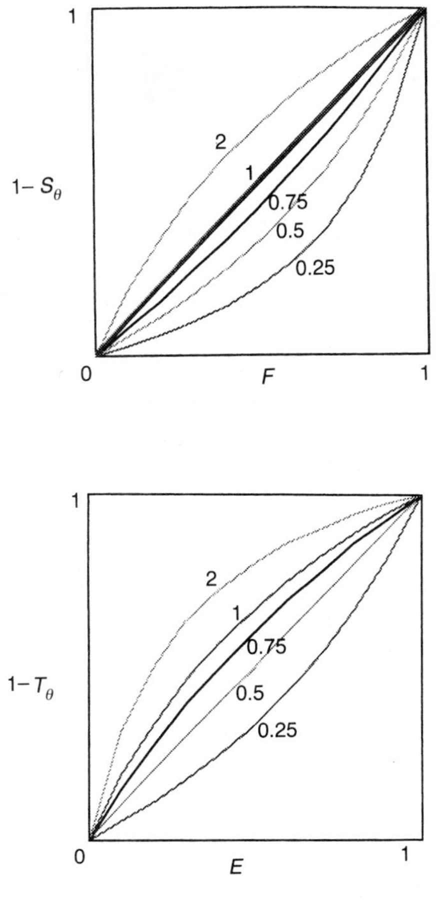

Fig. 4.1. *The curves of 1-S_θ against $F = pE/L$ and 1-T_θ against E, the extended matching coefficient, for a range of values of θ. The monotonicity of the curves with increasing θ and with increasing E and F is evident.*

a centroid of a set of points is within the convex hull of those points. However, the main interpretive tool for all these methods is that the ρ-dimensional display gives approximations to the distances between samples and that prediction depends on nearness to the CLPs.

The latter remark introduces the concept of **prediction regions**. A prediction region for some category-level, say red of *colour*, is the region of space that is nearer the CLP for red than it is to the CLP for any other colour. Thus, space is divided into disjoint convex neighbour-regions each of which identifies the nearest category-level. Any sample-point lying within a prediction region predicts the associated category-level for that sample. For categorical variables, this is the equivalent to the predictions given by the Eckart–Young theorem for quantitative variables. Just as every quantitative variable has an associated axis in the full space and a biplot axis in the approximation space, so every categorical variable has an associated set of prediction regions both in the full space and in the approximation space. The markers on linear biplot axes become the labels of the CLPs and instead of predicting the nearest scale marker on each axis, one predicts the nearest label associated with each categorical variable. The prediction regions in the approximation space are defined by the intersections of the approximation space with the prediction regions in the full space. Although the prediction-regions in the intersection space are disjoint for any one variable, they will overlap for more than one variable. Consequently, unlike in the quantitative case where all axes may be shown on a single diagram, if confusion is to be avoided, prediction regions for categorical variables will normally require a separate diagram for each variable. Prediction regions are discussed in more generality in Chapter 7 but they are shown in the examples at the end of this chapter without any further discussion at this stage as to how they are computed.

4.6 Examples of MCA

Table 4.6 shows the observations on five variables observed on twenty farms from the Dutch island of Terschelling. This table is reported by Jongman, ter Braak and van Tongeren (1987) and forms part of a much larger survey. It is concerned with environmental factors and different forms of farm management. Three variables may be regarded as categorical variables:

- *moisture class* at five levels (labelled M1,M2,M4,M5 in the diagrams – level 3 does not occur in the table);
- *grassland management type* at four levels (standard farming SF, biological farming BF, hobby farming HF and nature conservation management NM); and

Table 4.6 Data on five variables observed at 20 farms on the island of Terschelling (from Jongman, ter Braak and van Tongeren, 1987)

Farm number	*A1 horizon (cms)*	*Moisture class*	*Grassland management type*	*Grassland use*	*Manure class*
1	2.8	1	SF	2	4
2	3.5	1	BF	2	2
3	4.3	2	SF	2	4
4	4.2	2	SF	2	4
5	6.3	1	HF	1	2
6	4.3	1	HF	2	2
7	2.8	1	HF	3	3
8	4.2	5	HF	3	3
9	3.7	4	HF	1	1
10	3.3	2	BF	1	1
11	3.5	1	BF	3	1
12	5.8	4	SF	2	2
13	6.0	5	SF	2	3
14	9.3	5	NM	3	0
15	11.5	5	NM	2	0
16	5.7	5	SF	3	3
17	4.0	2	NM	1	0
18	4.6	1	NM	1	0
19	3.7	5	NM	1	0
20	3.5	5	NM	1	0

- *grassland use* at three levels (hay production U1, intermediate U2 and grazing U3).

In this chapter, we also assume that *manure class* (0, 1, 2, 3 and 4 labelled C0, C1, C2, C3 and C4) is categorical, although because '0' stands for zero use of manure and the other levels for increasing use, it is clear that this is at least an ordered categorical variable and possibly might be treated as a conventional quantitative variable. The soil-depth variable, *A1 horizon* (cms), is quantitative and is not considered until Chapter 7 where the same data is analysed by more general methods.

From Table 4.6, we derive the indicator matrix (Table 4.7) and the Burt matrix (Table 4.8). Figure 4.2 shows a two-dimensional MCA of the data of Table 4.6. In Fig. 4.2 farms are numbered and the category levels are labelled. The interpolation of the first sample is shown. The Burt matrix of Table 4.8 when inserted into (4.15), has the 11 non-trivial eigenvalues (one less than min $(L-p, n-1)$ because

Table 4.7 The indicator matrix **G** corresponding to Table 4.6

Farm	*Moisture*	*Management*	*Grass use*	*Manure*
1	1000	1000	010	00001
2	1000	0100	010	00100
3	0100	1000	010	00001
4	0100	1000	010	00001
5	1000	0010	100	00100
6	1000	0010	010	00100
7	1000	0010	001	00010
8	0010	0010	001	00010
9	0001	0010	100	01000
10	0100	0100	100	01000
11	1000	0100	001	01000
12	0010	1000	010	00100
13	0001	1000	010	00010
14	0001	0001	001	10000
15	0001	0001	010	10000
16	0001	1000	001	00010
17	0100	0001	100	10000
18	1000	0001	100	10000
19	0001	0001	100	10000
20	0001	0001	100	10000
Totals	7427	6356	785	63443

the columns for NM and CO are the same in Table 4.7)

Dimension	1	2	3	4	5	6	7	8	9	10	11
Eigenvalue	2.600	2.221	2.068	1.528	1.241	0.884	0.533	0.356	0.310	0.190	0.070

so that the two-dimensional percentage goodness of fit is

$$\frac{(2.600 + 2.221)}{12} \times 100 = 40.2\%$$

not exceptionally good but better than usual for MCA. These eigenvalues are p times those of (4.15) because the divisor p has been omitted and this makes the sum of the non-trivial eigenvalues equal to $L - p = 12$. Chapter 10 discusses the intricate issues surrounding the measurement of fit for MCA. An analysis of the extended matching coefficient of section 4.5 has also been done and is shown in Fig. 4.3; the sample-shifted representation (section 4.5) has been used.

As in Fig. 4.2, Fig. 4.3 farms are numbered and the category levels are labelled, and the interpolation of the first sample is shown. It

Table 4.8 The Burt matrix **B** derived from Table 4.5

Moisture				*Management*				*Grass use*			*Manure*				
1	*2*	*4*	*5*	*SF*	*BF*	*HF*	*NM*	*1*	*2*	*3*	*0*	*1*	*2*	*3*	*4*
7															
	4														
		2													
			7												
1	2	1	2	6											
2	1	0	0		3										
3	0	1	1			5									
1	1	0	4				6								
2	2	1	2	0	1	2	4	7							
3	2	1	2	5	1	1	1		8						
2	0	0	3	1	1	2	1			5					
1	1	0	4	0	0	0	6	4	1	1	6				
1	1	1	0	0	2	1	0	2	0	1		3			
3	0	1	0	1	1	2	0	1	3	0			4		
1	0	0	3	2	0	2	0	0	1	3				4	
1	2	0	0	3	0	0	0	0	3	0					3

Because the matrix is symmetric, only the lower-triangular part is shown. **L** appears down the diagonal and the off-diagonal is partitioned into its component 2 × 2 contingency tables

seems best to discuss the properties of the two figures jointly rather than separately. The non-trivial eigenvalues associated with the extended matching coefficient are as follows

Dimension	1	2	3	4	5	6	7	8	9	10	11
Eigenvalue	15.301	10.946	9.992	6.479	4.690	3.566	2.068	1.917	1.514	0.902	0.225

so that the two-dimensional percentage goodness of fit is

$$\frac{(15.301 + 10.946)}{57.6} \times 100 = 45.6\%$$

a slight improvement on MCA. Note that

$$57.6 = np - \frac{1}{n}\mathbf{1}'\mathbf{L}^2\mathbf{1}$$

Figures 4.2 and 4.3 have the same general characteristics. In both cases, the NM (nature management) farms are in the region of greatest moisture (M5), zero fertilizer (C0) and hay production

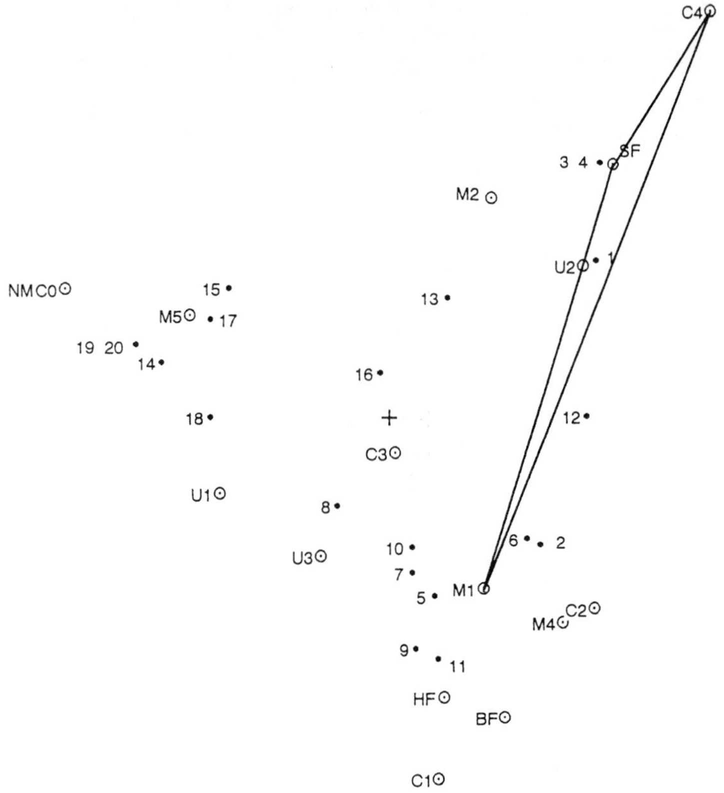

Fig. 4.2. *Two-dimensional MCA of the categorical variables shown in Table 4.6.*

(U1), showing no grazing. The SF (scientific management) farms are in the region of modest levels of moisture (M2), highest use of fertilizer (C5) and grassland management (U2) showing both grazing and hay production. The HF and BF (hobby and biological) farms are associated with extremes of moisture (M1 and M4), which denote very dry and rather wet, modest use of fertilizer (C1 and C2) and a tendency towards grazing (U3). It may be confirmed that these observations faithfully mirror the properties shown in Table 4.6. The differential weighting of variables given by the χ^2 ddistance of MCA compared with the use of the extended matching coefficient seems to have had very little effect.

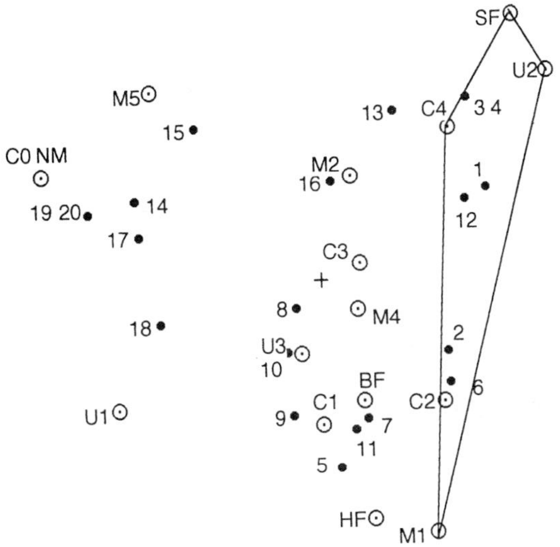

Fig. 4.3. *Two-dimensional analysis of the categorical variables shown in Table 4.6 based on the extended matching coefficient.*

Reference to Fig. 4.2 confirms that the centroids of each of the p sets of the projected CLPs, weighted by the category frequencies, are at the origin (section 4.3.3); in Fig. 4.3, this is also true because we have used the sample-shifted plot described in section 4.5. In both figures, the occurrence of a sample at the centroid of its CLPs is illustrated (for farm number 1); new samples may be interpolated in precisely the same way. No farm can occur outside the convex hull formed by its CLPs. Thus in Fig. 4.2, Farm 1 is the only one occuring within the convex hull shown in the figure so, provided that particular combination actually occurs is Table 4.6, then it must occur for Farm 1 and actually finding the centroid is superfluous. Other convex hulls will contain several farms (e.g. C4, SF, M1, U1). The centroid is somewhere between farms 12, 13 and in the vicinity of 16 but has little in common with these farms because the true interpolated position in $\mathscr{R}$ lies far from $\mathscr{L}$ so has a large residual. In Fig. 4.3, the illustrated convex hull contains several farms. Indeed, Farm 1 is at the centroid and Farms 2, 3, 4 and 6 are near the edges. These peripheral farms have at least two characteristics common to Farm 1.

Table 4.9 Predictions for the categorical variables for Table 4.6 given by a two-dimensional approximation from multiple correspondence analysis

Farm	*Moisture*			*Management*			*Grassland*			*Manuring*		
	T	*PR*	*PCLP*	*T*	*PR*	*PCLP*	*T*	*PR*	*PCLP*	*T*	*PR*	*PCLP*
1	1	2*	2*	SF	SF	SF	2	2	2	4	4	4
2	1	1	4*	BF	HF*	BF	2	2	3*	2	2	2
3	2	2	2	SF	SF	SF	2	2	2	4	4	4
4	2	2	2	SF	SF	SF	2	2	2	4	4	4
5	1	1	4*	HF	HF	BF*	1	1	3*	2	2	1*
6	1	1	4*	HF	HF	BF*	2	2	3*	2	2	2
7	1	1	4*	HF	HF	BF*	3	1*	3	3	2*	1*
8	5	1*	1*	HF	HF	HF	3	1*	3	3	0*	1*
9	4	1*	4	HF	HF	BF*	1	1	3*	1	1	1
10	2	1*	4*	BF	HF*	HF*	1	1	3*	1	2*	1
11	1	1	4*	BF	HF*	BF	3	1*	3	1	1	1
12	4	1*	4	SF	SF	SF	2	2	2	2	2	2
13	5	5	2*	SF	SF	SF	2	2	2	3	0*	4*
14	5	5	5	NM	NM	NM	3	1*	1*	0	0	0
15	5	5	5	NM	NM	NM	2	2	1*	0	0	0
16	5	5	2*	SF	SF	SF	3	2*	2*	3	0*	3
17	2	5*	5*	NM	NM	NM	1	1	1	0	0	0
18	1	5*	5*	NM	NM	NM	1	1	1	0	0	0
19	5	5	5	NM	NM	NM	1	1	1	0	0	0
20	5	5	5	NM	NM	NM	1	1	1	0	0	0
Errors	0	7	12	0	3	5	0	5	8	0	5	4

T denotes the true value given in Table 4.6, PR denotes the prediction-region prediction and PCLP denotes the prediction given by the nearest projected CLP. An asterisk denotes an incorrect prediction, the totals of which are given in the final row of the table

Figure 4.3 approximates **G** in terms of the inner product, or cosine formula, for all farm-CLP pairs. The most interesting of these are the zero values arising from right-angles subtended at the origin. Many of these are evident in Fig. 4.3 but the approximation to **G** is not good.

Table 4.9 shows the levels predicted by MCA for each variable for each farm, either by noting:

(a) the prediction region in which each farm lies, or
(b) the nearest projected CLP.

The prediction regions are shown in Fig. 4.4. The prediction regions give 20 incorrect predictions and the projected CLPs, 29 incorrect predictions; as expected, the prediction regions do better. The predicted values given in the PR column of Table 4.9 may be regarded as fitted values, $\hat{\mathbf{X}}$ corresponding to the optimal least-squares fitted values given by the Eckart–Young theorem. However, with categorical variables, there is no claim that the fitted values in ρ dimensions are optimal in any sense. It is an interesting, though unanswered, question as to what combination of distance and approximation predicts categorical levels with the fewest errors.

An advantage of prediction regions is that samples shown close to the boundaries between two regions, and hence likely to be correctly predicted with the greatest uncertainty, are easily ascertained by inspection. Thus Farm 16 in Figure 4.4 is only just predicted (correctly) as SF while Farm 13 is just within the region for zero manuring and very close to level 4, which is close to the true value of level 3. Incidentally, level 3 is never predicted, being hidden behind the levels 0, 1, 2 and 4 shown in the figure. Indeed, the Farms 7, 8, 13 and 16 with level 3 of manure lie central to the figure and behind the regions C0, C2 and C4. Similarly, a region for moisture level 4, which occurs only on Farms 9 and 12, does not appear in the plot for moisture, because it is hidden behind M1.

Table 4.10 gives the predictions both by prediction regions and nearness to the projected CLPs when using the extended matching coefficient. Fig. 4.5 shows the actual prediction regions for each variable, where the CLPs are those of PCA (i.e. using neither the shifted-sample nor shifted-CLP representations).

The prediction regions again give the fewest incorrect predictions – 23 compared to 28 for the projected CLPs. With these data, the prediction regions for MCA have done better than those for the

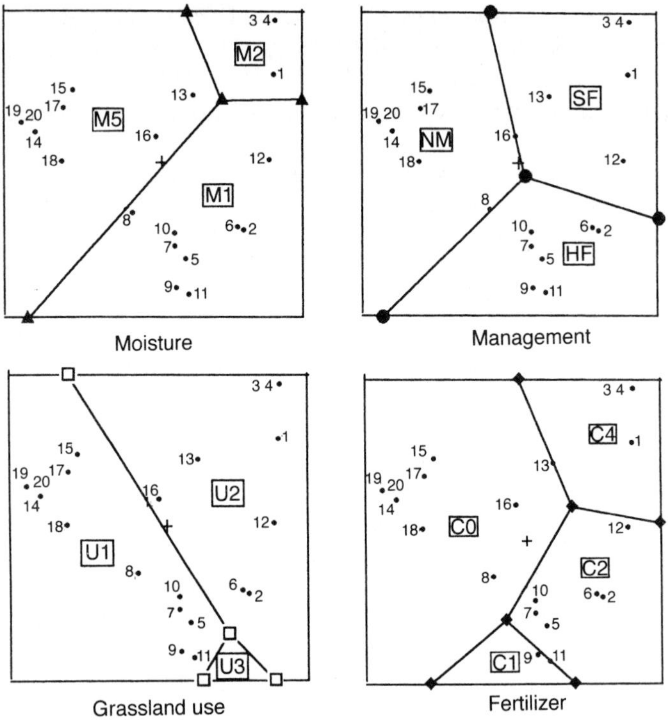

Fig. 4.4. *The prediction regions for each of the four variables, as given by MCA.*

extended matching coefficient but, as discussed above, we have to take into account the uncertainties of predictions made near the boundaries. In general, the prediction regions shown in Figs. 4.4 and 4.5 are very similar.

In Fig. 4.6, the MCA prediction regions for all four variables are superimposed; Fig. 4.7 shows the same thing for the extended matching coefficient. These displays are analogous to showing four (linear) biplot axes for continuous variables. With categorical variables, the display is more chaotic and with many variables would become unacceptable. Neverthless, in this case, the result is interesting and divides $\mathscr{L}$ into many mutually exclusive convex regions. There are 18 regions in Fig. 4.6 and 15 regions in Fig. 4.7

Table 4.10 The same as Table 4.9, but predictions arising from the analysis of the extended matching coefficient rather than χ^2 distance

Farm	*Moisture*			*Management*			*Grassland*			*Manuring*		
	T	*PR*	*PCLP*	*T*	*PR*	*PCLP*	*T*	*PR*	*PCLP*	*T*	*PR*	*PCLP*
1	1	2*	2*	SF	SF	SF	2	2	2	4	4	4
2	1	1	4*	BF	HF*	BF	2	2	3*	2	2	2
3	2	2	2	SF	SF	SF	2	2	2	4	4	4
4	2	2	2	SF	SF	SF	2	2	2	4	4	4
5	1	1	4*	HF	HF	HF	1	1	3*	2	2	1*
6	1	1	4*	HF	HF	BF*	2	2	3*	2	2	2
7	1	1	4*	HF	HF	HF	3	1*	3	3	2*	2*
8	5	5	1*	HF	NM*	BF*	3	1*	3	3	0*	3
9	4	1*	4	HF	HF	HF	1	1	3*	1	1	1
10	2	1*	4*	BF	HF*	BF	1	1	3*	1	0*	1
11	1	1	4*	BF	HF*	HF*	3	1*	3	1	2*	1
12	4	2*	4	SF	SF	SF	2	2	2	2	4*	4*
13	5	5	2*	SF	SF	SF	2	2	2	3	4*	4*
14	5	5	5	NM	NM	NM	3	1*	1*	0	0	0
15	5	5	5	NM	NM	NM	2	2	3*	0	0	0
16	5	5	2*	SF	SF	SF	3	2*	2*	3	4*	4*
17	2	5*	5*	NM	NM	NM	1	1	1	0	0	0
18	1	5*	5*	NM	NM	NM	1	1	1	0	0	0
19	5	5	5	NM	NM	NM	1	1	1	0	0	0
20	5	5	5	NM	NM	NM	1	1	1	0	0	0
Errors	0	6	12	0	4	3	0	6	8	0	7	5

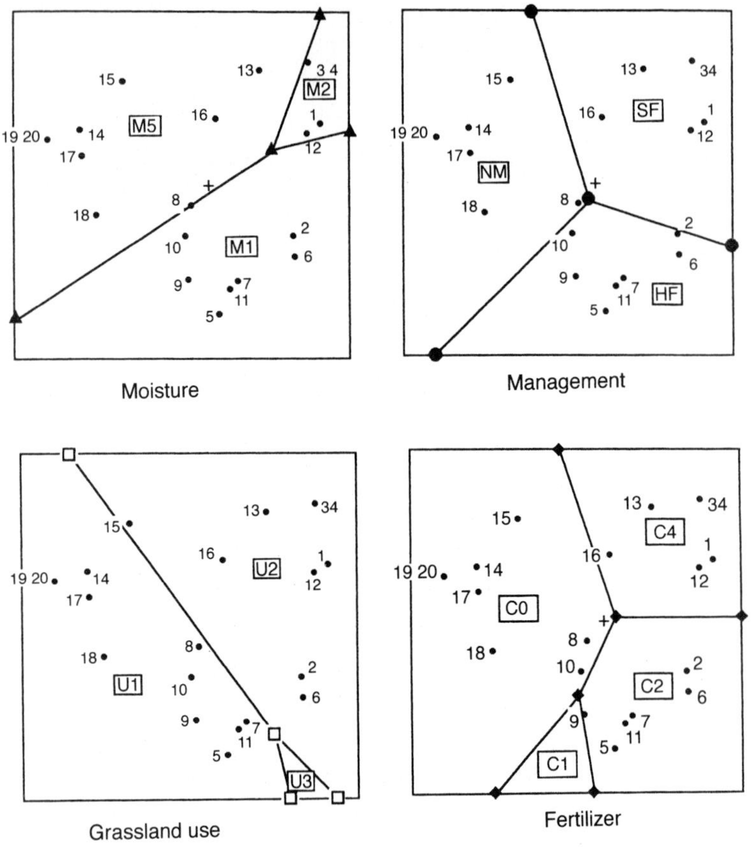

Fig. 4.5. *The prediction regions for each of the four variables, as given by the extended matching coefficient.*

but many of these are very small and others contain few or no examples. In both figures, three regions are fairly densely populated. These are:

MCA	(a)	M5, NM, U1, C0 containing Farms	14, 15, 17, 18, 19, 20
	(b)	M1, HF, U1, C1/2	5, 7, 9, 10, 11
	(c)	M2, SF, U2, C4	1, 3, 4
EMC	(a)	M5, NM, U1, C0 containing Farms	14, 17, 18, 19, 20
	(b)	M1, HF, U1, C2	5, 7, 9, 11
	(c)	M2, SF, U2, C4	1, 3, 4, 12

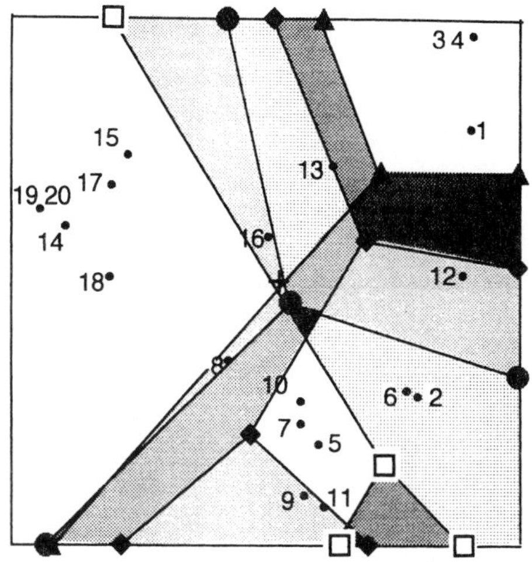

Fig. 4.6. *Combined prediction regions for MCA.*

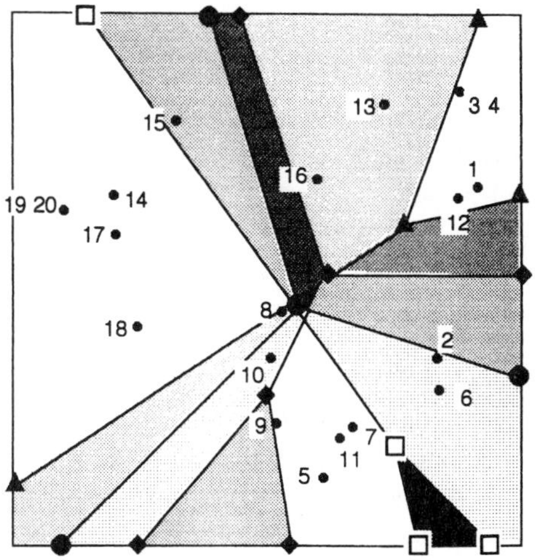

Fig. 4.7. *Combined prediction regions for the EMC.*

The category levels defining the groups are virtually the same in both cases. The farms included within the groups are virtually the same with MCA assigning two more farms than does EMC to the three groups.

CHAPTER 5

Canonical biplots

5.1 Introduction

This chapter is concerned with biplots in the context of canonical variate analysis (CVA) and therefore we begin by describing that method and in establishing notation. The distance used is Mahalanobis distance which differs fundamentally from those so far considered by not assuming that variables contribute independently to inter-sample ddistance. We shall see that, although both interpolative and predictive biplot axes exist and are linear, their directions differ; this is the most simple situation where this occurs.

5.2 Canonical variate analysis (CVA)

CVA is concerned with $\mathbf{X}(n \times p)$, assumed to be centred at the origin, partitioned into g groups of sizes given in the diagonal of the matrix $\mathbf{N}_g = \text{diag}(n_1, n_2, \ldots, n_g)$ from which can be derived a $g \times p$ matrix $\bar{\mathbf{X}}$ of group means. Thus, because of the centring, $\mathbf{1}'\mathbf{N}_g\bar{\mathbf{X}} = \mathbf{0}$. A between- and within-group analysis of variance can be constructed as in Table 5.1.

From Table 5.1, it can be seen that the between-and within-groups sums-of-squares-and-products matrices $\mathbf{B}$ and $\mathbf{W}$ are estimated in the usual way. CVA asks the question: 'What linear combination $\mathbf{Xv}$ of the p variables maximizes the between- to within-groups variance ratio?' Thus we want the vector $\mathbf{v}$ which maximizes

$$\frac{\mathbf{v}'\mathbf{B}\mathbf{v}}{\mathbf{v}'\mathbf{W}\mathbf{v}} \tag{5.1}$$

the solution of which may be regarded as providing a linear discriminator which is potentially useful for assigning to the

Table 5.1 Between and within groups Manova

	d.f.	*Sums of squares and products*
Between groups	$g-1$	$\mathbf{B}=\bar{\mathbf{X}}'\mathbf{N}_g\bar{\mathbf{X}}$
Within groups	$n-g$	$\mathbf{W}=\mathbf{X}'\mathbf{X}-\bar{\mathbf{X}}'\mathbf{N}_g\bar{\mathbf{X}}$
Total	$n-1$	$\mathbf{X}'\mathbf{X}$

groups. The scaling of $\mathbf{v}$ is arbitrary so, to obtain a unique result, we impose a constraint and choose $\mathbf{v}'\mathbf{W}\mathbf{v}=1$. The reason for this particular choice of constraint will soon become clear. Differentiating (5.1), with the Lagrangian term $\mu(\mathbf{v}'\mathbf{W}\mathbf{v}-1)$, yields after some simplification

$$\mathbf{B}\mathbf{v}=\lambda\mathbf{W}\mathbf{v} \tag{5.2}$$

where the constraint requires that $\mathbf{v}'\mathbf{B}\mathbf{v}=\lambda$, identifying λ as the value of the ratio (5.1). Equation (5.2) is the two-sided eigenvalue problem (Wilkinson, 1965; section A.2) which when $\mathbf{W}=\mathbf{I}$ becomes the classical eigenvalue problem (section A.1). Equation (5.1) is maximized when $\mathbf{v}=\mathbf{v}_1$, the eigenvector corresponding to the largest eigenvalue λ_1 of (5.2). Thus

$$\mathbf{y}=\mathbf{X}\mathbf{v}_1 \tag{5.3}$$

is the linear combination giving the maximal variance ratio. That solves the problem as posed and in a strict discriminant context may suffice, but it is well known that (5.2) has a total of p different solutions, $p-1$ of which give sub-optimal values of λ. What do these represent? All solutions of (5.2) may be gathered together in the form

$$\mathbf{B}\mathbf{V}=\mathbf{W}\mathbf{V}\mathbf{\Lambda} \tag{5.4}$$

where $\mathbf{V}=(\mathbf{v}_1,\mathbf{v}_2,\ldots,\mathbf{v}_p)$ and $\mathbf{\Lambda}=\operatorname{diag}(\lambda_1,\lambda_2,\ldots,\lambda_p)$, the eigenvalues being assumed to be given in non-ascending form. It is shown in section A.2 that the eigenvectors are orthogonal in the metric $\mathbf{W}$, i.e. $\mathbf{V}'\mathbf{W}\mathbf{V}$ is diagonal, and hence, with our chosen constraints, that

$$\mathbf{V}'\mathbf{W}\mathbf{V}=\mathbf{I} \tag{5.5}$$

Note that alternative forms of (5.5) are $\mathbf{V}\mathbf{V}'=\mathbf{W}^{-1}$ and $\mathbf{V}^{-1}=\mathbf{V}'\mathbf{W}$. Premultiplying (5.4) by $\mathbf{V}'$ gives

$$\mathbf{V}'\mathbf{B}\mathbf{V}=\mathbf{\Lambda} \tag{5.6}$$

All the linear combinations for a sample with values $\mathbf{x}$ are given by

$$\mathbf{y} = \mathbf{xV} \tag{5.7}$$

This gives a linear transformation from the original variables to a new set of variables, known as canonical variables. The group means of these canonical variables are termed canonical means and are given by

$$\mathbf{Y} = \bar{\mathbf{X}}\mathbf{V} \tag{5.8}$$

With the normalization (5.5), $\mathbf{YY}' = \bar{\mathbf{X}}\mathbf{W}^{-1}\bar{\mathbf{X}}'$, so that the distances between the canonical means are Mahalanobis distances. Thus, the Pythagorean distances between means in the canonical space are Mahalanobis distances in the original space and hence the canonical space is Euclidean. This space, which has min $(p, g-1)$ dimensions, is the counterpart of $\mathscr{R}$ of previous chapters, and will be so labelled here.

We shall need some further notation. The first ρ columns of $\mathbf{V}$ will be written $\mathbf{V}_\rho$ and the first ρ rows of $\mathbf{V}^{-1}$ will be written $\mathbf{V}^\rho$. The $p \times \rho$ matrix $\mathbf{J}_\rho$ has units on its diagonal but is otherwise zero; thus $\mathbf{VJ}_\rho = \mathbf{V}_\rho$ and $\mathbf{J}_\rho'\mathbf{V}^{-1} = \mathbf{V}^\rho$. Also, $\mathbf{V}^\rho\mathbf{V}_\rho$ is a $\rho \times \rho$ unit matrix and $\mathbf{J}_\rho\mathbf{J}_\rho'$ is a $p \times p$ matrix, zero everywhere except for units in the first ρ diagonal positions.

Our objective is to display the canonical means in only ρ dimensions, especially where $\rho = 2$, and canonical biplots are concerned with including information on the original variables in this ρ-dimensional canonical space which is labelled $\mathscr{L}$ in the following.

The approximation $\mathbf{Z}$ in $\mathscr{L}$ is simply obtained as the first ρ columns of $\mathbf{Y}$, i.e.

$$\mathbf{Z} = \bar{\mathbf{X}}\mathbf{V}_\rho = \bar{\mathbf{X}}\mathbf{VJ}_\rho$$

So far, the rationale of CVA has been presented as a concern with maximizing variance ratios. The approximation in $\mathscr{L}$ also approximates the Mahalanobis distances but it is not clear how, because the principal axes in $\mathscr{R}$ are obtained from $\mathbf{B}$ defined in weighted form in Table 5.1.

Thus $\mathscr{L}$ is not spanned by the first ρ principal axes of $\bar{\mathbf{X}}'\bar{\mathbf{X}}$, as it should be for a true principal components interpretation. Redefining $\mathbf{B} = \bar{\mathbf{X}}'\bar{\mathbf{X}}$ and substituting this definition into (5.2) gives new forms of $\mathbf{V}$ and the canonical means, but preserves the Mahalanobis distance property in the space $\mathscr{R}$. (Thus the new canonical axes in $\mathscr{R}$ are merely rotations of the old ones.) Then, because of the diagonality of (5.6), now with **B** unweighted, $\mathscr{L}$ is spanned by the ρ principal axes of the points that generate the canonical means. It follows (section 2.5) that the proportion of variance accounted for among the canonical means in $\mathscr{L}$ is

$$\frac{\sum_{i=1}^{\rho} \lambda_i}{\sum_{i=1}^{p} \lambda_i}$$

Gower (1966b) shows how this unweighted approach can be developed as the direct principal coordinates analysis (PCO) of the matrix of squared Mahalanobis distances. With $\mathscr{L}$ defined in terms of the principal axes of **Y**, we could use the methodology of Chapter 2 to obtain biplot axes. These axes would refer to the canonical variables and would be useful if the canonical axes were reified. However, our principal aim is to represent the *original* variables. In the remainder of this chapter, we see how to do this.

5.3 Interpolation

To interpolate a new sample $\mathbf{x} = (x_1, x_2, \ldots, x_p)$ into $\mathscr{R}$ requires its canonical representation

$$\mathbf{y} = \mathbf{x}\mathbf{V} \tag{5.9}$$

we may write $\mathbf{x} = x_1\mathbf{e}_1 + x_2\mathbf{e}_2 + \cdots + x_p\mathbf{e}_p$, where $\mathbf{e}_k$ is a unit vector on the kth Cartesian axis associated with **X**. Thus (5.9) becomes

$$\mathbf{y} = \sum_{k=1}^{p} x_k(\mathbf{e}_k\mathbf{V}) \tag{5.10}$$

To obtain the representation of the point **y** in the ρ dimensional space $\mathscr{L}$, **V** is replaced by $\mathbf{V}_\rho$ and (5.10) becomes

$$\mathbf{z} = \sum_{k=1}^{p} x_k(\mathbf{e}_k\mathbf{V}_\rho) \tag{5.11}$$

The quantities $\mathbf{e}_k\mathbf{V}_\rho$ $(k = 1, 2, \ldots, p)$ are the interpolants in $\mathscr{L}$ for unit values of the original variables, the coordinates of all p of which are given by the rows of $\mathbf{V}_\rho$, we refer to these points as **unit points**. For the kth variable, we may plot the coordinates given by the kth row of $\mu \mathbf{e}_k \mathbf{V}_\rho$, for a range of values of μ and mark these values of μ on the axis so obtained, to give a conventional linear coordinate axis with conveniently placed scale markers. The kth axis is the kth biplot axis for interpolation. Equation (5.11) shows that these axes may be used to interpolate graphically by finding the vector-sum of the markers for $x_1, x_2, \ldots, x_p$. Practically, this is more easily effected by finding the centroid of the markers and extending its position from the origin by multiplying by p, as was illustrated in section 2.3. Note that the means $\bar{\mathbf{X}}$ are interpolated into their correct positions in $\mathscr{L}$.

5.4 Prediction

In this section, prediction is first developed in a similar manner to that of Chapters 2 and 3; then it is given a characterization which links with the Eckart–Young rank ρ approximation of $\bar{\mathbf{X}}$. Finally the representation developed here is associated with a more conventional representation.

5.4.1 *Prediction in the canonical space*

As was seen in Chapter 1, prediction is the inverse of interpolation: given a point with coordinates $\mathbf{y}$ in the canonical space $\mathscr{R}$, predict the values $\mathbf{x}$ to be associated with its original variables. Inverting (5.9) gives

$$\mathbf{x} = \mathbf{y}\mathbf{V}^{-1} \tag{5.12}$$

All points that predict a value μ for the kth variable satisfy

$$\mu = \mathbf{y}\mathbf{V}^{-1}\mathbf{e}_k' \tag{5.13}$$

where $\mathbf{e}_k$ is a unit vector along the kth original axis. When $\mathbf{y}$ lies in $\mathscr{L}$, only its first ρ coordinate values, $\mathbf{z}$, can be non-zero, so we have

$$\mu = \mathbf{z}\mathbf{V}^{\rho}\mathbf{e}_k' \tag{5.14}$$

Thus these points lie in a $(\rho - 1)$-dimensional plane in $\mathscr{L}$ and as μ varies the resulting planes are parallel, all being normal to the direction $\mathbf{e}_k(\mathbf{V}^{\rho})'$. Graphically, the problem is to identify in which of these planes the point $\mathbf{z}$ lies and then to predict the corresponding value of μ. This is most simply accomplished by taking the line through the origin normal to all the parallel ρ-dimensional planes and to mark it with scale values for conveniently spaced values of the markers μ. Then subsequent orthogonal projection of any $\mathbf{z}$ onto this line gives the correct marker for prediction. Any point on this line has the form

$$\mathbf{z}_\mu = \sigma \mathbf{e}_k (\mathbf{V}^\rho)' \tag{5.15}$$

and substitution into (5.14) identifies σ as satisfying

$$\mu = \sigma \mathbf{e}_k (\mathbf{V}^\rho)' \mathbf{V}^\rho \mathbf{e}_k'$$

Thus, finally

$$\mathbf{z}_\mu = \frac{\mu}{\mathbf{e}_k (\mathbf{V}^\rho)' \mathbf{V}^\rho \mathbf{e}_k'} \mathbf{e}_k (\mathbf{V}^\rho)' \tag{5.16}$$

gives the position of the marker μ on the kth predictive biplot axis. It follows that the set of unit markers for all ρ variables are given by the columns of

$$\mathbf{V}^\rho \operatorname{diag}^{-1}\{(\mathbf{V}^\rho)'\mathbf{V}^\rho\} \tag{5.17}$$

Comparing (5.16) with the interpolation marker $\mu \mathbf{e}_k \mathbf{V}_\rho$ found in section 5.3 shows that the predictive and interpolative biplot axes for the kth variable are both linear but lie in different directions. Note that this remains true even in the full canonical space $\mathscr{R}$ where the unit points for interpolation are given by the rows of $\mathbf{V}$ and those for prediction are given by the rows of $(\mathbf{V}^{-1})'$.

The above development is similar to that of Chapters 2 and 3 but modified to take account of the fact that $\mathscr{R}$ now is not the space that holds $\mathbf{X}$ but is a space that holds the canonical variables. PCA corresponds to $\mathbf{W} = \mathbf{I}$, and with this setting, the results of this chapter should then agree with those of Chapter 2. That this is so can be checked by noting that when $\mathbf{W} = \mathbf{I}$ then $\mathbf{V}$ becomes an orthogonal matrix, ensuring that $(\mathbf{V}^{-1})' = \mathbf{V}$ and hence $(\mathbf{V}^\rho)' = \mathbf{V}_\rho$ so that the prediction and interpolation directions now coincide. The prediction unit point (5.15) becomes $\mathbf{z}_\mu = \sigma \mathbf{e}_k \mathbf{V}_\rho$, agreeing with the interpolation direction but not in the spacing of the scale markers; indeed, the unit markers for interpolation and prediction were shown to be inversely related (section 2.3) as is easily verified by evaluating the product of the lengths of the two markers for $\mu = 1$.

CVA provides the most simple example of where interpolative and predictive biplot axes are linear but with different sets of directions.

Returning to the canonical case, we can check that orthogonal projection onto the kth prediction biplot axis gives the correct prediction. The length u_k of a unit marker for the kth variable is given by

$$u_k^2 = \mathbf{z}_1\mathbf{z}_1' = \frac{1}{\mathbf{e}_k(\mathbf{V}^\rho)'\mathbf{V}^\rho\mathbf{e}'_k}$$

and hence for one unit, (5.16) becomes

$$\mathbf{z}_1 = u_k^2\mathbf{e}_k(\mathbf{V}^\rho)' \tag{5.18}$$

This gives the coordinates of the unit marker on the kth biplot axis in $\mathscr{L}$. The unit vector on this axis is $u_k\mathbf{e}_k(\mathbf{V}^\rho)'$so the length of the projection of any sample point $\mathbf{z}$ in $\mathscr{L}$ is $u_k\mathbf{z}\mathbf{V}^\rho e_k'$ and this represents a prediction $\hat{x}_k$ for the kth variable, given by

$$\hat{x}_k = \mathbf{z}\mathbf{V}^\rho e_k' \tag{5.19}$$

units and hence, in agreement with (5.14), this checks that (5.19) gives the correct prediction to be associated with $\mathbf{z}$ for the value of the kth variable.

Thus, (5.19) is the basic algebraic result for prediction; it has already been shown that this value is easily found graphically by reading off the marker value at the point of projection, where unit markers are given by (5.17). Of special interest is when $\mathbf{z} = \mathbf{x}\mathbf{V}_\rho$ arises from one of the original sample means. Then the prediction for all p variables is $\mathbf{x}\mathbf{V}_\rho\mathbf{V}^\rho$ and for all g means is

$$\hat{\mathbf{X}} = \bar{\mathbf{X}}\mathbf{V}_\rho\mathbf{V}^\rho \tag{5.20}$$

When $\rho = p$, then $\mathbf{V}_\rho = \mathbf{V}$ and $\mathbf{V}^\rho = \mathbf{V}^{-1}$, so that in the full canonical space, where there is no approximation, $\hat{\mathbf{X}}$ correctly becomes $\bar{\mathbf{X}}$. It will be seen below that (5.20) is a fundamental representation of a rank ρ approximation to $\bar{\mathbf{X}}$.

5.4.2 *An alternative characterization*

It was shown in Section 5.2 that the rows of $\bar{\mathbf{X}}\mathbf{V}$ generate Mahalanobis distances and that

$$\mathbf{Z} = \bar{\mathbf{X}}\mathbf{V}\mathbf{J}_\rho\mathbf{J}'_\rho$$

gives the best least-squares ρ-dimensional approximation to $\bar{\mathbf{X}}\mathbf{V}$, in the sense of PCA. Writing $\mathbf{Z}$ in this form (rather than $\bar{\mathbf{X}}\mathbf{V}\mathbf{J}_\rho$) ensures that it is referred to the p-dimensional space $\mathscr{R}$ rather than to the ρ-dimensional space $\mathscr{L}$.

Thus, $\| \bar{\mathbf{X}}\mathbf{V} - \mathbf{Z} \|$ is minimized over all rank ρ matrices $\mathbf{Z}$ when $\mathbf{Z} = \bar{\mathbf{X}}\mathbf{V}\mathbf{J}_\rho\mathbf{J}'_\rho$. Now

$$\|\bar{\mathbf{X}}\mathbf{V} - \bar{\mathbf{X}}\mathbf{V}\mathbf{J}_\rho\mathbf{J}'_\rho\| = \text{trace}\,(\bar{\mathbf{X}} - \mathbf{X}\mathbf{V}\mathbf{J}_\rho\mathbf{J}_\rho{}'\mathbf{V}^{-1})\,\mathbf{V}\mathbf{V}'(\bar{\mathbf{X}} - \bar{\mathbf{X}}\mathbf{V}\mathbf{J}_\rho\mathbf{J}'_\rho\mathbf{V}^{-1})'$$

$$= \text{trace}\,(\bar{\mathbf{X}} - \bar{\mathbf{X}}\mathbf{V}_\rho\mathbf{V}^\rho)\,\mathbf{W}^{-1}\,(\bar{\mathbf{X}} - \bar{\mathbf{X}}\mathbf{V}_\rho\mathbf{V}^\rho)' \qquad (5.21)$$

Equation (5.21) establishes that

$$\hat{\mathbf{X}} = \bar{\mathbf{X}}\mathbf{V}_\rho\mathbf{V}^\rho$$

given by (5.20) is the best ρ-dimensional weighted approximation to $\bar{\mathbf{X}}$, where the sum-of-squares of residuals is measured in the Mahalanobis metric. This implies that (5.20) must be a generalization of the Eckart–Young theorem, giving a rank ρ approximation to $\bar{\mathbf{X}}$ in the metric $\mathbf{W}$. The next section shows that (5.20) is consistent with a more conventional form expressed in terms of the singular value decomposition of $\bar{\mathbf{X}}$.

5.4.3 *Links with SVD and the Eckart–Young theorem*

By considering $\mathbf{V}_s$ the s columns of $\mathbf{V}$ corresponding to the non-zero eigenvalues of (5.4), the relationship

$$\mathbf{V}'\bar{\mathbf{X}}'\bar{\mathbf{X}}\mathbf{V} = \mathbf{\Lambda} = \mathbf{\Sigma}^2$$

(say), shows that there exists an orthonormal matrix $\mathbf{U}_s$ such that

$$\bar{\mathbf{X}}\mathbf{V}_s\mathbf{\Sigma}_s^{-1} = \mathbf{U}_s$$

This gives

$$\bar{\mathbf{X}}\mathbf{V}_s = \mathbf{U}_s\mathbf{\Sigma}_s$$

Augmenting $\mathbf{V}_s$ by the null eigenvectors of (5.4), $\mathbf{\Sigma}_s$ by corresponding zeros and $\mathbf{U}_s$ by any vectors orthogonal to each other and to $\mathbf{U}_s$, gives

$$\bar{\mathbf{X}}\mathbf{V} = \mathbf{U}\mathbf{\Sigma}$$

so that

$$\bar{\mathbf{X}} = \mathbf{U}\boldsymbol{\Sigma}\mathbf{V}^{-1} \tag{5.22}$$

where $\mathbf{U}$ is orthogonal and $\mathbf{V}'\mathbf{W}\mathbf{V} = \mathbf{I}$. Equation (5.22) is the general SVD for a matrix $\bar{\mathbf{X}}$ in terms of the metric $\mathbf{W}$ and this clearly takes the familiar form when $\mathbf{V}$ is orthogonal. From $\mathbf{V}'\mathbf{W}\mathbf{V} = \mathbf{I}$, (5.22) may be written as

$$\bar{\mathbf{X}} = \mathbf{U}\boldsymbol{\Sigma}\mathbf{V}'\mathbf{W} \tag{5.23}$$

and this is the form used by Gabriel (1972) in the first exposition of canonical biplots (see also Appendix A of Greenacre, 1984). Now, trivially, (5.22) may be written

$$\bar{\mathbf{X}} = (\bar{\mathbf{X}}\mathbf{V})\mathbf{V}^{-1}$$

which is in the same form as (5.20)

$$\hat{\mathbf{X}} = \bar{\mathbf{X}}\mathbf{V}_\rho\mathbf{V}^\rho$$

which (5.21) showed to give the best rank ρ approximation to $\bar{\mathbf{X}}$ in the metric $\mathbf{W}$. Equivalently, from the relationships

$$\mathbf{V}'\mathbf{W} = \mathbf{V}^{-1}, \mathbf{V}^\rho = \mathbf{J}_\rho'\mathbf{V}^{-1} = \mathbf{J}_\rho'\mathbf{V}'\mathbf{W} = (\mathbf{W}\mathbf{V}\mathbf{J}_\rho)' = (\mathbf{W}\mathbf{V}_\rho)'$$

corresponding to (5.23) we may write the approximation as

$$\hat{\mathbf{X}} = (\bar{\mathbf{X}}\mathbf{V}_\rho)(\mathbf{W}\mathbf{V}_\rho)'$$

The predictive canonical biplot is formed by plotting in $\mathscr{L}$ the usual approximation $\mathbf{X}\mathbf{V}_\rho$ to the canonical means and the columns of $\mathbf{V}^\rho$, or equivalently $\mathbf{V}_\rho'\mathbf{W}$, for the variables. Equation (5.16) gives the position of units on the kth axis, so these axes, although non-orthogonal, again may be used for prediction similarly to the use of Cartesian axes; interpolation axes are given by the rows of $\mathbf{V}_\rho$ (section 5.3)

In the above, $\mathbf{B}$ has been treated as an unweighted sums-of-squares-and-products matrix $\bar{\mathbf{X}}'\bar{\mathbf{X}}$. Using the weighted form $\mathbf{B} = \bar{\mathbf{X}}'\mathbf{N}_g\bar{\mathbf{X}}$, with centring $\mathbf{1}'\mathbf{N}_g\bar{\mathbf{X}} = \mathbf{0}$, rather than $\mathbf{1}'\bar{\mathbf{X}} = \mathbf{0}$, then (5.4) defines a new matrix $\mathbf{V}$ and (5.22) now becomes

$$\bar{\mathbf{X}} = \mathbf{N}_g^{-1/2}\mathbf{U}\boldsymbol{\Sigma}\mathbf{V}^{-1}$$

However, we still have $\hat{\mathbf{X}} = \bar{\mathbf{X}}\mathbf{V}_\rho\mathbf{V}^\rho$ with unit points given by (5.16) and interpolation $\bar{\mathbf{X}}\mathbf{V}_\rho$ with unit points given by $\mathbf{V}_\rho$. By regarding

$$\hat{\mathbf{X}} = \bar{\mathbf{X}}\mathbf{V}_\rho\mathbf{V}^\rho$$

as the primary representation of the Eckart–Young rank ρ approximation to $\bar{\mathbf{X}}$ (rather than $\mathbf{U\Sigma V}'$ or (5.23)), all cases are covered and interpolation $\bar{\mathbf{X}}\mathbf{V}_\rho$ with the 'inversion' $\mathbf{V}^\rho$ for prediction are nicely encapsulated in a particularly simple and transparent formula. In addition, variants of the singular value decomposition can be avoided and therefore the same computer program can encompass the conventional Eckart–Young approximation and its generalized forms using either weighted or unweighted means. Finally, we note that (5.20) may be interpreted as giving the projection in the metric $\mathbf{W}$ of $\bar{\mathbf{X}}$ onto the ρ-dimensional space spanned by the columns of $\mathbf{V}_\rho$.

5.5 Canonical correlation biplots

It is well known that CVA may be formulated as a special case of canonical correlation analysis in which the variables of one set, representing the group structure, are dummies (Gower, 1989). The close relationship of the two methods extends to their biplots. We give the main results in this section, but leave it to the reader to justify them by making the necessary slight modifications to the developments of sections 5.3 and 5.4. ter Braak (1990b) discusses biplots in the context of canonical correlation.

In canonical correlation, we are given two data matrices $\mathbf{X}$ and $\mathbf{Y}$ which refer to the same n samples, presented in the same order, but to two different sets of variables, one of size p and the other of size q. The objective is to find linear combinations $\mathbf{Xl}$ and $\mathbf{Ym}$ that have maximum correlation. Writing $\mathbf{X}'\mathbf{X} = \mathbf{P}$, $\mathbf{Y}'\mathbf{Y} = \mathbf{Q}$ and $\mathbf{X}'\mathbf{Y} = \mathbf{R}$ these are given by solutions to the following equations:

$$\mathbf{RM} = \mathbf{PL\Sigma} \tag{5.24}$$

$$\mathbf{R}'\mathbf{L} = \mathbf{QM\Sigma}'$$

In the usual way, the equations (5.24) may be written as the two-sided eigenvalue problems:

$$(\mathbf{RQ}^{-1}\mathbf{R}')\mathbf{L} = \mathbf{PL\Sigma\Sigma}'$$

$$(\mathbf{R}'\mathbf{P}^{-1}\mathbf{R})\mathbf{M} = \mathbf{QM\Sigma}'\mathbf{\Sigma} \tag{5.25}$$

The vectors $\mathbf{l}$ and $\mathbf{m}$ are those that correspond to the maximum eigenvalues of (5.25), but the complete canonical set of solutions is of interest or, at least, the ρ most important solutions. Because we are maximizing a correlation, which is a ratio, the scaling of the vectors is entirely arbitrary but to fix on a unique solution it is usual to choose the scalings

$$\mathbf{L}'\mathbf{PL} = \mathbf{I} \quad \text{and} \quad \mathbf{M}'\mathbf{QM} = \mathbf{I} \tag{5.26}$$

which when combined with (5.24) gives

$$\mathbf{L}'\mathbf{RM} = \Sigma \tag{5.27}$$

A ρ-dimensional plot is obtained by plotting the two sets of canonical variables

$$\mathbf{XL}_\rho \quad \text{and} \quad \mathbf{YM}_\rho \tag{5.28}$$

and these have unit markers for interpolation given by the rows of

$$\mathbf{L}_\rho \quad \text{and} \quad \mathbf{M}_\rho \tag{5.29}$$

and for prediction (see (5.17)) by the columns of

$$\mathbf{L}^\rho \text{diag}^{-1}\{(\mathbf{L}^\rho)'\mathbf{L}^\rho\} \quad \text{and } \mathbf{M}^\rho \text{diag}^{-1}\{(\mathbf{M}^\rho)'\mathbf{M}^\rho\} \tag{5.30}$$

Distances between points of the first (second) set are of the Mahalanobis type in the metric $\mathbf{P}$ ($\mathbf{Q}$) and in ρ-dimensional approximations, the canonical axes are principal axes.

The angles between the biplot axes are of some interest. Before approximation, the direction of the kth interpolative axis is given by the vector $\mathbf{e}_k\mathbf{L}$ so the angle θ_{hk} between axis h and axis k is given by

$$\begin{aligned} \cos(\theta_{hk}) &= \mathbf{e}_k\mathbf{L}\mathbf{L}'\mathbf{e}_h' / \{(\mathbf{e}_k\mathbf{L}\mathbf{L}'\mathbf{e}_k')^{1/2}(\mathbf{e}_h\mathbf{L}\mathbf{L}'\mathbf{e}_h')^{1/2}\} \\ &= \frac{p^{hk}}{(p^{hh}p^{kk})^{1/2}} \end{aligned} \tag{5.31}$$

where p^{hk} is the (h, k)th element of $\mathbf{P}^{-1}$.

Writing

$$\mathbf{p} = \text{diag}^{-1}((\mathbf{L}')^{-1}\mathbf{L}^{-1}) = \text{diag}(p_{11}^{-1}, p_{11}^{-1}, \ldots, p_{pp}^{-1})$$

the direction of the kth predictive axis is given by the vector $\mathbf{L}^{-1}\mathbf{pe}_k'$ so the angle ϕ_{hk} between axis h and axis k is given by:

$$\cos(\phi_{hk}) = \frac{\mathbf{e}_k\,\mathbf{p}(\mathbf{L}')^{-1}\mathbf{L}^{-1}\mathbf{pe}_h'}{\{(\mathbf{e}_k\,\mathbf{p}(\mathbf{L}')^{-1}\mathbf{L}^{-1}\mathbf{pe}_k)^{1/2}(\mathbf{e}_h\,\mathbf{p}(\mathbf{L}')^{-1}\mathbf{L}^{-1}\mathbf{pe}_h)^{1/2}\}}$$

$$= \frac{p_{hk}}{(p_{hh} p_{kk})^{1/2}} \tag{5.32}$$

which is the correlation betwen $\mathbf{x}_h$ and $\mathbf{x}_k$.

Similar results to (5.31) and (5.32) apply to the second set of variables. Of course, in ρ-dimensional approximations these results will also be approximate. Equation (5.32) is useful but it seems that (5.31) has little interpretive interest. When two variables are completely correlated, ϕ_{hk} is zero and this remains true in approximations. Thus, predictive axes that have small angles in an approximation suggest high correlations between the corresponding variables.

Finally, we note the links with generalized SVDs. The most simple form derives directly from (5.27) to give

$$\mathbf{P}^{-1}\mathbf{R}\mathbf{Q}^{-1} = \mathbf{L}\Sigma\mathbf{M}' \quad \text{with} \quad \mathbf{L}'\mathbf{P}\mathbf{L} = \mathbf{I} \text{ and } \mathbf{M}'\mathbf{Q}\mathbf{M} = \mathbf{I} \tag{5.33}$$

It is easy to verify that (5.33) satisfies (5.24). The generalized singular value decomposition consistent with (A.22) and (A.23) is

$$\mathbf{R} = \mathbf{S}\Sigma\mathbf{T}^{-1} \quad \text{with} \quad \mathbf{S}'\mathbf{P}^{-1}\mathbf{S} = \mathbf{I} \quad \text{and} \quad \mathbf{T}'\mathbf{Q}\mathbf{T} = \mathbf{I} \tag{5.34}$$

where $\mathbf{L} = \mathbf{P}^{-1}\mathbf{S}$ and $\mathbf{M} = \mathbf{Q}^{-1}(\mathbf{T}^{-1})'$. Approximations in ρ dimensions are given by the generalized Eckart–Young theorem (A.40) with its alternative forms (A.41).

5.6 Example

We give an example of the use of biplots in CVA; with canonical correlation analysis, both sets of variables can be handled in precisely the same ways that they are in the following and no further example is necessary. This example uses the meteorological data published in Table 2 of Gabriel (1972) concerning a cloud-seeding experiment.

The experimental units are days, falling into one of $g = 5$ groups. On each day, the rainfall is measured at $p = 4$ sites:

(a) north,
(b) buffer (lying between (a) and (c)),
(c) centre and
(d) south.

They are allocated to groups according to a cross-classification of [Site examined for suitable cloud conditions] and [Whether seeding was done]. Thus:

G1 North site examined, seeding suitable and done
G2 North site examined, seeding unsuitable and not done
G3 Centre site examined, seeding suitable and done
G4 Centre site examined, seeding unsuitable and not done

and there was a control group, G5, measured before the experiment started.

Table 5.2 shows the number of days in each group, together with the average values of daily rainfall at each of the four sites in each of the five groups of days. The labelling of variables and groups given in Table 5.2 is used in the following discussion. Table 5.3 shows the within-sites dispersion matrix **W** for the four experimental groups (i.e. excluding G5, for which only mean values are available).

The analysis was done, both in the weighted and unweighted forms, using (5.1) and (5.8), providing interpolative axes with units $\mathbf{e}_k\mathbf{V}_\rho$ and predictive axes with units given by (5.17). There is little difference between the two analyses. The weighted analysis is what

Table 5.2 Number of observation-days and average daily rainfall (mm) at four sites and five groups – from Gabriel (1972)

Group	*Days*	*Site*			
		a	*b*	*c*	*d*
G1	119	13.442	11.526	10.364	8.409
G2	89	2.424	1.893	1.960	1.549
G3	97	11.393	13.704	12.422	7.724
G4	86	3.217	2.912	2.558	1.729
G5	846	6.780	7.659	6.721	4.234

Table 5.3 The within-group dispersion matrix **W**

Site (a)	93.90			
Site (b)	95.33	165.56		
Site (c)	68.94	89.63	111.72	
Site (d)	25.69	27.76	56.27	70.32

is essentially Gabriel's (1972) two-dimensional predictive biplot and his interpretation remains valid in the unweighted form. Therefore, there is little to add concerning the interpretation of this particular set of data except to discuss the relationship between interpolation and prediction. Because the unweighted analysis gives the better approximation to Mahalanobis distances, this is used as the basis for the following discussion.

Figure 5.1 shows the unweighted analysis; Gabriel's confidence circles are omitted but could easily be included, if desired. This figure has been constructed using the horizontal and vertical canonical coordinate axes but these are not shown, because it is the biplot axes which refer to the original variables that are used for interpretation. The two-dimensional approximation is essentially exact, with eigenvalues (1.248, 0.238, 0.002, 0.000). The interpolative axes are shown in Fig. 5.1(a) and the predictive axes in Fig. 5.1(b), both equipped with scale markers. The two are shown separately, partially to avoid confusion but also because interpolation and prediction are likely to be independent processes. Figures 5.1 confirms that, unlike with classical biplots based on PCA, the predictive and interpolative axes differ not only in scale but also in direction. Figs. 5.1(a) and 5.1(b) share the same canonical means (G1, G2, G3, G4) and are also linked by a point labelled P which is discussed below. The interpolation of the set of means (13.442, 11.526, 10.364, 8.409) mm of rainfall observed for G1 is obtained by finding the centroid C1 of the markers (indicated by arrowed vectors along the biplot axes) for these values on the four interpolation axes and multiplying by $p = 4$, as is shown in Fig. 5.1(a). As expected, the position representing the canonical mean of G1 is recovered. The predictions for this group, read from projection onto the prediction axes of Fig. 5.1(b), are (13.4, 11.4, 10.3, 8.4), agreeing well with the original figures because the two-dimensional approximation is essentially exact; with a less accurate approximation, the agreement would be less good. The same technique can be used to interpolate new samples to give new points in the display and to predict the values to be associated with a given point. Thus we can interpolate the values (11, 16, 11, 12), whose markers are indicated by the pointing arrows, to give the point P shown in both figures. The predicted values for P obtained from Fig. 5.1(b) are about (13.6, 15.0, 13.4, 9.0) which do not agree well with the original values because the new sample lies far outside the two-dimensional display. The point for G5 has been interpolated in the same way.

From the near coincidence of b, and c, whose directions are indistinguishable in Fig. 5.1(b), it is clear that the predicted values for these two variables are nearly linearly related and hence that they have very high intergroup correlation. A similar remark applies to variables a and d, but the correlation is less strong. In general, correlation decreases as angles increase or, as it is usually put, the cosine of the correlation approximates correlation. However, the precise nature of this approximation needs examination.

a)

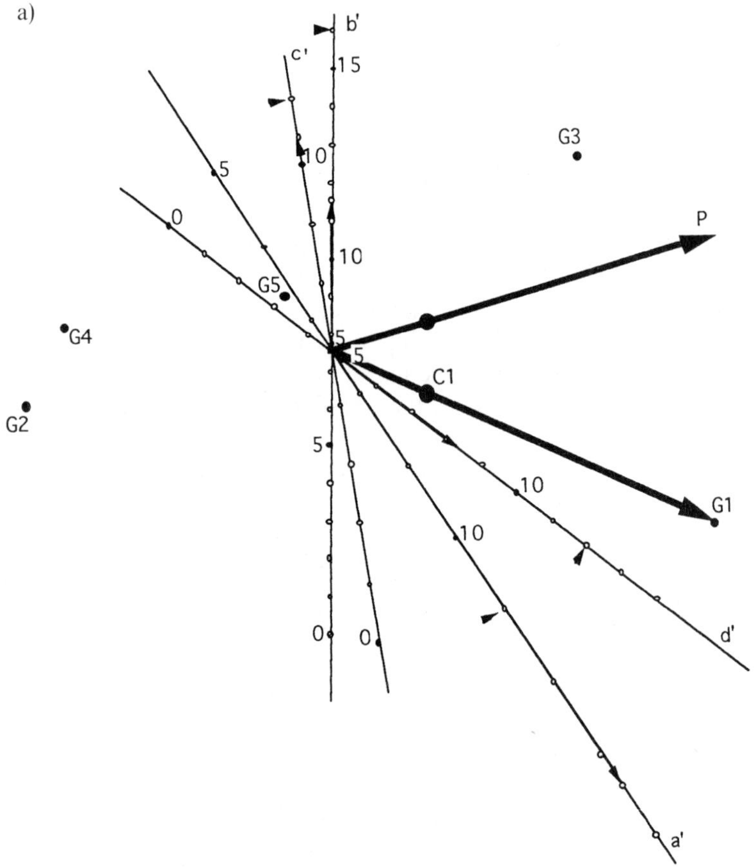

b)

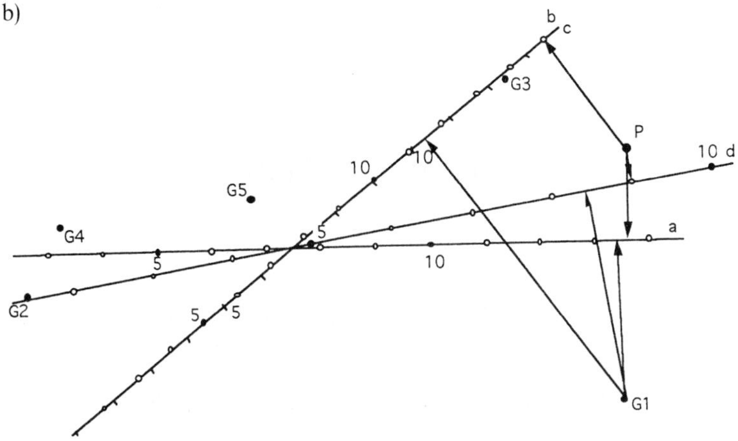

Fig. 5.1. *The two-dimensional approximation of the canonical space of the five groups of Table 5.1: (a) the interpolation biplot axes are calibrated in mean daily rainfall (mm) for the four regions labelled a′, b′, c′ and d′; (b) the prediction biplot axes, labelled a, b, c, d.*

Even in the classical case ($\mathbf{W} = \mathbf{I}$), one would need to plot $\mathbf{L}_\rho \mathbf{\Lambda}_\rho^{-1/2}$ rather than $\mathbf{L}_\rho$ to give a true correlational interpretation. When visualization of either inter- or intra-group correlations is required, it seems better to use the methods advocated by Hills (1969).

CHAPTER 6

Nonlinear biplots

6.1 Introduction

In previous chapters, we have met several specific examples of biplots. These are characterized by a data-matrix **X**, a choice of distance – Pythagorean, Mahalanobis, χ^2 – and a method of MDS. The general aim is to represent the inter-sample distances in a ρ-dimensional space $\mathscr{L}$, preferably of few dimensions, and then to equip $\mathscr{L}$ with axes representing the original p variables. These axes can be used for interpolation or prediction. We have already mentioned that any method of metric MDS can be used to approximate the distances and have shown (in Chapter 3) that linear predictive biplot axes can then be superimposed when distance is Pythagorean. Some difficulties have been encountered:

- linear axes are not always available except as approximations;
- with categorical variables prediction regions replace axes.

Further, all the distances so far encountered are Euclidean, so that points representing the given rows of **X** and any future samples that might be interpolated can be represented in the finite-dimensional space $\mathscr{R}$ (p dimensions for PCA and CVA but L dimensions for MCA). Because the previously defined matrices **V**, **W** and **L** pertain to the original form of **X**, interpolation is not the same as doing the analysis *ab initio* with **X** augmented by the extra sample. Nevertheless, even if the augmented form were used, only p dimensions would be required, although the positions of the points in $\mathscr{R}$ would change. Indeed, except with Pythagorean distance, because **W** and **L** change then so do the inter-sample distances in $\mathscr{R}$. The asumption is that the differences between complete and augmented analyses would be minor, but if the new sample differed greatly from those of **X** itself, then this could have a major effect on the approximation of inter-sample distances in $\mathscr{L}$.

With non-Euclidean distances, commonly used with general forms of MDS, additional complications occur. In this chapter, we discuss some generalizations that extend biplot methodology to cover these cases. Initially, we confine our attention to Euclidean embeddable distances. For example

$$d_{ij} = \left(\sum_{k=1}^{n} |x_{ik} - x_{jk}| \right)^{1/2} \tag{6.1}$$

is embeddable but its square is not. This means that all the distances $\{d_{ij}\}$ derived between the rows of **X** can be generated in a Euclidean space $\mathscr{R}$ by the distances between points whose coordinates are given as the rows of a matrix **Y**. The dimensionality of $\mathscr{R}$ is now written as m; in previous chapters, $m = p$ or $m = L$ but now $m = n - 1$. Many ways of deriving Euclidean embeddable distances from **X** are discussed by Gower and Legendre (1986), and a detailed presentation is not given here. Attention is restricted to quantitative variables, which we show necessarily lead to the consideration of nonlinear biplot axes. Later, in section 6.4, we relax the restriction to include distances which are not Euclidean embeddable. Chapter 7 extends the ideas developed here to include categorical variables.

The basic properties of Euclidean embeddable distance matrices that we shall need are given in section A.5, where it is shown to be convenient to consider the distances to be given in a matrix

$$\mathbf{D} = \{ -\tfrac{1}{2} d_{ij}^2 \}$$

The generating coordinates **Y** are given by any decomposition satisfying

$$\mathbf{B} = (\mathbf{I} - \mathbf{1s}')\mathbf{D}(\mathbf{I} - \mathbf{s1}') = \mathbf{YY}' \tag{6.2}$$

where **s** is any vector satisfying $\mathbf{1}'\mathbf{s} = 1$ implying that $\mathbf{s}'\mathbf{Y} = 0$. Thus there are many representations of the generating coordinates but these are all equivalent, differing only in translation (given by the choice of **s**) and orientation (given by the different decompositions). For several reasons, which will become evident, we choose $\mathbf{s} = \mathbf{1}/n$ and choose **Y** to be given by the spectral decomposition of **B**, so that $\mathbf{YY}' = \boldsymbol{\Lambda}$, the diagonal matrix of the eigenvalues, asumed to be given in non-increasing order, of **B**. With this choice, **Y** is centered at its centroid and is referred to its principal axes. It so happens that this choice is also that used in the metric scaling approximation method of principal coordinates analysis (PCO)/classical scaling, which we shall discuss below in section 6.3, but it is important to realize here that the representation of $\mathbf{Y}$ in $\mathscr{R}$ is exact and that considerations of approximation are not relevant. Thus

$$\mathbf{B} = (\mathbf{I} - \mathbf{N})\mathbf{D}(\mathbf{I} - \mathbf{N}) = \mathbf{YY}' \tag{6.3}$$

where

$$\mathbf{N} = \mathbf{11}'/n \quad \text{and} \quad \mathbf{YY}' = \mathbf{\Lambda}$$

$\mathbf{Y}$ is referred to its orthogonal principal axes as coordinate axes. In the following, these principal axes are used as the main tool for plotting but we regard them as having little substantive interest. We think of them as a kind of scaffolding to be used to support information which is of substantive interest (see Chapter 2 for some comments on reification):

The first thing we want to do is to endow $\mathscr{R}$ with coordinate axes that relate to the original variables. In the classical Cartesian case, the axis for the kth variable is the locus of $\mu\mathbf{e}_k$ as μ varies. It is therefore natural to consider the interpolation of $\mu\mathbf{e}_k$ in $\mathscr{R}$. $\mu\mathbf{e}_k$ represents a sample with value μ for the kth variable and zero for all other variables; recall that zero represents the mean. We term such a 'sample' a **pseudosample** for the kth variable; for certain values of μ, these pseudosamples will be most unusual. The formula for interpolating any sample into $\mathscr{R}$ is given in (A.67) as

$$\mathbf{y} = \mathbf{\Lambda}^{-1}\mathbf{Y}'(\mathbf{d}_{n+1} - \tfrac{1}{n}\mathbf{D1}) \tag{6.4}$$

where $\mathbf{d}_{n+1} = \{-\frac{1}{2}d_{n+1}^2\}$ is the vector giving the ddistances of the new sample (the pseudosample) from each of the n original samples. These ddistances are measured in whatever is the chosen form – (e.g.(6.1)) – of defining d_{ij}. Becuase we are dealing with embedding, the new sample requires an $(m + 1)$th dimension whose coordinate y_{m+1} is given by (A.65) as

$$y_{m+1}^2 = \frac{1}{n}\mathbf{1}'\mathbf{D1} - \frac{2}{n}\mathbf{1}'\mathbf{d}_{n+1} - \mathbf{y}'\mathbf{y} \tag{6.5}$$

We shall say that y_{m+1} gives a coordinate in a residual dimension. As μ varies, (6.4) and (6.5) trace out a locus in $\mathscr{R}$, which we label ξ_k, and provisionally identify as representing the kth variable. This locus is normally nonlinear, so we term ξ_k a **trajectory** rather than an axis but it can be marked with a scale of markers in the usual way, with $\mu = 1$ coresponding to the marker for one unit of the variable; we term this the **unit marker** for the kth variable. Each variable generates one such trajectory; $\xi_1, \xi_2, \ldots, \xi_p$ share the value $\mu = 0$ and hence are concurrent at a point, which we label O, corresponding to the mean of **X**. In previous chapters, the linear axes were concurrent at the centroid G of the samples and G and O coincided, but this is no longer so, because G continues to be the centroid of **Y**, where $\mathbf{1}'\mathbf{Y} = 0$, which differs from O.

The residual dimension requires special consideration because, properly speaking, each value of μ creates a new dimension giving an infinite continuum of dimensions beyond the m dimensions of $\mathscr{R}$ that contain **Y**. Fortunately, for many purposes, it will be seen that we can proceed as if there were only one residual dimension and this augmented version of $\mathscr{R}$ wil be written $\mathscr{R}^+$. The key thing to observe is that ddistances between any point of $\mathscr{R}$ and any point of $\mathscr{R}^+$ are always correct (section A.12) but ddistances between points that are both in $\mathscr{R}^+$ are not; the latter are not required.

It is one thing to define these trajectories, but another to show that they are useful. Rather than consider the full generality, we shall first discuss the important simplification of additive ddistance, which contains most of the distances of the preceding chapters as special cases.

6.2 Additive ddistance

Pythagorean ddistance and χ^2 ddistance are additive in the contributions from each variable; each variable contributes independently of the others. This is not true of Mahalanobis ddistance where the correlations between the variables are taken into acount. When independence holds, we have

$$d_{ij}^2 = \sum_{k=1}^{p} d_k(x_{ik}, x_{jk}) \tag{6.6}$$

where $d_k(x_{ik}, x_{jk})$ gives the contribution of the kth variable to the ddistance between the ith and jth samples. The suffix k allows a different function to pertain to each variable but in this chapter, normally the same distance-function will be used for every variable and the suffix may be dropped. When ddistances are additive as in (6.6) then

$$\mathbf{D} = \sum_{k=1}^{p} \mathbf{D}_k \tag{6.7}$$

and (6.2) becomes

$$\mathbf{B} = \sum_{k=1}^{p} \mathbf{B}_k \tag{6.8}$$

where

$$\mathbf{D}_k = \left\{ -\frac{1}{2} d_k^2(x_{ik}, x_{jk}) \right\} \quad \text{and} \quad \mathbf{B}_k = (\mathbf{I} - \mathbf{N})\mathbf{D}_k(\mathbf{I} - \mathbf{N})$$

If we take a pseudosample $\mu\mathbf{e}_k$ then its ddistances from the original n samples are given by

$$\mathbf{d}_{n+1} = \sum_{h=1}^{p} \{d_h(x_{ih}, 0)\} + \{d_k(x_{ik}, \mu)\} - \{d_k(x_{ik}, 0)\} \tag{6.9}$$

The only terms which vary with μ are members of the vector $\{d_k(x_{ik}, \mu)\}$ and, because these are ddistances, $d_k(x_{ik}, \mu) = 0$ only when $\mu = x_{ik}$. What this means is that the point on the kth trajectory, nearest to the ith sample point of $\mathbf{Y}$, bears the scale-marker x_{ik}, for $i = 1, 2, \ldots, n$. This result is valid for the rows of $\mathbf{Y}$, and we assume its validity for other points of $\mathscr{R}$. This is an important result, for it shows that the kth trajectory is acting like an ordinary coordinate axis in that, given a sample point in $\mathscr{R}$, then its correct values are obtained by projecting normally onto the p trajectories and reading off the markers. Thus, all points that predict μ lie in the subspace that is normal to the trajectory at the marker μ.

In one respect, the embedding differs from the Euclidean framework of previous chapters. If a new sample is interpolated into $\mathscr{R}$ then this requires the extra residual dimension and so lies in $\mathscr{R}^+$. Because the markers on the trajectories are also in $\mathscr{R}^+$, the ddistances from the interpolated values to the trajectories are not correct (section 6.1). We have already seen (section 2.3) that even in classical biplots in the context of PCA, points interpolated into the approximation space $\mathscr{L}$ do not normally predict their original values; in this more general situation, this property extends to $\mathscr{R}$ itself.

6.3 Approximation

Approximation is given by any form of MDS that yields a configuration of points occupying a ρ-dimensional space $\mathscr{L}$ and whose coordinates are given by the rows of a matrix **Z**. An introduction to MDS has been given in section 3.1. In Chapter 3, it was assumed that the set of distances $\{d_{ij}\}$ are Pythagorean but in this chapter we are allowing $\{d_{ij}\}$ to represent any Euclidean embeddable distances in $\mathscr{R}$; we shall continue to assume that the transformation τ of (3.1) is the identity. The linear coordinate axes of previous chapters have now become nonlinear trajectories in $\mathscr{R}^+$. We are concerned with how to approximate these axes in $\mathscr{L}$ to aid interpolation and prediction. To make further progress, we must consider the actual method of MDS.

6.3.1 Interpolation with Euclidean embedding

In Chapter 3, we saw that, even with Pythagorean distance, for most methods of metric scaling, graphical interpolation is available, at best, as an approximation. The position is precisely the same with Euclidean embeddable distances and the basic ideas of section 3.4 which do not depend critically on the Pythagorean assumption remain valid. However, explicit results can be obtained for principal coordinates analysis (PCO) and these are derived in below. With Pythagorean distances, PCO is identical to PCA and the methods of Chapter 2 become a special case of what follows. With Euclidean embeddable distances, PCO defines the approximation **Z** to be the first ρ columns of **Y** of (6.3) and hence $\mathscr{L}$ is the space spanned by these columns. The choice of the spectral decomposition of **B** ensures that **Z** is obtained from the first ρ principal components of **Y**. Thus $\mathscr{L}$ is a subspace of $\mathscr{R}$ obtained by orthogonal projection.

Interpolation in PCO

Interpolation of a sample $\mathbf{x} = (x_1, x_2, \ldots, x_p)$ into $\mathscr{R}^+$ is given by (6.4), (6.5) and (A.64), (A.65) with the vector $\mathbf{d}_{n+1}$ now defined as giving the ddistances from **x** rather than from a pseudosample. Its projection **z** onto $\mathscr{L}$ is obtained as the first ρ columns of **y** and (6.5) can be ignored because the residual dimension is orthogonal to $\mathscr{R}$ and hence has a null projection onto the subspace $\mathscr{L}$. The ddistance of the pseudosample $x_k\mathbf{e}_k$ from the original samples is given by (6.9) as

$$\mathbf{d}_{n+1}(x_k)=\sum_{h=1}^{p}\{d_h(x_{ih},0)\}+\{d_k(x_{ik},x_k)\}-\{d_k(x_{ik},0)\}$$

It follows that for the sample

$$\mathbf{x}=\sum_{k=1}^{p}x_k\mathbf{e}_k$$

we have

$$\mathbf{d}_{n+1}(\mathbf{x})=\sum_{h=1}^{p}\{d_k(x_{ik},x_k)\}=\sum_{k=1}^{p}\mathbf{d}_{n+1}(x_k)-(p-1)\sum_{h=1}^{p}\{d_h(x_{ih},0)\}$$

Substitution into (6.4) gives

$$\mathbf{y}=\mathbf{\Lambda}^{-1}\mathbf{Y}'\left[\sum_{k=1}^{p}(\mathbf{d}_{n+1}(x_k)-\tfrac{1}{n}\mathbf{D1})-\left(\sum_{h=1}^{p}\{d_h(x_{ih},0\}-\tfrac{1}{n}\mathbf{D1}\right)\right] \quad (6.10)$$

The first summation gives the vector-sum for the markers $x_1,x_2,\ldots,x_p$ of the pseudosamples; the second summation gives the interpolant of the mean $(0,0,\ldots,0)$, which we have seen to be the point of concurrency O of the trajectories. Thus, graphical interpolation is obtained as the vector-sum of the appropriate markers relative to the point of concurrency. Because the centroid of a set of points, here representing pseudosamples, projects into the centroid of the projections, the interpolate $\mathbf{z}$ in $\mathscr{L}$ remains a vector-sum of the markers on the trajectories in $\mathscr{R}$ (or, equivalently, $\mathscr{R}^+$) projected onto $\mathscr{L}$. Despite the nonlinear nature of the trajectories, the independence assumption (6.6) ensures that interpolation can continue to be expressed as a vector-sum and so admits simple graphical usage. Figure 6.1 illustrates the procedure for graphical interpolation. H is the centroid of the appropriate markers on the biplot trajectories which when extended three times ($p=3$) gives the interpolant P which is the vector-sum of the arrowed vectors at O, the point of concurrency of the trajectories. Also shown in Figure 6.1 is the centroid G of the coordinates $\mathbf{z}$ of the two-dimensional PCO approximation.

This method is a natural extension of interpolation for Pythagorean distance which has been shown in Chapters 2 and 3 to give vector-sums relative to the centroid of the points with

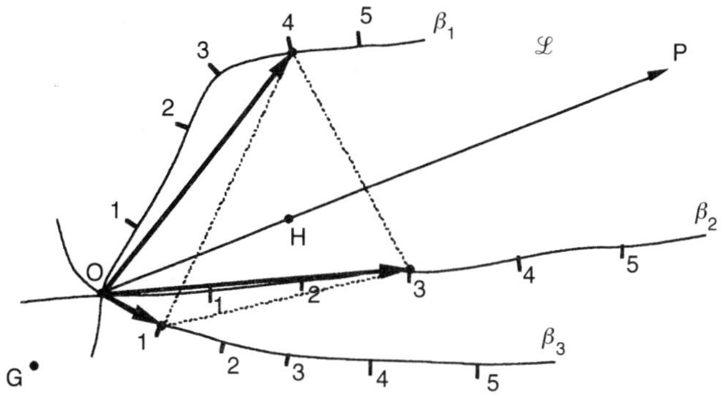

Fig. 6.1. *The procedure for graphical interpolation of the point P(4, 3, 1).*

coordinates **X**. Any original sample will automatically interpolate into its correct position in the PCO approximation. Note that now the centroid G of **Y** differs from the point of concurrency which we have labelled O.

When **x** is one of original sample values, say $\mathbf{x} = x_{ik}\mathbf{e}_k$, then $\mathbf{d}_{n+1}$ becomes a column of $\mathbf{D}_k$ and this can be used to obtain explicit coordinates for the n points on each trajectory that correspond to the actual data values given in **X**. We term these the **basic points** of the kth variable. The results are not particularly interesting but the corresponding results for generalized biplots (Chapter 7) are, and throw light on the results for nonlinear biplots; therefore, further discussion is postponed until section 7.3.1.

6.3.2 *Prediction with Euclidean embedding*

Prediction in the context of Euclidean embedding proceeds in essentially the same way as was described in Chapter 3 but with changes to cope with the nonlinearity of the trajectories. Figure 6.2 is the counterpart of Figure 3.1. ξ is a trajectory in $\mathscr{R}^+$ for some variable; two markers ξ_1 and ξ_2 are shown.

With PCO, $\mathscr{L}$ is already a subspace of $\mathscr{R}$ but, with other methods of metric scaling, $\mathscr{L}$ must first be embedded by the Procrustean method given in section 3.3.1. If ξ_1 is a marker on the trajectory ξ in $\mathscr{R}^+$ then all points in the space $\mathscr{N}_1$ normal to ξ at ξ_1 will predict the value ξ_1 for the variable under consideration. In particular, all points in $\mathscr{L}$ that predict ξ_1 lie in $\mathscr{N}_1 \cap \mathscr{L}$ which,

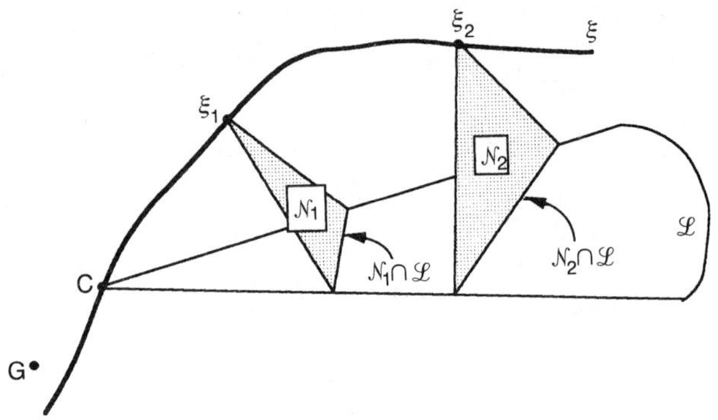

Fig. 6.2. *The basis of prediction.*

despite the nonlinearity of ξ, is a linear subspace of $\mathscr{L}$ and hence of $\mathscr{R}$. Figure 6.2 illustrates the geometry and also shows a second marker ξ_2 with its resulting intersection space $\mathscr{N}_2 \cap \mathscr{L}$. A departure from the properties of the linear case is that now the intersection spaces are not parallel and consequently a linear biplot axis that is simultaneously orthogonal to all the intersection spaces is not available. It follows that the intersection spaces themselves will usually intersect, so that for some points of $\mathscr{L}$ prediction will not be unique; with a good approximation, such points should be outside the area of the data. Recall that prediction of the values to be associated with a point P in $\mathscr{L}$ is a matter of deciding in which intersection space P lies and then predicting its associated marker. Some new graphical method is needed to accomplish this.

Figure 6.3 illustrates the procedure for constructing nonlinear biplot trajectories and how they are to be used. We define the nonlinear biplot trajectory β to be the locus of the orthogonal projections of O onto the intersection spaces. In the linear case, parallelism ensures that this process gives the linear biplot axes of previous chapters. Figure 6.3 shows two points on the locus with their markers ξ_1 and ξ_2. The figure also shows a point P, which happens to lie on $\mathscr{N}_1 \cap \mathscr{L}$. How is this fact ascertained? The answer is to construct a circle on OP as diameter and predict the value of the marker at its point of intersection with β. This works because diameters of circles subtend right angles at the circumference. Thus, orthogonal projection associated with Pythagorean

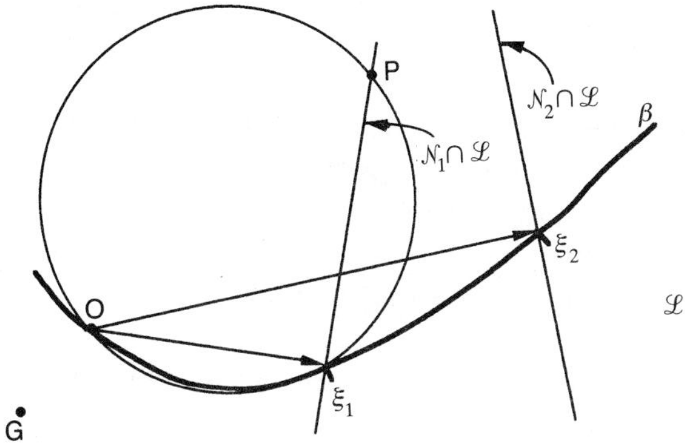

Fig. 6.3. *Construction and use of a nonlinear predictive biplot trajectory β. Prediction for a point P is given by circular projection.*

distance, is for Euclidean embeddable distances replaced by what we shall term **circular projection**. Note that when β is linear through O, circular projection is the same as orthogonal projection.

In the construction of this form of predictive biplot trajectory β, there is some arbitrariness in the choice of O as origin. Any point in $\mathscr{L}$ could have been chosen as origin, each generating a different β, but all would give the same predictions. Each variable could be chosen to have a different origin but it is convenient to choose a common origin so that all p trajectories are concurrent at that origin. O has the advantage of being associated with definite values (the means) of the original variables but has the disadvantage that no simple algorithm seems to be available for computing the coordinates of O. G is an obvious rival to O but, although trivial to compute, does not normally correspond to 'nice' values of the variables. Another possibility is to choose an origin that corresponds to some integer values (or whatever are convenient units) of the variables; this will simplify the calibration of the trajectories but needs sophisticated algorithms to identify, automatically, a suitable origin and scale-interval from the raw data.

The coordinates of the predictive trajectories may be obtained from the basic equations (6.4) and (6.5) which determine the kth trajectory as the locus of $\mathbf{y}$ as μ varies. To make progress, the parametric equation of each trajectory is needed in a more tractable form and the following discussion is based on the additivity assumption (6.6) but, as has already been shown with canonical biplots (Chapter 5), progress can also be made with other assumptions. When (6.6) is valid, $\mathbf{d}_{n+1}$ in (6.4) is replaced by (6.9) and differentiation with respect to μ immediately gives the direction of the tangent at μ in the first m dimensions of $\mathscr{R}$ as

$$\frac{\partial \mathbf{y}}{\partial \mu} = \boldsymbol{\Lambda}^{-1}\mathbf{Y}'\left\{\frac{\partial d_k(x_{ik}, \mu)}{\partial \mu}\right\} \tag{6.11}$$

The direction in the residual dimension is obtained by differentiating (6.5) to give

$$2y_{m+1}\frac{\partial y_{m+1}}{\partial \mu} = -\frac{\partial(\mathbf{y}'\mathbf{y} + \frac{2}{n}\mathbf{1}'\mathbf{d}_{n+1})}{\partial \mu}$$

Writing

$$\mathbf{y}'\mathbf{y} = (\mathbf{d}_{n+1} - \tfrac{1}{n}\mathbf{D1})'\mathbf{B}^{-}(\mathbf{d}_{n+1} - \tfrac{1}{n}\mathbf{D1})$$

yields by routine methods

$$\frac{\partial y_{m+1}}{\partial \mu} = (y_{m+1})^{-1}\left[\tfrac{1}{n}\mathbf{1}'\mathbf{DB}^{-} - \tfrac{1}{n}\mathbf{1}' - \mathbf{d}'_{n+1}\mathbf{B}^{-}\right]\left\{\frac{\partial d_k(x_{ik}, \mu)}{\partial \mu}\right\} \tag{6.12}$$

Thus, (6.11) and (6.12) give the directions of the tangent and hence the plane $\mathscr{N}$ normal to the trajectory at μ. We shall write this plane as $\mathbf{t}'\mathbf{y} = \mathbf{t}'\mathbf{y}_\mu$ where $\mathbf{y}_\mu$ are the coordinates at μ given by (6.4) and (6.5) and $\mathbf{t} = (t_1, t_2, \ldots, t_{m+1})$ is given by (6.11) and (6.12). The intersection space is given by

$$t_1 y_1 + t_2 y_2 + \cdots + t_\rho y_\rho = \mathbf{t}'\mathbf{y}_\mu \tag{6.13}$$

and we require the point $\mathbf{z}_\rho$ in this space that is nearest to a chosen origin $\mathbf{g} = (g_1, g_2, \ldots, g_\rho)$, say. Thus, we must minimize $(\mathbf{z}_\mu - \mathbf{g})(\mathbf{z}_\mu - \mathbf{g})'$ subject to (6.13), i.e. $\mathbf{t}'_\rho\mathbf{z}_\mu = \mathbf{t}'\mathbf{y}_\mu$. This gives

$$\mathbf{z}_\mu = \mathbf{g} + \frac{(\mathbf{t}'\mathbf{y}_\mu - \mathbf{t}'_\rho\mathbf{g})}{(\mathbf{t}'_\rho\mathbf{t}_\rho)} \tag{6.14}$$

Therefore, as μ varies (6.14) traces out the prediction biplot trajectory for the kth variable; note that the coefficients of $\mathbf{t}$ are themselves functions of μ and hence determine the nonlinearity of the trajectory.

6.4 The non-Euclidean case

In the above, we have assumed Euclidean embeddability. The steps of the procedures have been:

(i) from **X** generate the distance matrix **D**;
(ii) find coordinates **Y** in an m-dimensional Euclidean space $\mathscr{R}$ that generate **D** via Pythagorean ddistances (the embedding step);
(iii) place p axes/trajectories (the reference system χ) in $\mathscr{R}$ that can be used for interpolating or predicting values in the measurement units of **X**; this may need extending $\mathscr{R}$ to $\mathscr{R}^+$ (the reference step);
(iv) approximate **D** by Δ generated by coordinates **Z** in a ρ-dimensional space $\mathscr{L}$ (MDS step);
(v) if $\mathscr{L}$ is not already a subspace of $\mathscr{R}$, embed it in $\mathscr{R}$ by the Procrustes method of Chapter 3; and
(vi) approximate χ in $\mathscr{R}$ by $\mathscr{B}_I$ (for interpolation) and $\mathscr{B}_P$ (for prediction) which form two separate reference systems in $\mathscr{L}$.

The assumption of Euclidean embeddability is crucial in step (ii). Many methods of metric scaling, including those discused in Chapter 3, need no Euclidean assumption. Thus **D** need not be embeddable, although the (usual) aim is to find a Euclidean approximation Δ. Without Euclidean embeddability, there is no $\mathscr{R}$, no **Y** and no χ, and hence no approximation to them. All that survives is the MDS step (iv). Even step (i) is not essential, as most MDS methods operate directly on **D** which may be directly observed, i.e. **D** is the data and there is no **X**. Of course, if **X** does not exist, then there is no question of superimposing information on variables on the MDS, so we disregard this possibility here apart from noting the possibility of attempting reification of directions, trajectories or regions in $\mathscr{L}$. Useful as has been the above framework, it is useless for the generalizations under discussion. How then may we proceed?

When **D** is Euclidean embeddable and the MDS method is PCO, including its special cases such as PCA and MCA, and the additive ddistance (6.6) assumption is valid, then we arrive back at the methods of section 6.3 with their Euclidean underpinning. When **D** is nearly Euclidean, then we can expect the previous methods to remain approximately valid. By near Euclidean, we mean that there are points in $\mathscr{R}$ that generate Euclidean distances $\mathbf{D}_M$ that are close to those of $\mathbf{D}$ as judged by the chosen MDS criterion S_M (3.1). Indeed, we can define $\mathbf{D}_M$ to be generated by the points that minimize S_M; they are the best Euclidean fit to $\mathbf{D}$. Bailey and Gower (1990) show that the dimensionality of the points that generate the best $\mathbf{D}_M$ cannot exceed the **positivity** of $\mathbf{D}$; the most simple definition of positivity is the number of positive eigenvalues of (6.3). Thus, for a good Euclidean approximation to $\mathbf{D}$, we require that its negative eigenvalues be small relative to its positive eigenvalues. Having defined $\mathbf{D}_M$ and $\mathscr{R}$, we may proceed as before.

When there is no good Euclidean approximation to $\mathbf{D}$ then, as was noted by Gower and Harding (1988), it is always possible to superimpose pseudosamples $\mu\mathbf{e}_k$ directly onto the MDS in $\mathscr{L}$. The process has already been illustrated in section 3.2, for least-squares and least-squares squared scaling. In general, if $\mathbf{d}_{n+1}$ gives the distances of a new sample with values $\mathbf{x}_{n+1}$ from the original n samples $\mathbf{X}$, then superimposition amounts to finding the coordinates of a point $\mathbf{z}_\mu$ in $\mathscr{L}$ with distances δ_{n+1} from the MDS coordinates $\mathbf{Z}$, regarded as given, that minimizes the special form of (3.1)

$$S(\tau(d_{n+1,i}), \delta_{n+1,i})$$

This guarantees that when $\mathbf{x}_{n+1}$ has the values of one of the original samples then the superimposed point will coincide with its proper position in the MDS; even the validity of the additive ddistance assumption is unnecessary. The transformation $\tau(.)$ will often be the identity transformation but, if not, it is given or, at least, it has already been determined by the MDS. The locus as μ varies will trace out a trajectory for the kth variable, to which markers may be attached as previously.

Thus, there is no difficulty in principle, in providing biplot trajectories in $\mathscr{L}$; the problem is how they are to be interpreted. Different interpolative and predictive trajectories are not available, but because the original samples are properly interpolated, we can expect the trajectories to have something to do with interpolation. In Chapter 3, we saw that reliable graphical interpolation was not available even with Pythagorean ddistance and that predictive and approximate interpolative biplots depended on the Procrustean embedding of $\mathscr{L}$ and $\mathscr{R}$. Without embeddability, it seems that the trajectories can be interpreted only in a qualitative way. Perhaps, nearness of a sample-point to higher, rather than lower, values of a variable may be meaningful, although precise numerical interpolations and predictions are not. The root of the problem seems to be that it is questionable to try to fit a Euclidean representation to a highly non-Euclidean distance matrix and then expect Euclidean interpretations that link information on the variables with Euclidean positions of the samples to remain valid. A discussion of non-metric scaling is given in section 11.2.

The regression method described in sections 3.3.2 and 3.4.3 may always be used to give linear biplots. Then the directions of predictive and interpolative axes coincide but, as usual, different scales may be marked for the two purposes. It has already been pointed out that interpretation with this method is not entirely clear, even with Pythagorean distance and becomes more questionable in the context of nonlinearity and non-Euclideanarity. One possibility is to try some form of curvilinear regression, such as polynomial regression, but the details have to be developed.

6.5 Relaxing the additive ddistance assumption

The additive ddistance assumption (6.6) has been central to obtaining many detailed results concerning graphical interpolation by using vector-sums in the better-known methods of MDS. However, in Chapter 5, it was seen that vector-sum interpolation was available with canonical biplots even though Mahalanobis D^2, the ddistance concerned, does not satisfy (6.6). The essential requirements for interpolation in terms of vector-sums are that:

(a) an interpolant can be expressed algebraically as a linear combination of basic interpolants in $\mathscr{R}$ – the interpolated pseudo-samples and CLPs; and
(b) their orthogonal projection into $\mathscr{L}$ is relevant for the MDS method that is being used.

The key result is that the projection of a sum of a set of vectors is the vector-sum of their projections. These requirements are satisfied for PCA, CVA, MCA and, provided (6.6) is valid, for PCO. In other situations, vector-sum interpolation is not available, as we have already seen in Chapter 3, where algebraic interpolation is feasible but does not translate into a graphical method although graphical approximations to the optimal algebraic interpolants may be considered (section 3.5.1); this is the general position when the requirements are not satisfied.

Predictive biplots depend on finding the nearest CLP or marker on a trajectory in $\mathscr{R}$ and it has been shown in (section 6.2) that the validity of (6.6) is sufficient to justify this usage. Again CVA, discussed in Chapter 5, shows that (6.6) is not a necessary assumption. However, the existence of the reference system χ in $\mathscr{R}$ is essential. The concept of nearness, as developed in this book, is meaningless without the reference system. If we attempt to interpret **nearness** in a similar way to that of the Eckart–Young theorem, this would require the rank ρ matrix $\mathbf{Z}$ which minimizes (3.1), which is precisely the definition of MDS. If now $\mathbf{z}$ is a general point of $\mathscr{L}$, rather than one of the samples that have generated $\mathscr{L}$, we can ask what value of $\mathbf{x}$ is nearest in terms of the original variables and the criterion $S_M(.,.)$? Unfortunately, this question is not well posed. Certainly there is no unique answer. We cannot regard as satisfactory, the answer which says that we predict for $\mathbf{z}$ the value of $\mathbf{x}$ which interpolates at $\mathbf{z}$. Even in the linear case and with Pythagorean ddistance there is no unique $\mathbf{x}$, because many values of $\mathbf{x}$ will project into $\mathbf{z}$. Eckart and Young choose the unique solution given by the intersection of the projection with $\mathscr{L}$ and this is possible because of the embedding of $\mathscr{L}$ in $\mathscr{R}$, thus linking the MDS, including $\mathbf{z}$, with the coordinate system for $\mathbf{X}$. This linkage can be imposed by Procrustean embedding as in section 3.4. Alternatively, noting that $\mathbf{Z}$ predicts $\mathbf{X}$, we may set up an appropriate regression and use it to predict $\mathbf{x}$ from $\mathbf{z}$ (section 3.3.2). However, we have already seen that in nonlinear situations there is even more reason to question the use of linear regression than there was with the linear biplot axes associated with Pythagorean ddistance of Chapter 3.

6.6 Examples

Figure 6.4 shows examples of nonlinear interpolative biplot trajectories given by Gower and Harding (1988) for data which consists of 12 variables on the logarithms of parts per million of trace elements for soil sampled at 15 sites in Glamorganshire. Gower and Harding give that data and its source.

The Euclidean embeddable ddistance used it that defined in (6.1). The interesting thing about (6.1) is that it generates polygonal

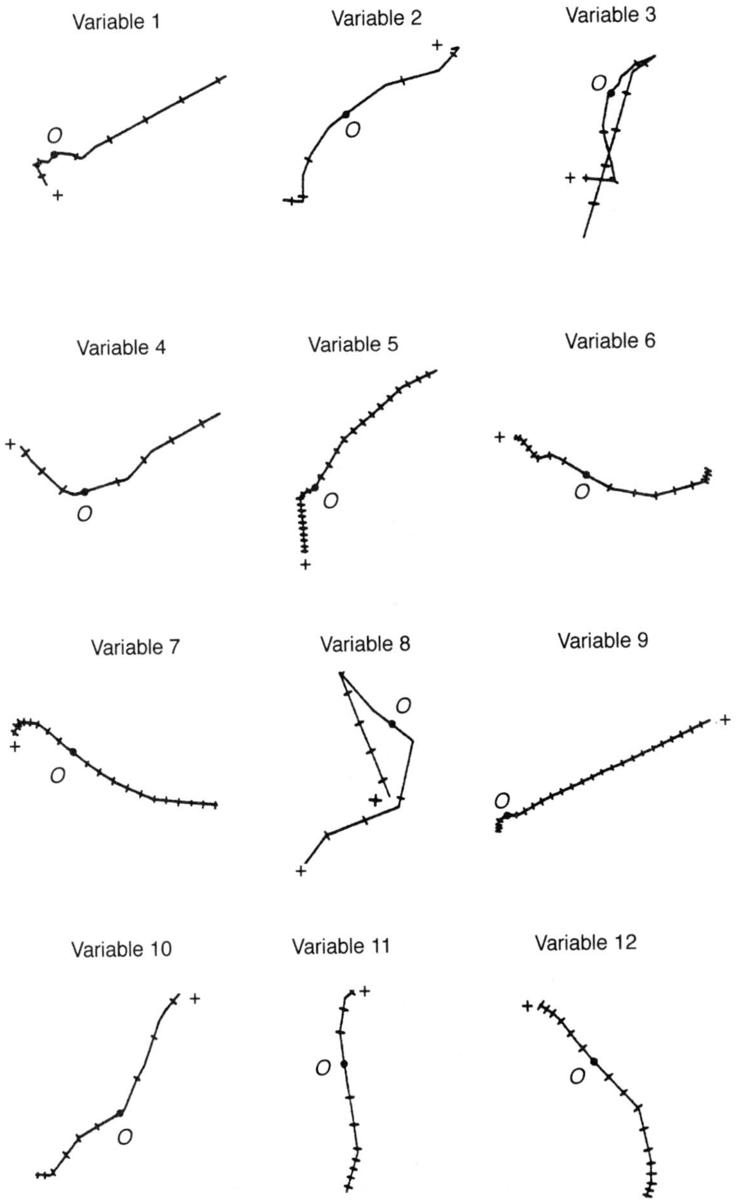

Fig. 6.4. *Examples of nonlinear interpolative biplot trajectories for 12 commensurable (logarithm of parts per million of trace elements from soil samples) variables from Gower and Harding (1988).*

trajectories with corners at every basic point. The reason is evident from (6.9) where, if we assume that the values $x_{1k}, x_{2k}, \ldots, x_{nk}$ for the kth variable are in non-decreasing order, then the elements of the vector $\{d_k(x_{1k}, \mu)\}$ decrease for x_{1k} in the interval $(-\infty, \mu)$ and increase in the interval (μ, ∞). Between basic points, the changes in the vector are linear with μ, generating a linear kth trajectory. As μ passes through a basic point, there is an abrupt change in the pattern of the vector and a corresponding change in the associated trajectory. The scales are linear between basic points but with unequal separation of scale points in different edges of the polygon. This effect is clear from Fig. 6.4 where it is also clear that the trajectories have finite length, being bounded by the smallest and biggest basic points. The finiteness is not merely because each trajectory has not been drawn outside these bounds, but is more fundamental. Consider any value $\mu \leqslant x_{1k}$ then $\{d_k(x_{1k}, \mu)\} = \mathbf{x}_k - \mu\mathbf{1}$. When substituted into (6.9) and (6.4) the constant vector $\mu\mathbf{1}$ gives a zero contribution to the coordinates of the pseudosample and therefore, in this range, the trajectory remains fixed at the point given by $\mu = x_{1k}$. Similarly, when $\mu \geqslant x_{pk}$ the trajectory remains fixed at the point given by $\mu = x_{pk}$. This is what happens in $\mathscr{R}$; in $\mathscr{R}^+$ the trajectory is linear in these ranges. Thus, what look like finite trajectories in Fig. (6.4), in reality, continue but are orthogonal to $\mathscr{L}$. This artefact is only a problem when we are interested in values of the variables beyond the range of the data. Just as with conventional coordinate axes, it is good practice not to plot biplot axes for values much beyond the range of observed values. We note that when $0 < d_{ij} < 1$, as happens with distances based on dissimilarity, we do obtain genuine finite trajectories.

Scale markers at equal intervals are shown in Fig. 6.4 on either side of O the point representing the mean of each variable, which is the point of concurrency for all variables. In Fig. 6.4, the positive end of each scale is also indicated. From Fig. 6.4, we see that some trajectories are very convoluted, even intersecting themselves, and that parts of the scales can be very distorted. Intersections cannot occur in $\mathscr{R}$ and the observed distortions in $\mathscr{L}$ are largely induced by projection. Interpolation that uses distorted regions of the biplot axes is likely to be very unreliable. However, the trajectories in Fig. 6.4 are drawn to viewing scales that help inspection of the features discussed above; in reality, they are of very different sizes as can be verified by examining Fig. 3 of Gower and Harding

(1988). A short trajectory indicates either small values, incommensurable with the values of other variables included in the study, in which case transformation might be considered, or a trajectory that is nearly orthogonal to $\mathscr{L}$ and hence is of limited value for interpolation and should be discarded; this remark applies equally to linear biplots which are short within the range of the basic points. Ideally, good interactive computing facilities are needed to facilitate inspection of MDS and its associated reference system, so that decisions can be taken on which trajectories to include and which to reject.

Inspection shows that the only variables that contribute significantly to interpolation are numbers 5, 7, 9 and 12 and Fig. 6.4 shows that these are all reasonably well behaved but with 5, 7 having compressed scales at the positive end of their ranges and with 9 having a compressed scale at the negative end of its range. Incommensurability is not a problem with the data discussed here so there is no need to consider transformations of the rejected variables. Figure 6.5 shows a nonlinear biplot using these four variables, numbered 5, 7, 9 and 12 being a selection from 12 original variables.

Figure 6.5 is from Gower and Harding (1988) but modified:

(i) by removing the ineffective trajectories; and
(ii) by scaling all trajectories from O by a factor of four ($p = 4$, the number of variables shown).

The latter improves the relative sizes of samples and trajectories and allows the convenience of effecting vector-sum interpolation as the centroid of the points to be interpolated, eliminating the extension by p. The combined effects of (i) and (ii) is to provide a less cluttered and more balanced view of the biplot. The separation of the sample centroid G from the point of concurrency O is evident. With only four of twelve variables shown, the original samples will no longer interpolate exactly to the sample positions. Nevertheless, the general features of the data remain clear. Sample (site) g is an obvious outlier which can only interpolate to the position shown for a high positive value of variable 9 and high negative values of the other variables. Reference to the original data shows that this site has the extreme values for all these variables, except variable 5 which has a modest negative value. The temptation to use this biplot for prediction should be avoided as can be illustrated for site

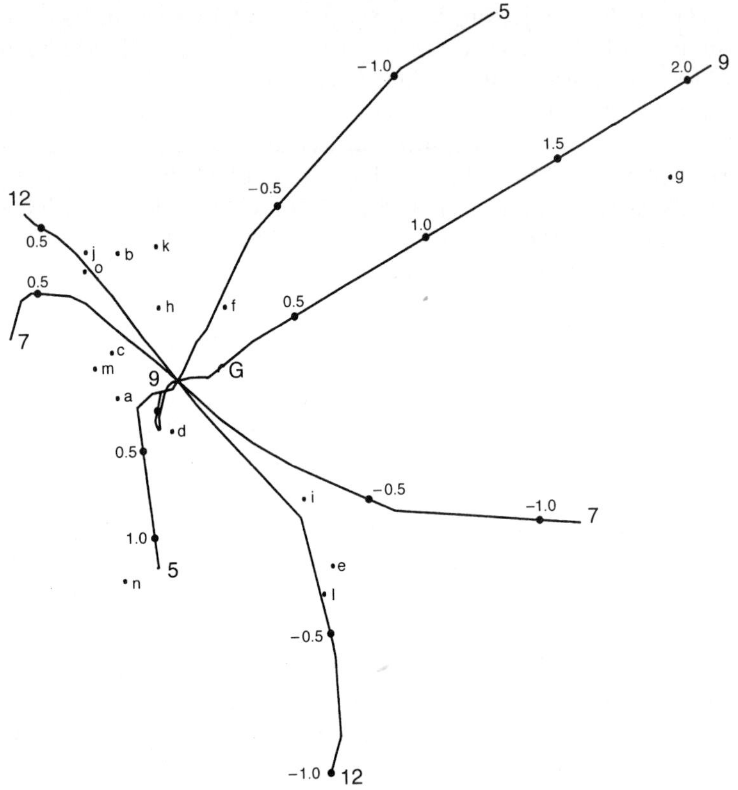

Fig. 6.5. *PCO of 15 Glamorganshire soils (lower case letters) based on the square root of Minkowski L1-distance (6.1). This figure is a modified version of Fig. 2 of Gower and Harding (1988).*

n which interpolates quite well but gives unsatisfactory predictions. This should be no surprise, given the enormous differences between the prediction and interpolation evident even in the linear case treated in section 5.6 where the two-dimensional fit is very much better than that of the example discused here.

We do not yet have an example of predictive nonlinear biplots using the methodology of section 6.3. Another example of a nonlinear interpolative biplot is included in the example of Chapter 7 (section 7.4).

CHAPTER 7

Generalized biplots

7.1 Introduction

Nonlinear biplots (Chapter 6) give a fairly general methodology for providing biplots for quantitative variables in association with any form of metric scaling (section 11.2 for non-metric scaling). The theory is most complete for predictive biplots but simple interpolative graphical biplots have been found for PCO, especially when the additive assumption (6.6) is valid. Most of the methods discussed in previous chapters are special cases of nonlinear biplots but MCA and other methods that admit categorical variables are exceptions. The pseudosample $\mu\mathbf{e}_k$, that is the basis of nonlinear biplots, cannot handle categorical variables because it depends on zero representing the mean of a variable and categorical variables do not have zero means. Hence, if the theory is to be generalized to handle categories, some new pseudosample must be found and preferably one that can also include quantitative variables. Ddistances that may include both categorical and quantitative variables are available; see, for example, Gower (1968), which discusses a flexible family of (dis)similarity coefficients satisfying (6.6). Further, it would be very desirable if the generalization could subsume nonlinear biplots as a special case. This chapter is concerned with such developments.

7.2 Pseudosamples for categorical variables

We adopt an approach in which the kth variable takes the value μ while leaving the values of the other variables unchanged. Thus we choose pseudosamples, the ith of which becomes $(x_{i1}, x_{i2}, \ldots, x_{ik-1}, \mu, x_{ik+1}, \ldots, x_{ip})$ and this can be superimposed on $\mathscr{R}$ using the formulae (A.64) and (A.65) of the appendix. This may be done for all n samples $(i = 1, 2, \ldots, n)$ to give n superimposed points. The locus of the centroid of these points, as μ varies, defines the kth biplot trajectory. For quantitative variables, a true trajectory is generated and we shall shortly investigate how this trajectory relates to the corresponding nonlinear biplot trajectory. When the kth variable is categorical, μ can take only L_k distinct values, the number of levels of the kth variable, so rather than generating a trajectory, this process generates L_k points, known as the **category-level-points** (CLPs) a special case of which we have already met in MCA (Chapter 4). To avoid referring to trajectories and CLPs individually, we shall term the combined concepts a **reference system**.

The previous paragraph gives a very general operational algorithm for constructing a reference system for both quantitative and categorical variables, but when the additivity assumption (6.6) is valid, explicit formulae may be obtained. The ddistances between the n samples and n pseudosamples for the kth variable may be written as

$$\begin{pmatrix} \mathbf{D} & \mathbf{D} - \mathbf{D}_k + \{d(\mu, x_{ik})\}\mathbf{1}' \\ \mathbf{D} - \mathbf{D}_k + \mathbf{1}\{d(\mu, x_{ik}\}' & \mathbf{D} - \mathbf{D}_k \end{pmatrix} \tag{7.1}$$

Where $\{d(\mu, x_{ik})\}$ includes the usual factor of $-\frac{1}{2}$.

From (7.1), (A.62) gives the ddistances $\mathbf{d}_{n+1}$ of the centroid of the pseudosamples from each of the n samples as

$$\mathbf{d}_{n+1} = -\frac{1}{2}\left(\frac{\mathbf{1}'(\mathbf{D} - \mathbf{D}_k)\mathbf{1}}{n^2}\mathbf{I} + (\mathbf{D} - \mathbf{D}_k)\right)\frac{\mathbf{1}}{n} + \{d(\mu, x_{ik})\} \tag{7.2}$$

This is the value of $\mathbf{d}_{n+1}$ to be substituted into (A.64) to obtain the coordinates $\mathbf{y}$ in $\mathscr{R}$ of the centroid of the pseudosamples, giving

$$\mathbf{y} = \mathbf{\Lambda}^{-1}\mathbf{Y}'\left(\{d(\mu, x_{ik})\} - \mathbf{D}_k\frac{\mathbf{1}}{n}\right) \tag{7.3}$$

Also, applying (A.62) to (7.1) we can evaluate the ddistance between the centroid of the pseudosamples and the centroid of the samples $\mathbf{Y}$, which is our origin, as

$$\frac{1}{n^2}\mathbf{1}'\mathbf{D}_k\mathbf{1} - \frac{2}{n}\mathbf{1}'\{d(\mu, x_{ik})\}$$

which combined with (7.3) gives the coordinate of the centroid of the pseudosamples in the $(m+1)$th dimension of $\mathscr{R}^+$ as

$$y_{m+1}^2 = \frac{1}{n^2}\mathbf{1}'\mathbf{D}_k\mathbf{1} - \frac{2}{n}\mathbf{1}'\{d(\mu, x_{ik})\} - \mathbf{y}'\mathbf{y} \tag{7.4}$$

Thus (7.3) and (7.4) define a coordinate on the kth trajectory or, when the variable is categorical, a CLP.

From (7.2), we see that because $\{d(\mu, x_{ik})\}$ is the only term involving μ then just as for nonlinear biplots (section 6.2), each sample point in $\mathscr{R}$ is nearest its correct marker on the trajectory or CLP. Thus the generalized biplot pseudosample also leads to a reference system with the nearness properties of conventional coordinate axes. Indeed, comparing (7.3) with (6.9) shows that in both the only term that varies with μ is $\{d(\mu, x_{ik})\}$ so that for continuous variables the nonlinear trajectories are parallel to the generalized biplot trajectories. The nature of the generalized biplot pseudosample implies that $\mu = 0$ is not a point common to all trajectories. We shall show (section 7.3.1) that when interpolative biplots are available, an adjustment is permissible that restores the concurrent nonlinear biplot trajectories; we have already seen that, for predictive biplots, the choice of origin is arbitrary and so the concurrency property is then of less importance, although there is some loss of simplicity as a consequence of non-concurrency. Of course, the concept of concurrency has no relevance for categorical variables; a set of CLPs cannot be concurrent.

7.2.1 *Coordinates of the basic points*

The representation in $\mathscr{R}$ of the actual sample values in $\mathbf{X}$ will be termed the *basic points* of $\mathscr{R}$. For categorical variables, the basic points are the CLPs and so have special interest. For quantitative variables, the basic points are an arbitrary set of points, depending on the vagaries of the sample, lying on the trajectories and have less interest. From (7.3), we see that the coordinates of the basic points of the kth variables are given by

$$\mathbf{Z}'_k = \mathbf{\Lambda}^{-1}\mathbf{Y}'\mathbf{D}_k(\mathbf{I} - \mathbf{N})$$

The transpose is used here for consistency with our notation where we prefer that the coordinates of points be given as row-vectors.

Because $\mathbf{Y}'\mathbf{1} = \mathbf{0}$, the above may be written as

$$\mathbf{Z}_k = \mathbf{B}_k\mathbf{Y}\boldsymbol{\Lambda}^{-1} \tag{7.5}$$

This result ignores the residuals (7.4) and hence applies only to $\mathscr{R}$, or what amounts to the same thing, the projection from $\mathscr{R}^+$ onto $\mathscr{R}$. Summing (7.5) yields

$$\sum_{k=1}^{p}\mathbf{Z}_k = \sum_{k=1}^{p}\mathbf{B}_k\mathbf{Y}\boldsymbol{\Lambda}^{-1} = (\mathbf{Y}\mathbf{Y}')\mathbf{Y}\boldsymbol{\Lambda}^{-1} = \mathbf{Y}\boldsymbol{\Lambda}\boldsymbol{\Lambda}^{-1} = \mathbf{Y} \tag{7.6}$$

The results (7.5) and (7.6) express, respectively (i) the coordinates in the reference system in terms of the coordinates of the samples and (ii) the coordinates of the samples in terms of the coordinates in the reference system. These are a generalization of the transition formulae (4.12) of MCA. A further simple result that follows from (7.5) is that

$$\mathbf{1}'\mathbf{Z}_k = 0 \tag{7.7}$$

This is of special interest when all the variables are categorical. Then, (7.5) has only L_k distinct columns giving the coordinates of the L_k distinct CLPs for the kth variable. Each coordinate is repeated with the frequency of the corresponding category-level, so that the mean of all the CLPs of a variable, weighted by the category frequencies, is at the centroid of the samples. This is a property often quoted for MCA, but here we see that it remains true for any method of metric MDS of categorical variables when the independence assumption (6.6) holds, as it certainly does for χ^2 distance and the extended matching dissimilarity (4.5). Actually, these centroid properties are less remarkable than they may seem. We have already pointed out that with any method of MDS based on categorical variables, CLPs may be defined by placing them at the centroids of those sample-points with the same category-levels. With the notation of Chapter 4 for categorical variables, the means of all samples for each of the L_k levels of the kth variable are given by

$$\mathbf{Z}_k = \mathbf{L}_k^{-1}\mathbf{G}_k'\mathbf{Y} \tag{7.8}$$

In deriving (7.8), we note that the hth row of $\mathbf{G}_k'\mathbf{Y}$ gives the sum of all samples with the hth level of the kth variable and this is divided by its frequency given by the hth, diagonal, element of $\mathbf{L}_k$. Similarly, the means over all p variables may be written

$$\mathbf{Z} = \mathbf{L}^{-1}\mathbf{G}'\mathbf{Y} \tag{7.9}$$

The rows of $\mathbf{Z}$ give the coordinates of the L centroids of those samples which have each of the L category levels. These formulae give the centroids in $\mathscr{R}$; approximation is discussed in section 7.3 but we note here that for many applications ρ-dimensional approximations may be obtained by replacing $\mathbf{Y}$ by $\mathbf{Y}_\rho$.

Special cases

In this section, we look at some instances of (7.5) and, for the most part, confirm that this general result is indeed consistent with results found previously as special cases.

(*a*) *Pythagorean distance.* Pythagorean distance is the choice of classical biplots and PCA discussed in Chapter 2. It follows that $\mathbf{B}_k = \mathbf{x}_k\mathbf{x}_k'$ so that substituting into (7.5) we have

$$\mathbf{Z}_k = \mathbf{x}_k\mathbf{x}_k'\mathbf{Y}\mathbf{\Lambda}^{-1}$$

which has rank one so represents a set of collinear points. Substitution of $\mathbf{Y} = \mathbf{X}\mathbf{V}$, the PCA interpolation formula, confirms that $\mathbf{Z}_k = \mathbf{x}_k\mathbf{e}_k'\mathbf{V}_\rho$, the correct result for interpolating the basic points for the kth variable by projection onto the ρ dimensions of $\mathscr{L}$.

(*b*) *The extended matching coefficient.* For the extended matching coefficient we have

$$-2\mathbf{D}_k = \mathbf{1}\mathbf{1}' - \mathbf{G}_k\mathbf{G}_k'$$

giving unit distance for a mismatch and zero distance for a match. Hence

$$\begin{aligned}\mathbf{B}_k &= -\frac{1}{2}(\mathbf{I}-\mathbf{N})(\mathbf{1}\mathbf{1}' - \mathbf{G}_k\mathbf{G}_k')(\mathbf{I}-\mathbf{N})\\ &= (\mathbf{I}-\mathbf{N})\left(\frac{1}{2}\mathbf{G}_k\mathbf{G}_k'\right)(\mathbf{I}-\mathbf{N})\end{aligned}$$

Substitution into (7.5) gives

$$\mathbf{Z}_k = (\mathbf{I}-\mathbf{N})\left(\frac{1}{2}\mathbf{G}_k\mathbf{G}_k'\right)(\mathbf{I}-\mathbf{N})\mathbf{Y}\mathbf{\Lambda}^{-1} = \frac{1}{2}(\mathbf{I}-\mathbf{N})(\mathbf{G}_k\mathbf{G}_k')\mathbf{Y}\mathbf{\Lambda}^{-1}$$

Using (4.2) this becomes

$$\mathbf{Z}_k = \frac{1}{2}\left(\mathbf{G}_k - \frac{1}{n}\mathbf{1}\mathbf{1}'_{L_k}\mathbf{L}_k\right)\mathbf{G}'_k\mathbf{Y}\boldsymbol{\Lambda}^{-1} \tag{7.10}$$

The coordinates for each level of the kth categorical variable will be repeated as many times as its frequency in $\mathbf{X}$ so that $\mathbf{Z}_k$ has only L_k distinct columns. The formula corresponding to (7.10) but giving each level only once is

$$\mathbf{Z}_k^* = \frac{1}{2}\left(\mathbf{I}_{L_k} - \frac{1}{n}\mathbf{1}_{L_k}\mathbf{1}'_{L_k}\mathbf{L}_k\right)\mathbf{G}'_k\mathbf{Y}\boldsymbol{\Lambda}^{-1} \tag{7.11}$$

and the basic points for all variables may be gathered into the matrix

$$\mathbf{Z}^* = \frac{1}{2}\left(\mathbf{I} - \frac{1}{n}\mathbf{J}\mathbf{L}\right)\mathbf{G}'\mathbf{Y}\boldsymbol{\Lambda}^{-1} \tag{7.12}$$

where $\mathbf{J} = \text{diag}\,(\mathbf{1}_{L_1}\mathbf{1}'_{L_1}, \mathbf{1}_{L_2}\mathbf{1}'_{L_2}, \ldots, \mathbf{1}_{L_p}\mathbf{1}'_{L_p})$. The matrix $\mathbf{I} - 1/n\ \mathbf{J}\mathbf{L}$ is idempotent.

(*c*) *χ^2 distance.* We have from (4.3) and (4.4) that for χ^2 distance, the centred inner product matrix for the kth variable is:

$$\mathbf{B}_k = \frac{1}{p}(\mathbf{I} - \mathbf{N})\mathbf{G}_k\mathbf{L}_k^{-1}\mathbf{G}'_k(\mathbf{I} - \mathbf{N})$$

which on substitution into (7.5), making the simplifications as in (b), above, gives:

$$\mathbf{Z}_k = \frac{1}{p}\left(\mathbf{G}_k - \frac{1}{n}\mathbf{1}\mathbf{1}'_{L_k}\mathbf{L}_k\right)\mathbf{L}_k^{-1}\mathbf{G}'_k\mathbf{Y}\boldsymbol{\Lambda}^{-1}$$

Thus:

$$\mathbf{Z}_k = \frac{1}{p}\mathbf{G}_k\mathbf{L}_k^{-1}\mathbf{G}'_k\mathbf{Y}\boldsymbol{\Lambda}^{-1} \tag{7.13}$$

and avoiding repetitions of columns

$$\mathbf{Z}_k^* = \frac{1}{p}\mathbf{L}_k^{-1}\mathbf{G}'_k\mathbf{Y}\boldsymbol{\Lambda}^{-1} \tag{7.14}$$

All basic points for χ^2 distance are therefore given by

$$\mathbf{Z}^* = \frac{1}{p}\mathbf{L}^{-1}\mathbf{G}'\mathbf{Y}\boldsymbol{\Lambda}^{-1} \tag{7.15}$$

This result agrees precisely with (4.12) giving the CLPs of MCA; the eigenvalue form $\mathbf{\Lambda}^{-1}$ replaces the singular value form Σ^{-2} but this is a standard relationship. If other methods of scaling the dimensions are used, as is sometimes the case with MCA (Chapter 4) and CA (Chapter 9) then there would be only partial agreement. We have that $p\mathbf{Z}^*\mathbf{\Lambda} = \mathbf{L}^{-1}\mathbf{G}'\mathbf{Y}$, which is (7.9), showing that a scaled form of the CLPs of MCA gives the centroids of the samples with the corresponding category levels. For the extended matching coefficient (7.12) gives a similar but less obvious relationship.

(*d*) *Combining the extended matching coefficient with Pythagorean distance.* A particularly simple special case with mixed types of variables is when in (6.6) Pythagorean distance is used for all quantitative variables and the extended matching dissimilarity (section 4.4.5) is used for all categorical variables. For commensurability, the quantitative variables should first be normalized in some way, say, to have unit sum or sum-of-squares. When (6.6) holds, the ordering of the variables is irrelevant so, without loss of generality, we may assume that the quantitative variables preceed the categorical variables. With this understanding, we may write $(\mathbf{X}, \mathbf{G})$ to represent a partition of the normalized data into its quantitative and categorical components. The contribution to total ddistance generated by the Lcategorical variables is

$$\{d_{ii}^2\} = L\mathbf{1}\mathbf{1}' - \mathbf{G}\mathbf{G}'$$

and, on including the factor $-\frac{1}{2}$ of $\mathbf{D}$ in section A.5, as in (a) above, the contribution to the total centred inner-product matrix is

$$(\mathbf{I} - \mathbf{N})\left(\frac{1}{2}\mathbf{G}\mathbf{G}'\right)(\mathbf{I} - \mathbf{N})$$

Including the continuous variables, we have that $\mathbf{Y}$ is given by the eigenvectors of the centred matrix

$$(\mathbf{I} - \mathbf{N})\left(\mathbf{X}\mathbf{X}' + \frac{1}{2}\mathbf{G}\mathbf{G}'\right)(\mathbf{I} - \mathbf{N})$$

scaled as described in section A.5. With this setting of $\mathbf{Y}$, (7.5) gives the CLPs of the EMC for categorical variables, as in (b) above, and concurrent linear biplots for the quantitative variables, as in (a) above. An example of this kind of biplot is shown in section 11.3, Fig. 11.2.

Distances between sample points and basic points, including CLPs.

The coordinates given by (7.3) and (7.5) may be used to derive an $n(p+1) \times n(p+1)$ matrix giving the ddistances between all basic points of the reference system and those of the original samples generated by the matrices $\mathbf{Y}, \mathbf{Z}_1, \ldots, \mathbf{Z}_p$. Because we have that

$$\mathbf{Z}_h \mathbf{Z}_k' = \mathbf{B}_h \mathbf{Y} \mathbf{\Lambda}^{-2} \mathbf{Y}' \mathbf{B}_k = \mathbf{B}_h \mathbf{B}^{-1} \mathbf{B}_k$$

(see (A.57)) and

$$\mathbf{Y}\mathbf{Z}_k' = \mathbf{Y}\mathbf{\Lambda}^{-1}\mathbf{Y}'\mathbf{B}_k = \mathbf{B}_k$$

(see (A.59)) the inner-product form $\mathbf{\Pi}$ of this distance matrix is

$$\mathbf{\Pi} = \begin{pmatrix} \mathbf{B} & \mathbf{B}_1 & \mathbf{B}_2 & \cdots & \mathbf{B}_p \\ \mathbf{B}_1 & \mathbf{B}_1\mathbf{B}^-\mathbf{B}_1 & \mathbf{B}_1\mathbf{B}^-\mathbf{B}_2 & \cdots & \mathbf{B}_1\mathbf{B}^-\mathbf{B}_p \\ \mathbf{B}_2 & \mathbf{B}_2\mathbf{B}^-\mathbf{B}_1 & \mathbf{B}_2\mathbf{B}^-\mathbf{B}_2 & \cdots & \mathbf{B}_2\mathbf{B}^-\mathbf{B}_p \\ & & & \cdots & \\ \mathbf{B}_p & \mathbf{B}_p\mathbf{B}^-\mathbf{B}_1 & \mathbf{B}_p\mathbf{B}^-\mathbf{B}_2 & \cdots & \mathbf{B}_p\mathbf{B}^-\mathbf{B}_p \end{pmatrix} \tag{7.16}$$

Because the row and column sums are all zero, (7.16) shows that $\mathbf{\Pi}$ is in double-centred form. Thus, the ddistance in $\mathscr{R}$ between the ith level of variable h and the jth level of variable k is, by (A.51), given by the Mahalanobis-like ddistance:

$$(\mathbf{b}_{hi} - \mathbf{b}_{kj})'\mathbf{B}^-(\mathbf{b}_{hi} - \mathbf{b}_{kj}) \tag{7.17}$$

where $\mathbf{b}_{hi}$ is the ith column of $\mathbf{B}_h$. By defining $\mathbf{B}_0$ to be a synonym of $\mathbf{B}$, (7.17) includes inter-sample ddistances (i.e. d_{ij}^2) and ddistances between a sample and a level of any variable. We have the following interpretations for ddistances in $\mathscr{R}$:

(i) The ddistances generated by $\mathbf{B}_0$ (i.e. $\mathbf{B}$) are d_{ij}^2.
(ii) The ddistances generated by $\mathbf{B}_k$ ($k \neq 0$) are given by (7.17) with $h = k$. In $\mathscr{R}^+$, these inter-level distances are defined by the choice of CLPs. Thus, for MCA, they depend on the frequencies in the sample of the category-levels in such a way that the rarer levels are closer together than the more common ones. With the extended matching coefficient, the CLPs are equidistant, forming a regular L_k simplex. In $\mathscr{R}$, these relationships are approximate. Even when exactly known, these properties, at best, are interesting only for comparing frequencies. The main value of their approximation is their potential for detecting

departures from a known pattern, which can be used to judge the adequacy of the approximation – but see (iii) below.

(iii) The ddistances generated between levels of $\mathbf{B}_h$ and $\mathbf{B}_k$ ($h \neq k$, h, $k \neq 0$) in $\mathscr{R}^+$ also seem to be of little interest. However, in $\mathscr{R}$ interesting things happen. When h and k refer to two categorical variables which happen to have the same profile for all samples, then $\mathbf{B}_h = \mathbf{B}_k$ so (7.17) vanishes and both sets of CLPs coincide. Of course, for this to occur, we must also have $L_h = L_k$. When the patterns do not match exactly but nevertheless have a similar structure for the ith level of h and jth level of k, then $(\mathbf{b}_{hi} - \mathbf{b}_{kj})$ will have some common values and will tend to be small, thus generating close CLPs. Similar remarks pertain to quantitative variables but with less force. Surprisingly, the ddistances in $\mathscr{R}$ are more useful than those in $\mathscr{R}^+$.

(iv) The ddistances generated between levels of $\mathbf{B}_0$ and $\mathbf{B}_k$ ($k \neq 0$) are between samples and levels of variables and therefore have special interest. Rather than examine ddistances, which from (7.17) are seen to be hard to interpret, we shall concentrate attention on the inner product. This corresponds to the inner-product interpretations based on SVD of earlier chapters. The inner-product between sample coordinates $\mathbf{Y}$ and the basic points given by (7.5) is, as is shown in (7.16)

$$\mathbf{Y}\mathbf{Z}_k' = \mathbf{Y}\mathbf{\Lambda}^{-1}\mathbf{Y}'\mathbf{B}_k = \mathbf{B}_k \tag{7.18}$$

It follows that if one plots the axis joining the origin to the point whose coordinates are given by the ith row of $\mathbf{Z}_k$, then the projections of the samples with coordinates $\mathbf{Y}$ onto this line gives the ith row of $\mathbf{B}_k$. This is similar to regression biplots (section 3.3.2) but rather than predict $\mathbf{X}$, we now predict $\mathbf{B}_k$ whose elements are generally nonlinear transformations of the kth variable $\mathbf{x}_k$. Indeed, this interpretation is reinforced by observing that on replacing $\mathbf{\Lambda}$ by $\mathbf{Y}'\mathbf{Y}$, then (7.18) is recognizable as a multivariate multiple regression for predicting $\mathbf{B}_k$ from $\mathbf{Y}$, with regression coefficients $\mathbf{Z}_k'$. Thus, every variable gives rise to n axes but these need not all be distinct. For the kth categorical variable, there are L_k such axes and with Pythagorean distance, and hence PCA, all axes coincide with the usual unique biplot axis – see (a) above. In general, the interpretation for quantitative variables is that the n projections from $\mathbf{Y}$ are roughly ordered according to the size of the values of the kth variable but the situation is complicated because the ordering is replicated, with variations, n times – except with Pythagorean distance. With categorical variables, there are only two possibilities; either a sample has the category level associated with the axis of projection or it does not. How this manifests itself depends on the distance used. With χ^2 distance, the ith row of $\mathbf{B}_k$ becomes

$$\frac{1}{l_i} - \frac{1}{n}$$

for every sample with the same category level of the kth variable as does the ith sample, and $-1/n$ for all other samples. Here, l_i is the frequency of the category level of the variable occuring in the ith sample. Thus, in $\mathscr{R}$, it is easy to detect the level of any categorical variable associated with a sample, by inspecting its projections onto the vectors joining the CLPs to the origin. With the extended matching coefficient, the position is more complicated but essentially the same geometry applies. All these results apply in $\mathscr{R}$ and are only approximately valid in $\mathscr{L}$ (section 7.3).

7.2.2 *Joint MDS*

Because (7.16) is in centred form, it may be used directly to give a joint PCO of all the samples and all the basic points. Thus, the coordinates of all $n(p+1)$ points are given by $\mathbf{U}$ in the spectral decomposition

$$\mathbf{\Pi} = \mathbf{U}\mathbf{U}' \quad \mathbf{U}'\mathbf{U} = \mathbf{\Gamma}. \tag{7.19}$$

where $\mathbf{\Gamma}$ is the diagonal matrix of non-increasing eigenvalues of $\mathbf{\Pi}$. Further, the inner-product $\mathbf{\Pi}$ may be replaced by its distance form $\mathbf{\Delta}$, a typical element of which is given by (7.17), and then any method of metric or non-metric scaling may be used to give a global MDS. We term this procedure joint MDS. Unfortunately, both $\mathbf{\Pi}$ and $\mathbf{\Delta}$ are normally very large matrices and hence most methods of MDS are inefficient and may exceed the bounds of computational practicability. The problem is ameliorated for categorical variables by noting that the matrices repeat the same row and column according to the frequency of each category-level. Each matrix $\mathbf{B}_h\mathbf{B}^-\mathbf{B}_k$ $(h, k = 1, 2, \ldots, p)$ may therefore be reduced from size $n \times n$ to one of size $L_h \times L_k$ so that operations are required on a new form of $\mathbf{\Pi}$ of order only $(n+L) \times (n+L)$, corresponding to an analysis unweighted by category frequencies (compare the relationship between χ^2 distance and the extended matching coefficient).

The above outlines the general approach to joint MDS. With PCO, the computations can be reduced to solving an eigen-problem for an $n \times n$ matrix. This follows from noting that (7.16) may be written as

$$\mathbf{\Pi} = \mathbf{P}'\mathbf{B}^{-}\mathbf{P} \tag{7.20}$$

where $\mathbf{P} = (\mathbf{B}, \mathbf{B}_1, \mathbf{B}_2, \ldots, \mathbf{B}_p)$. Because $\mathbf{B}^{-} = \mathbf{Y}\mathbf{\Lambda}^{-2}\mathbf{Y}'$, we may define $\mathbf{Q} = \mathbf{P}'\mathbf{Y}\mathbf{\Lambda}^{-1}$ and write (7.20) as

$$\mathbf{\Pi} = \mathbf{Q}\mathbf{Q}' \tag{7.21}$$

The spectral decomposition required by PCO then becomes

$$\mathbf{\Pi}\mathbf{U} = \mathbf{Q}\mathbf{Q}'\mathbf{U} = \mathbf{U}\mathbf{\Gamma}$$

which may be rewritten as the eigen-problem

$$\mathbf{Q}'\mathbf{Q}(\mathbf{Q}'\mathbf{U}) = (\mathbf{Q}'\mathbf{U})\mathbf{\Gamma} \tag{7.22}$$

(7.22) shows that $\mathbf{W} = \mathbf{Q}'\mathbf{U}$ are eigenvectors of $\mathbf{Q}'\mathbf{Q}$ corresponding to the eigenvalues given in $\mathbf{\Gamma}^2$. The normalization of the vectors $\mathbf{W}$ is given by

$$\mathbf{W}'\mathbf{W} = \mathbf{U}'\mathbf{Q}\mathbf{Q}'\mathbf{U} = \mathbf{U}'\mathbf{\Pi}\mathbf{U} = \mathbf{\Gamma}.$$

The advantage of solving (7.22) rather than (7.19) is that $\mathbf{Q}'\mathbf{Q}$ is of order only q. Indeed, we may write

$$\mathbf{Q}'\mathbf{Q} = \mathbf{B}^2 + \mathbf{B}_1^2 + \mathbf{B}_2^2 + \cdots + \mathbf{B}_p^2 \tag{7.23}$$

To avoid the repeated columns associated with categorical variables, we may define $\mathbf{B}_k^*$ to be the matrix of the L_k different columns of $\mathbf{B}_k$. Then, in (7.23), $\mathbf{B}_k^2$ may be replaced by $\mathbf{B}_k^*\mathbf{B}_k^{*\prime}$ ($k = 1, 2, \ldots, p$).

Alternatively, noting the Mahalanobis nature of (7.17), we are led to the methods of Chapter 5, where the two-sided eigenvalue problem (5.4) is replaced by

$$\mathbf{P}\mathbf{P}'\mathbf{V} = \mathbf{B}\mathbf{V}\mathbf{\Gamma}$$

with the normalization $\mathbf{V}'\mathbf{B}\mathbf{V} = \mathbf{I}$. From the eigenvectors $\mathbf{V}$, we derive $\mathbf{U} = \mathbf{P}'\mathbf{V}$. (Note that the symbols are defined as in this chapter and differ from those of Chapter 5.) To get an unweighted analysis we may replace $\mathbf{P}$ by $\mathbf{P}^* = (\mathbf{B}, \mathbf{B}_1^*, \mathbf{B}_2^*, \ldots, \mathbf{B}_p^*)$. An intriguing consequence is that, using the methods of Chapter 5, we arrive at interpolative and predictive linear biplots for 'variables' with values given by the rows of $\mathbf{P}$ which are themselves the rows, with repetitions, of $\mathbf{B}$. Thus we have an axis for each variable, including the categorical variables.

The above gives the methodology for constructing a joint MDS. The approach hitherto has been to concentrate on the accuracy of the inter-sample distances and to add information on the variables as a secondary objective. With the joint form of analysis, both types of information are given parity, which must mean that the inter-sample distances are likely to be less well represented than by more conventional methods but the CLPs should be better represented. Gower and Dijksterhuis (1994) give an example of joint MDS in which the representation of the CLPs is indeed improved.

Joint MDS and components analysis

It is interesting to investigate the effect of joint PCO analysis in the context of classical linear biplots and components analysis. It is possible to proceed as above but in this simple case it is more direct to operate on the matrix:

$$\begin{pmatrix} & & \mathbf{X} & & \\ \mathbf{x}_1 & 0 & 0 & \cdots & 0 \\ 0 & \mathbf{x}_2 & 0 & \cdots & 0 \\ 0 & 0 & \mathbf{x}_3 & \cdots & 0 \\ . & . & . & . & . \\ 0 & 0 & 0 & \cdots & \mathbf{x}_p \end{pmatrix} \tag{7.24}$$

where $\mathbf{x}_k$ is the kth column of $\mathbf{X}$. This matrix gives the original sample values augmented by the basic points and hence, in this case, a PCA of (7.24) will give the same results as the more general theory. The column totals of (7.24) remain zero so we may proceed immediately by evaluating its sums-of-squares-and-products matrix as

$$\mathbf{X}'\mathbf{X} + \text{diag}\,(\mathbf{X}'\mathbf{X}) \tag{7.25}$$

When $\mathbf{X}$ is normalized so that $\mathbf{X}'\mathbf{X} = \mathbf{R}$, a correlation matrix, then (7.25) becomes $\mathbf{R} + \mathbf{I}$ and has the same eigenvectors as $\mathbf{R}$ itself. It follows that the rows of (7.24) project precisely as they do in PCA and, in particular, the basic points project into their corresponding positions on the linear biplot axes. The only thing that changes is that all the eigenvalues are increased by unity. Further, we saw in Chapter 4 that MCA can be treated as the PCA of the scaled (normalized) indicator matrix $\mathbf{GC}^{-1/2}$ and hence if this replaces $\mathbf{X}$ in (7.24), with corresponding changes in the basic points, the joint PCA analysis, ignoring deviations from the means as before, will give both the sample points and the CLPs of a conventional MCA. The CLPs will be repeated according to their frequencies and this can be avoided by including each basic point only once so that (7.24) has only $n + L$ rows but then the resulting joint MDS is not a conventional MCA. Similar remarks pertain to the use of the extended matching coefficient of section 4.4. When $\mathbf{X}$ is not normalized, a PCA of (7.25) will not repeat the conventional analysis and the difference between the two will depend on the size of the variation among the diagonal elements of $\mathbf{X}'\mathbf{X}$. This difference is related to the problem of commensurability of variables discussed in section 2.5.

The above shows that the joint MDS can give sensible results, at least when $\mathscr{R}$ and $\mathscr{R}^+$ coincide. Because the residual dimension of $\mathscr{R}^+$ is not included in (7.5), the ddistances between pairs of CLPs are not complete but can be made so by direct evaluation and then included in $\mathbf{\Delta}$, thus extending joint MDS to any method involving Euclidean embeddable distances.

7.3 Approximation

For quantitative variables, approximation, both for interpolation and for prediction, proceeds much as in Chapter 6. Thus, graphical interpolation is available only in the context of PCO and when the additive assumption (6.6) is valid. For categorical variables, interpolation remains similar to that for quantitative variables but we shall see that prediction is very different.

7.3.1 *Interpolation*

The interpolation of a sample $(x_1, x_2, \ldots, x_p)$ remains precisely as in (6.10) although now some, or all, of the sample values may be categorical. This interpolant must now be interpreted in terms of the generalized biplot pseudosamples rather than those of nonlinear biplots. Using the general interpolation formula (A.67) immediately gives the interpolant

$$\mathbf{y} = \mathbf{\Lambda}^{-1}\mathbf{Y}'\left[\sum_{h=1}^{p}\{d_h(x_{ih}, x_h)\} - \frac{1}{n}\mathbf{D1}\right] \tag{7.26}$$

or, from (6.7)

$$\mathbf{y} = \mathbf{\Lambda}^{-1}\mathbf{Y}'\sum_{h=1}^{p}\left[\{d_h(x_{ih}, x_h)\} - \frac{1}{n}\mathbf{D}_h\mathbf{1}\right] \tag{7.27}$$

which is merely the vector-sum of the markers $\mu = x_1, x_2, \ldots, x_p$ for the generalized biplot pseudosample interpolants (7.3). In (7.27), it makes no difference whether an x_h refers to a categorical or quantitative variable. Further this process remains valid when the approximation space $\mathscr{L}$ is obtained by projection from $\mathscr{R}^+$. Thus interpolation remains merely a matter of finding vector-sums as previously.

The only point that needs consideration is the non-concurrency of the trajectories, if any, pertaining to quantitative variables. Clearly, if these trajectories are translated in such a way that the vector-sum of the translations is null, then this has no effect on the interpolant. Hence, we seek a set of translations of this kind that achieves concurrency. Consider the point O_k on the kth trajectory that corresponds to the mean of that variable. Then (7.3) shows that the coordinates of O_k are

$$\mathbf{y} = \Lambda^{-1}\mathbf{Y}'\left[\{d_k(x_{ik}, 0)\} - \frac{1}{n}\mathbf{D}_k\mathbf{1}\right] \tag{7.28}$$

The most simple choice of translation is to move all the trajectories to be concurrent at $\bar{O}$, the centroid of all O_k (i.e. for all the quantitative variables). Every point on the kth trajectory is translated by the vector $O_k - \bar{O}$ so the new position of the trajectory is parallel to the old position; the CLPs are unchanged. Then, interpolation is obtained as the vector-sum, relative to G, of the appropriate markers on the concurrent trajectories and of the appropriate CLPs. As in section 2.3, the easist way of doing this is to find the centroid H, say, of the markers and extend it p times from G. Figure 7.1 illustrates the use of the method in conjunction with categorical variables. The CLPs are shown for two variables: Sex (black squares) and Hair Colour (white squares). The markers for the sample values are joined by a dotted line and H is their centroid. The interpolated value is at P and is five ($p = 5$) times GH.

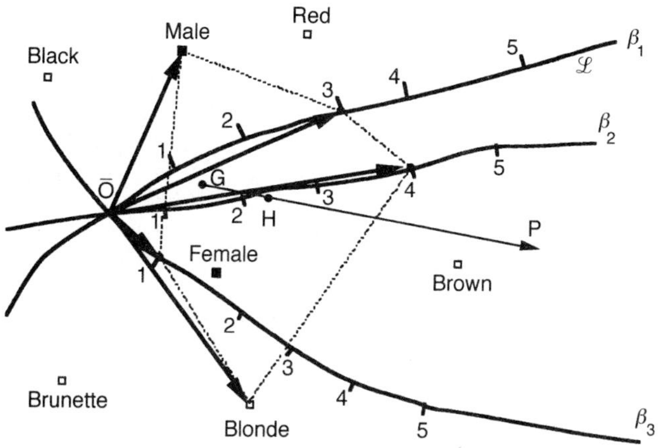

Fig. 7.1. *Interpolation of the sample (3, 4, 1, Male, Blonde).*

In analogy with the vector-sum interpretation for nonlinear biplots (section 6.3.1), we show in the figure arrowed lines with origin $\bar{O}$ that pick out the points for interpolation. This device is merely for comparison and as shown, the interpolated point P is not the vector-sum of these arrowed lines but it is the sum of vectors pointing to the same markers but with origin G. When all the variables are quantative, this method of interpolation may be compared with that described in section 6.3.1 for nonlinear biplots. There the trajectories are concurrent at O, which corresponds to the sample means, and vector-sums are taken relative to O rather than to G. Summing (7.28) for all variables shows that $O = p\bar{O}$. Figure 7.2 shows the relationships between the two methods.

Figure 7.2 shows the equivalence of extension from O (nonlinear biplots) to extension from G (adjusted generalized biplots). It also shows that in generalized biplots, one may extend from $\bar{O}$ rather than G, provided that the vector $\bar{O}O$ is added in at the end, but this would be inconvenient when all the variables are quantitative, in which case the nonlinear biplot form may as well be used. In Figure 7.2 a marker for the same value is shown on each trajectory. Parallelism is indicated by the congruence of the dotted triangle connecting the makers with the triangle $O\bar{O}O_2$; the same is true of

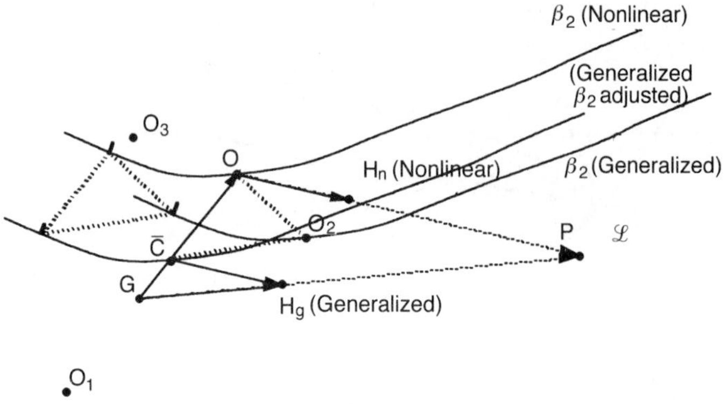

Fig. 7.2. *Comparison of interpolation for variant forms of trajectory for, it is assumed, the second variable.*

triangles connecting any equal markers. H_n is the centroid of markers interpolating some sample using the nonlinear biplot markers; H_g is the same thing for adjusted generalized biplots. The extension of H_n from O gives the same interpolant P as the extension of H_g from G; the extension factor shown is for $p = 3$.

The important thing is that all methods yield the same results for quantitative variables; they are merely graphical variants for evaluating the same interpolation formula (A.64). Thus the generalized interpolative biplot subsumes the nonlinear interpolative biplot but the generalized form must be used when some, or all, variables are categorical.

7.3.2 *Prediction*

For quantitative variables, the generalized biplot is almost identical to its nonlinear counterpart described in section 6.3.2. Because of the parallelism properties, even the differential coefficients remain the same. The different translations of the trajectories give a displacement of the intersection space $\mathscr{N} \cap \mathscr{L}$ but the methodology of Chapter 6 needs no further adaptation. In particular, the choice of origin remains arbitrary and can be chosen as O when all variables are quantitative or as G or $\bar{O}$ when some variables are categorical.

Prediction of categorical variables raises problems which were touched upon in Chapter 4 in the context of MCA. Recall that the key to prediction for the value of variable k to be associated with any point in $\mathscr{L}$ is given by the nearest marker on the kth trajectory or the nearest CLP in $\mathscr{R}^+$. Just as all the points in $\mathscr{R}^+$ that predict a value x for the kth quantitative variable lie in the sub-space $\mathscr{N}$ that is normal to the kth trajectory at the marker x, so all the points in $\mathscr{R}^+$ that predict a category-level l for the kth categorical variable lie in the sub-space that contains all points that are nearer the CLP for l than they are to any other of the remaining $L_k - 1$ CLPs associated with the kth variable. Thus the whole of $\mathscr{R}^+$ is divided into L_k distinct disjoint regions $\mathscr{F}_1, \mathscr{F}_2, \ldots, \mathscr{F}_{L_k}$ one for each of the categories. These are termed **neighbour-regions** and each region is notionally labelled with the name of the category-level which it predicts. The boundary between two neighbour-regions, say $\mathscr{F}_h, \mathscr{F}_k$, is a hyperplane that bisects the join of the corresponding CLPs. Thus, the boundaries of the neighbour-regions themselves are at the intersection of such disecting planes and therefore define convex regions with linear boundaries.

For quantitative variables, the essence of a biplot is to provide information on the exact trajectories (or axes) in $\mathscr{R}^+$ in an approximation space $\mathscr{L}$ and this involves an investigation of $\mathscr{N} \cap \mathscr{L}$. For categorical variables, the equivalent is to investigate the intersections $\mathscr{F}_h \cap \mathscr{L}$ $(h = 1, 2, \ldots, L_k)$. These regions in $\mathscr{L}$ are just a slice through the global neighbour-regions and therefore are also convex regions with linear boundaries. In two dimensions, the linear boundaries become straight lines. The intersections of $\mathscr{L}$ with the neighbour-regions are termed **prediction regions**.

Figure 7.3 shows the prediction-regions in $\mathscr{L}$ of a categorical variable, *colour*, with five levels. Thus, any sample point lying in the region labelled '*red*' is predicted to have the value red, and so on. Many prediction-region diagrams relate to variables with only two or three levels, so are more simple than is Fig. 7.3. It may happen that $\mathscr{L}$ does not intersect with all the neighbour-regions of a categorical variable; in that case, certain categories are never predicted and do not appear in the prediction-region diagram. A prediction-region diagram pertains to one categorical variable and, therefore, is equivalent to a single biplot axis. With continuous variables, many different axes or trajectories can be assimilated into a single diagram but prediction-regions are not so conveniently

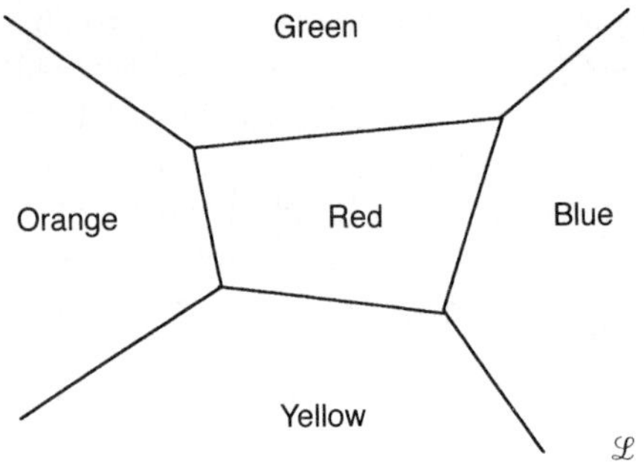

Fig. 7.3. *Five prediction regions in $\mathscr{L}$ associated with a categorical variable for colour.*

superimposed. Generally, separate prediction-regions will be produced for each categorical variable, although there may be times when it is interesting to see if two or more sets of prediction-regions may occupy similar areas of $\mathscr{L}$. Similar prediction-regions suggest correlation between the two categorical variables concerned, just as coincident axes or trajectories provide prima-facie evidence of correlation. Figures 4.4 and 4.5 show prediction regions for MCA and a PCA of the extended matching coefficient; Figures 4.6 and 4.7 show the combined prediction regions for the four variables of these analyses. The prediction regions for generalized biplots would have the same forms so we do not give a separate example.

It remains to discuss how to compute prediction-regions. Section A.11 in the Appendix outlines a detailed algorithm for the most general case (Gower 1993). However when $\rho = 2$, which is usual, a very simple algorithm is as follows. Regard $\mathscr{L}$ as a computer screen, which it very well may be, equipped with pixels. For the kth categorical variable, evaluate which of the associated CLPs is nearest each pixel and label accordingly. Labelling might be a matter of colouring the pixel or giving it some shade of grey. When all pixels are labelled, the computer screen will show the required prediction-regions. This is the method that has been used for the examples of section 4.6.

An alternative scheme is to project the CLPs onto $\mathscr{L}$ (just as for interpolation section 7.2.1) and then append a number giving the distance or ddistance of each projected point from its true position in $\mathscr{R}^+$. This provides all the information needed to calculate the neighbour-distances but it is cumbersome to use. This method does provide a feasible means for combining the neighbour information for all variables into a single diagram.

7.4 An example

This example uses the data of Table 4.6. In section 4.6, we have given an MCA and an analysis based on the extended matching coefficient ignoring the quantitative variable depth of soil and treating the remaining four variables as categorical. We here include soil depth and treat the levels of fertilizer C0, C1, C2, C3 and C4 as equidistant quantitative values. As in section 7.2.1, we treat the quantitative variables as Pythagorean and the categorical variables by the extended matching coefficient. It follows that the quantitative variables give linear biplots with a PCO analysis. Figure 7.4 shows the PCO with linear interpolative axes plus the projected CLPs for the other variables.

The basic points are shown rather than equidistant scale points for the quantitative variables. This method of labelling the axes differs from that demonstrated in previous chapters because, by definition, the basic points include the minimum and maximum values of the variables, which limit the length of the axes; the usual equal steps in the scale of a variable are not presented. For C0, C1, C2, C3 and C4, the basic points are collinear, which, because they are grouped into only five levels, are analogous to the CLPs for ordered categorical variables, and do give an equally spaced scale. For soil depth, the basic points are ungrouped and are clearly skewed towards the lower depths, a feature of the data that is brought to notice when plotting basic points that is unlikely to be noticed with more conventionally scaled axes. Both axes pass through the centroid of $\mathbf{Y}$. Comparing with Fig. 4.3, we see that the CLPs are in very similar positions. The main differences are the obvious ones that there is an additional variable (soil depth), which has induced slight differences in the positions of the farms, and that the fertilizer levels are now shown linearly. Interpolation is as described in section 7.3.1. Predictive scales for the quantitative variables could be shown as the complement of the interpolative scales; for the categorical variables we require prediction regions.

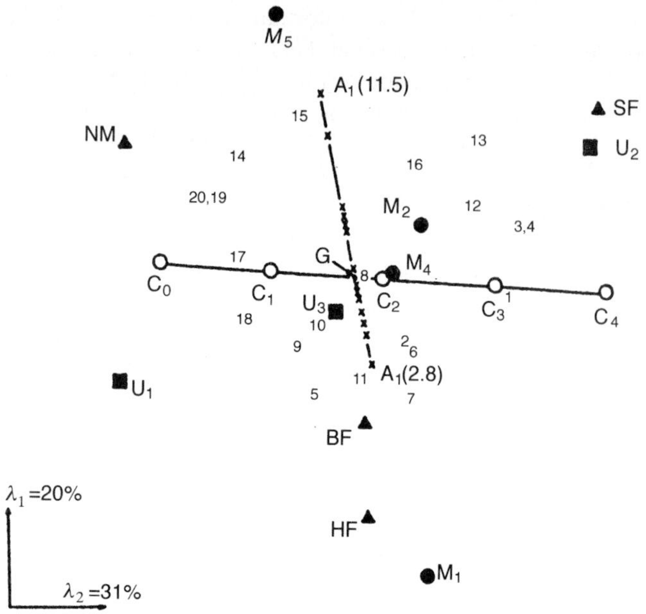

Fig. 7.4. *PCO of Table 4.6 using the extended matching coefficient for categorical variables and Pythagorean distance for the quantitative variables.*

Figure 7.5 shows the same data but using the extended matching coefficient for categorical variables and the square root of Minkowski $L1$-distance for the quantitative variables. The basic points are shown rather than equidistant scale points, supplemented by points O_c for mean fertilizer level and O_a for mean soil depth. Thus, in (6.6) the contribution to distances for quantitative variables are defined by

$$d_k^2(x_{ik}, x_{jk}) = |x_{ik} - x_{jk}|$$

where we have normalized so that

$$\max_k |x_{ik} - x_{jk}| = 1$$

Gower (1971) showed that such a distance is Euclidean embeddable. As explained in Chapter 6, the trajectories are polygons with corners at the basic points where at the end points the trajectories

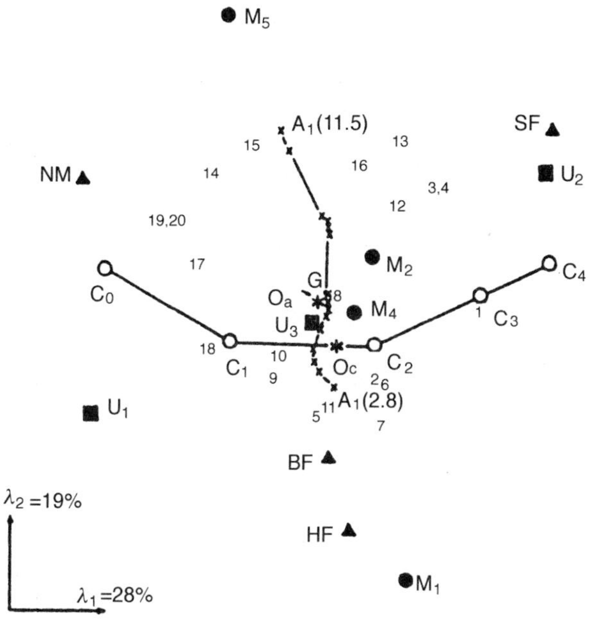

Fig. 7.5. *PCO of Table 4.6 using the extended matching coefficient for categorical variables and the square root of Minkowski L1-distance for the quantitative variables.*

continue orthogonal to $\mathscr{L}$. The two trajectories do not meet at the centroid of **Y**. If there were more than two quantitative trajectories, they could be translated to concurrency at the centroid of the points representing the means of the quantitative variables; section 7.3.1. In Fig. 7.5, the means O_c and O_a are shown, but with only two quantitative variables, there is no point in using these for translation.

CHAPTER 8

Biadditive models

8.1 Introduction

In this chapter and the next, we continue to be concerned with a two-way array **X** but it now refers to a two-way table rather than to a data-matrix. In this chapter, the table contains quantitative values which may be regarded as the values of a response variable such as yield of an agricultural crop or strength of a material. Unlike a data-matrix where the rows refer to samples and the columns to variables, and hence have different status, the row and column classisfiers of a two-way table have equal status, for example referring to two treatments. To emphasize this difference, we shall use the notation that **X** has p rows and q columns and set $n = pq$. Table 8.1 shows such a table from a plant breeding experiment where the response variable is yield and the columns refer to different environments and the rows to different genotypes.

Here $p = 21$, $q = 7$ and $n = 147$. To exhibit the relationship between a two-way table and a data-matrix, Table 8.2 shows Table 8.1 represented as a data-matrix with $n = 147$ rows and three columns which refer to the classifying row and column categorical variables and the response variable which is quantitative. After normalizing the response variable so that it does not dominate the analysis, Table 8.2 could be exhibited as a generalized biplot using the methods of Chapter 7. The next section examines this approach.

8.1.1 The generalized biplot approach

We now explore how the methods of previous chapters would perform when operating on a two-way table expressed as a data-matrix, such as that shown in Table 8.2. We shall write **z** for the

Table 8.1 Strength (average score) of 21 genotypes of perennial rye grass at seven locations in western France (from Denis and Gower, 1995)

Genotypes	*Locations*						
	L1	*L2*	*L3*	*L4*	*L5*	*L6*	*L7*
G1	266.0	420.7	337.5	86.0	398.4	406.0	780.0
G2	359.0	431.8	385.6	384.4	391.8	357.6	479.4
G3	321.7	399.4	365.0	288.9	312.0	331.4	552.2
G4	321.7	469.5	351.2	318.9	276.0	392.8	472.3
G5	310.1	434.4	426.8	295.0	380.8	350.2	643.1
G6	193.0	393.5	413.3	319.5	345.2	342.8	494.9
G7	253.2	485.2	383.8	233.0	346.4	310.0	458.1
G8	349.2	443.4	378.7	179.0	425.0	272.0	672.0
G9	367.1	385.9	456.7	503.0	428.9	347.2	556.7
G10	376.7	416.5	475.0	413.7	439.2	323.9	562.5
G11	381.3	492.0	389.7	385.2	484.8	479.3	689.0
G12	397.3	458.8	337.5	434.1	405.0	489.4	562.5
G13	288.8	335.9	331.4	358.5	365.2	353.3	605.9
G14	360.0	294.6	310.0	276.3	385.1	403.3	636.5
G15	385.6	333.8	349.4	313.6	408.4	417.0	584.2
G16	425.5	285.0	330.6	360.1	438.3	472.5	618.6
G17	374.2	349.3	409.7	373.4	385.1	536.0	615.6
G18	372.3	319.9	356.5	336.8	477.0	451.2	602.3
G19	305.9	314.3	416.8	413.4	484.8	541.0	642.2
G20	306.8	294.2	371.9	406.6	431.6	458.3	694.0
G21	419.3	393.5	389.7	289.7	448.2	375.0	694.0

third column of the stretched out version of $\mathbf{X}$ and shall assume that $\mathbf{z}$ is normalized to unit range, to make it comparable with the categorical variables, and zero mean. We shall write $z = \mathbf{z}'\mathbf{z}$ and the row and column means of $\mathbf{X}$ as column-vectors $\mathbf{x}$ and $\mathbf{y}$ of length p and q, respectively. The inner-product matrix derived from Table 8.2 corresponds to the extended matching coefficient for the row and column categorical variables supplemented by the quantitative variable $\mathbf{z}$ and is then

$$\begin{pmatrix} q\mathbf{I} & \mathbf{11}' & \mathbf{x} \\ \mathbf{11}' & p\mathbf{I} & \mathbf{y} \\ \mathbf{x}' & \mathbf{y}' & z \end{pmatrix}$$

which has eigenvalues and transposed eigenvectors as follows:

- $\lambda = q$; $(\mathbf{u}', \mathbf{0}', 0)$ where $\mathbf{1}'\mathbf{u} = \mathbf{x}'\mathbf{u} = 0$, repeated $p - 2$ times. The vectors $\mathbf{u}$ may be taken as any $p - 2$ columns of $\mathbf{I} - \mathbf{11}'/p - \mathbf{xx}'/(\mathbf{x}'\mathbf{x})$.

Table 8.2 Part of the information of Table 8.1 exhibited as a data-matrix of 147 samples and three variables

Genotype	*Location*	*Strength*
G1	L1	266.0
G1	L2	420.7
G1	L3	337.5
G1	L4	86.0
G1	L5	398.4
G1	L6	406.0
G1	L7	780.0
G2	L1	359.0
G2	L2	431.8
G2	L3	385.6
G2	L4	384.4
G2	L5	391.8
G2	L6	357.6
G2	L7	497.4
G3	L1	321.7
G3	L2	399.4
G3	L3	365.0
G3	L4	288.9
G3	L5	312.0
G3	L6	331.4
G3	L7	552.2
⋮	⋮	⋮
G20	L1	306.8
G20	L2	294.2
G20	L3	371.9
G20	L4	406.6
G20	L5	431.6
G20	L6	458.3
G20	L7	694.0
G21	L1	419.3
G21	L2	393.5
G21	L3	389.7
G21	L4	289.7
G21	L5	448.2
G21	L6	375.0
G21	L7	694.0

- $\lambda = p$; $(\mathbf{0}', \mathbf{v}', 0)$ where $\mathbf{1}'\mathbf{v} = \mathbf{y}'\mathbf{v} = 0$, repeated $q - 2$ times. The vectors $\mathbf{v}$ may be taken as any $q - 2$ columns of $\mathbf{I} - \mathbf{11}'/q - \mathbf{yy}'/(\mathbf{y}'\mathbf{y})$.
- $\lambda = 0$; $(\mathbf{1}', -\mathbf{1}', 0)$
- $\lambda = p + q$; $(q\mathbf{1}', p\mathbf{1}', 0)$
- $\lambda = \lambda_k$ $(k = 1, 2, 3)$; $(\mathbf{x}'/(\lambda_k - q), \mathbf{y}'/(\lambda_k - p), 1)$ where $(\lambda - p)$ $(\lambda - q)$ $(\lambda - z) - (\lambda - p)\mathbf{x}'\mathbf{x} - (\lambda - q)\mathbf{y}'\mathbf{y} = 0$.

If one also removes the column-means from the categorical variables, the only change is that the vector $(q\mathbf{1}', p\mathbf{1}', 0)$ now also corresponds to a zero eigenvalue. It is easy to show that, irrespective of the value of z, then

$$\lambda_1 \geqslant \max(p, q) \geqslant \lambda_2 \geqslant \min(p, q) \geqslant \lambda_3 \geqslant 0$$

Thus in a biplot, the first dimension will be based on λ_1 but the second and subsequent dimensions will be based on either $(\mathbf{u}', \mathbf{0}', 0)$ or $(\mathbf{0}', \mathbf{v}', 0)$. These two sets of vectors at least plot, i.e. interpolate, scores u_i, v_j for all samples in row i and column j, respectively, but this is not exactly informative; the predicted value for the sample in row i and column j given by the eigenvector corresponding to λ_k is

$$\frac{x_i}{\lambda_k - q} + \frac{y_j}{\lambda_k - p} + x_{ij}$$

which is a strange concoction, of little help in interpretation. One notes that because all row-column combinations occur once and only once and induce many regular simplices, so the distribution of the $n = pq$ points in the full $p + q - 1$ dimensions is unlikely to admit to good representations in few dimensions.

The position is not much better if one analyses a modified form of Table 8.2 by MCA. A modification is required to transform $\mathbf{z}$ into categorical form. Thus we first group the values of $\mathbf{z}$ (and hence of the quantitative part of $\mathbf{X}$) into a convenient number r, say, of categories. How this is done is arbitrary but provided pathological groupings are avoided, MCA seems not to be sensitive to the precise groupings that are chosen. Gower (1990b) has analysed the effects of this process, using the notation of section 4.2.3, in which the Burt matrix $\mathbf{B}$ is defined by $\mathbf{B} = \{\mathbf{B}_{ij}\} = \{\mathbf{G}_i'\mathbf{G}_j\}$ with $\mathbf{B}_{ii}' = \mathbf{C}_i$. Note that the 2×2 leading blocks of the previous matrix give the Burt matrix for the classification variables of $\mathbf{X}$, to which are now appended $\mathbf{B}_{13}, \mathbf{B}_{23}$ and $\mathbf{C}_3$. It turns out that the eigenvalues and transposed eigenvectors of $\mathbf{B}(\mathbf{u}', \mathbf{v}', \mathbf{w}')' = \lambda\mathbf{C}(\mathbf{u}', \mathbf{v}', \mathbf{w}')'$ (section 4.2.3) are

- $\lambda = 0; (\mathbf{1}', -\mathbf{1}', \mathbf{0})$ and $(\mathbf{1}', \mathbf{1}', -2\mathbf{1}')$
- $\lambda = 3; (\mathbf{1}', \mathbf{1}', \mathbf{1}')$
- $\lambda = 1;$ $(\mathbf{u}', \mathbf{v}', \mathbf{0})$ where $\mathbf{1}'\mathbf{u} = \mathbf{1}'\mathbf{v} = \mathbf{u}'\mathbf{B}_{13} + \mathbf{v}'\mathbf{B}_{23} = 0$, repeated $p + q - r - 1$ times

- $\lambda = \lambda_k, 2-\lambda_k$ $(k=1,\ldots,r-1)$; if $(\mathbf{u}'_k, \mathbf{v}'_k, \mathbf{w}'_k)$ is the vector corresponding to λ_k then $(-\mathbf{u}'_k, -\mathbf{v}'_k, \mathbf{w}'_k)$ is the vector corresponding to $2-\lambda_k$.

The vectors corresponding to $\lambda = 1$ satisfy $(\mathbf{B}_{31}, \mathbf{B}_{32})\ (\mathbf{u}', \mathbf{v}')' = 0$ so lie in the null-space of the $r \times (p+q)$ matrix of adjoined contingency tables $\mathbf{B}_{31}, \mathbf{B}_{32}$. Provided $r < p+q$ the rank of the adjoined matrices is normally r so the null-space has dimension $p+q-r$ but one of the null vectors is $(\mathbf{1}', -\mathbf{1}', \mathbf{0})$ which has already been accounted for, leaving an additional $p+q-r-1$ vector for which $\lambda = 1$.

The vectors corresponding to $\lambda = \lambda_k$, $2-\lambda_k$ satisfy

$$\mathbf{u}_k = \frac{1}{(\lambda_k - 1)} \frac{\mathbf{B}_{13}\mathbf{w}_k}{q}$$

$$\mathbf{v}_k = \frac{1}{(\lambda_k - 1)} \frac{\mathbf{B}_{23}\mathbf{w}_k}{p}$$

$$(\mathbf{C}_3 - \mathbf{B}_{31}\mathbf{B}_{13} - \mathbf{B}_{32}\ \mathbf{B}_{23})\mathbf{w}_k = \lambda(2-\lambda)\mathbf{C}_3\mathbf{w}_k \tag{8.1}$$

which gives a two-sided eigenvalue equation for these eigenvalues and vectors. Thus in this special structured case, MCA can be completed by solving the eigenvalue equation of the small matrix of (8.1) which has order r. Indeed, Gower (1990b) showed that the solution to (8.1) is essentially given by Fisher's optimal scores and therefore that the values of $\mathbf{w}_k$, corresponding to the biggest eigenvalue, are the scores for fitting to the grouped values of $\mathbf{z}$ in $\mathbf{X}$ an additive model that maximizes the ratio of between-group sums-of-squares to within-group sums-of-squares. This is very satisfactory if one wishes to fit optimal additive scores but viewed as an MCA things are more complicated. Discarding $\lambda = 3$, because of its uninteresting vector, attention will focus on the next highest eigenvalues. Because all eigenvalues in an MCA are non-negative and $\lambda_k + (2-\lambda_k) = 2$, it follows that we may assume that $1 \leqslant \lambda_k \leqslant 2$, in which case $0 \leqslant 2-\lambda_k \leqslant 1$. Thus those eigenvalues for which $\lambda_k \geqslant 1$ now form the basis of a biplot (or, in this case, an MCA) and, unlike the other vectors, these contain a component that pertains to the response variable. However, the plotted coordinates for the sample in the ith row and jth column of $\mathbf{X}$ is at least as complicated and as unhelpful as in the previous case. Furthermore, because of the balanced nature of the two-way table, the χ^2 distances induce

regular simplices and do not admit acceptable low-dimensional approximations. Other methods of analysis need to be exploited. An obvious approach is to fit an additive model *ab initio*, with or without recourse to optimal scores. We shall see in the following that the biplot approach has a role to play but in a different context from that of preceding chapters.

8.1.2 *Introduction to biadditive models*

The previous section has brought to light a flaw in the generalized biplot approach, and of all the methods discussed so far, in that the distinction between response variables and others is ignored; this is the same distinction as that between dependent and independent variables in models, especially linear models. We have already mentioned additive models; here we are concerned with biadditive models, i.e. models, in which parameters occur as products, such as

$$x_{ij} = m + a_i + b_j + c_i d_j \tag{8.2}$$

which becomes a simple additive model if either the suffix i or j is kept fixed. We shall find it convenient to write (8.2) in matrix form:

$$\mathbf{X} = m\mathbf{11}' + \mathbf{a1}' + \mathbf{1b}' + \mathbf{cd}' \tag{8.3}$$

In the usual way, the parameters of (8.2) require constraints for identifiability as it can easily be seen that adding any constant to any of the parameters does not change the form of the model. Accordingly, we apply the usual identification constraints:

$$\mathbf{1}'\mathbf{a} = \mathbf{1}'\mathbf{b} = \mathbf{1}'\mathbf{c} = \mathbf{1}'\mathbf{d} = 0 \tag{8.4}$$

Further identification constraints on the multiplicative parameters are discussed below. Equation (8.2) may be fitted by least-squares to give normal equations whose solutions for the additional parameters are precisely the same as those for the additive model that excludes the multiplicative parameters:

$$\left.\begin{aligned} m &= x_{..} \\ a_i &= x_{i.} - x_{..} \\ b_j &= x_{.j} - x_{..} \end{aligned}\right\} \tag{8.5}$$

Here a dot implies that the mean is taken over the unlabelled parameter.

The multiplicative parameters are obtained from $\mathbf{Z} = \{z_{ij}\}$ where

$$z_{ij} = x_{ij} - x_{i.} - x_{.j} + x_{..}$$

the matrix of residuals after fitting the additive part of the model. Because $\mathbf{cd}' = (\lambda\mathbf{c})(\lambda^{-1}\mathbf{d})'$ there is an indeterminacy in the relative scaling of the multiplicative parameters. This is resolved by giving both sets of parameters equal scaling, by imposing the further identification constraint $\mathbf{c}'\mathbf{c} = \mathbf{d}'\mathbf{d}$. Using a Lagrangian term $\sigma(\mathbf{c}'\mathbf{c} - \mathbf{d}'\mathbf{d})$, the normal equations then become

$$\left.\begin{aligned}\mathbf{Zd} &= \sigma\mathbf{c}\\ \mathbf{Z}'\mathbf{c} &= \sigma\mathbf{d}\end{aligned}\right\} \tag{8.6}$$

where σ is to be determined. From (8.6) we have

$$\mathbf{Z}'\mathbf{Zd} = \sigma^2\mathbf{d} \tag{8.7}$$

showing that $\mathbf{d}$ is an eigenvector of $\mathbf{Z}'\mathbf{Z}$ corresponding to the eigenvalue σ^2. Similarly

$$\mathbf{ZZ}'\mathbf{c} = \sigma^2\mathbf{c} \tag{8.8}$$

To decide which eigenvalue to choose, we note from (8.3) that the residual sum-of-squares is $\|\mathbf{Z} - \mathbf{cd}'\|$. Writing $\mathbf{c}'\mathbf{c} = \mathbf{d}'\mathbf{d} = C$ and using (8.6) gives

$$\begin{aligned}\|\mathbf{Z} - \mathbf{cd}'\| &= \text{trace } (\mathbf{Z} - \mathbf{cd}')(\mathbf{Z} - \mathbf{cd}')'\\ &= \text{trace } (\mathbf{ZZ}') - 2\mathbf{c}'\mathbf{Zd} + C^2\\ &= \text{trace } (\mathbf{ZZ}') - 2\sigma C + C^2\\ &= \text{trace } (\mathbf{ZZ}') + (C - \sigma)^2 - \sigma^2\end{aligned} \tag{8.9}$$

For any choice of σ, the residual sum-of-squares (8.9) is minimized by choosing the scaling $C = \sigma$ and we have a global minimum for the largest value of σ^2. Thus we select the largest eigenvalue of (8.7) and (8.8) and scale the eigenvectors such that $\mathbf{c}'\mathbf{c} = \mathbf{d}'\mathbf{d} = C = \sigma$. Another way of putting this is that the solution is given by the largest singular value and associated vectors of

$$\mathbf{Z} = \mathbf{U\Sigma V}' = \mathbf{CD}' \tag{8.10}$$

where

$$\mathbf{C} = \mathbf{U}\mathbf{\Sigma}^{1/2}, \quad \text{and} \quad \mathbf{D} = \mathbf{V}\mathbf{\Sigma}^{1/2}$$

The model (8.3) may be generalized to give

$$\mathbf{X} = m\mathbf{1}\mathbf{1}' + \mathbf{a}\mathbf{1}' + \mathbf{1}\mathbf{b}' + \sum_{k=1}^{\rho} \mathbf{c}_k\mathbf{d}_k' \tag{8.11}$$

The estimates of the additive parameters remain as given by (8.5) and those of the multiplicative parameters are given by the ρ biggest singular values of (8.10) and their corresponding vectors in the SVD.

Variant models of the same general type, include

$$\begin{aligned}
\mathbf{X} &= m\mathbf{1}\mathbf{1}' + \mathbf{a}\mathbf{1}' + \mathbf{1}\mathbf{b}' + \mathbf{c}\mathbf{b}' && \text{(i)} \\
\mathbf{X} &= m\mathbf{1}\mathbf{1}' + \mathbf{a}\mathbf{1}' + \mathbf{1}\mathbf{b}' + \mathbf{a}\mathbf{d}' && \text{(ii)} \\
\mathbf{X} &= m\mathbf{1}\mathbf{1}' + \mathbf{a}\mathbf{1}' + \mathbf{1}\mathbf{b}' + \lambda\mathbf{a}\mathbf{b}' && \text{(iii)}
\end{aligned} \tag{8.12}$$

which are better reparameterized to

$$\begin{aligned}
\mathbf{X} &= m\mathbf{1}\mathbf{1}' + \mathbf{a}\mathbf{1}' + \mathbf{c}\mathbf{d}' && \text{(i)} \\
\mathbf{X} &= m\mathbf{1}\mathbf{1}' + \mathbf{1}\mathbf{b}' + \mathbf{c}\mathbf{d}' && \text{(ii)} \\
\mathbf{X} &= m\mathbf{1}\mathbf{1}' + \mathbf{c}\mathbf{d}' && \text{(iii)} \\
\mathbf{X} &= \mathbf{c}\mathbf{d}' && \text{(iv)}
\end{aligned} \tag{8.13}$$

where (8.13 (iv)) has been added for completeness.

Equation (8.13 (i)) is often termed the rows-regression model, (8.13 (ii)) the columns regression model, (8.13 (iii)) is the model associated with Tukey's one degree of freedom for non-additivity and (8.13 (iv)), the first of this class of models to be discussed in the literature, is due to Fisher and Mackenzie (1923). The term row-regression arises from noting that if the row-suffix i is fixed, then the ith row of (8.13 (i)) defines a linear regression with slope c_i and intercept $m + a_i$; similarly for column-regression. The Tukey model postulates the same intercept m for all regressions, so sometimes this is referred to as the concurrent regression model. In the parameterization (8.13), the identification constraints appropriate to (8.11) need modification. This is evident for (8.13 (iii)) and (8.13 (iv)) which require no constraints; indeed any constraints imposed would not merely determine identifiability but would change the nature of the models. Similarly, in (8.13 (i)), **c** should not be constrained and, in (8.13 (ii)), **d** should not be constrained but the other parameters remain identified as in (8.11).

Finally, the models (8.13) can be generalized, in a way that is not evident from (8.12), to give

$$\begin{aligned}
&\mathbf{X} = m\mathbf{11}' + \mathbf{a1}' + \sum_{k=1}^{\rho} \mathbf{c}_k \mathbf{d}_k' && \text{(i)} \\
&\mathbf{X} = m\mathbf{11}' + \mathbf{1b}' + \sum_{k=1}^{\rho} \mathbf{c}_k \mathbf{d}_k' && \text{(ii)} \\
&\mathbf{X} = m\mathbf{11}' + \sum_{k=1}^{\rho} \mathbf{c}_k \mathbf{d}_k' && \text{(iii)} \qquad (8.14) \\
&\mathbf{X} = \sum_{k=1}^{\rho} \mathbf{c}_k \mathbf{d}_k' && \text{(iv)}
\end{aligned}$$

Estimation of parameters of (8.14 (i) and (ii)) proceeds as before with the usual least-squares estimates of **a** and **b** and with $\mathbf{c}_k$ and $\mathbf{d}_k$ estimated by the kth singular vectors of the appropriately defined residual matrix **Z**. Similarly, the parameters of (8.14 (iv)) are given by the first ρ singular vectors of **X** itself, which may be recognized as the Eckart–Young theorem of (A.32). The generalized Tukey model (8.14 (iii)) has no closed form solution and requires a special alternating least-squares algorithm for its solution. There is a vast literature on these models; here we are concerned only with associated biplots. For a full discussion, see Gauch (1992), Denis and Gower (1995) or Kempton and Talbot (1988).

8.1.3 Confidence regions for the parameters of biadditive models

The asymptotic covariance matrix for the parameters of biadditive models is available for computing confidence regions that may be superimposed on biplot diagrams. These results assume independent and identically distributed residuals. For completeness, in the remainder of this section, we summarize the results of Denis and

Gower (1994). The basic covariance matrix is

$$\sigma^2 \left(\begin{array}{ccc|cccc}
m & \mathbf{a} & \mathbf{b} & \mathbf{c}_1,\mathbf{d}_1 & \mathbf{c}_2,\mathbf{d}_2 & & \mathbf{c}_\rho,\mathbf{d}_\rho \\
\hline
r & \mathbf{0} & \mathbf{0} & -r\mathbf{f}'_1 & -r\mathbf{f}'_2 & \cdots & -r\mathbf{f}'_\rho \\
\mathbf{0} & \frac{1}{q}\mathbf{P} & \mathbf{0} & \mathbf{0} & \mathbf{0} & \cdots & \mathbf{0} \\
\mathbf{0} & \mathbf{0} & \frac{1}{p}\mathbf{Q} & \mathbf{0} & \mathbf{0} & \cdots & \mathbf{0} \\
\hline
-r\mathbf{f}_1 & \mathbf{0} & \mathbf{0} & \mathbf{E}_{11}+\mathbf{F}_{11} & \mathbf{E}_{12}+\mathbf{F}_{12} & \cdots & \mathbf{E}_{1\rho}+\mathbf{F}_{1\rho} \\
-r\mathbf{f}_2 & \mathbf{0} & \mathbf{0} & \mathbf{E}_{21}+\mathbf{F}_{21} & \mathbf{E}_{22}+\mathbf{F}_{22} & \cdots & \mathbf{E}_{2\rho}+\mathbf{F}_{2\rho} \\
\vdots & \vdots & \vdots & \vdots & \vdots & \ddots & \vdots \\
-r\mathbf{f}_\rho & \mathbf{0} & \mathbf{0} & \mathbf{E}_{\rho 1}+\mathbf{F}_{\rho 1} & \mathbf{E}_{\rho 2}+\mathbf{F}_{\rho 2} & \cdots & \mathbf{E}_{\rho\rho}+\mathbf{F}_{\rho\rho}
\end{array}\right)$$

where σ^2 is an estimate of error variance, derived in the usual way from replication or possibly from high order interactions. The other symbols fall into two classes:

(i) those that are fixed; and
(ii) those that depend on the details of the model.

Not all parameters occur in all models; in the above covariance matrix, the rows and columns corresponding to any unwanted parameters are to be ignored. The fixed parameters are

$$\mathbf{P} = \mathbf{I} - \frac{1}{p}\mathbf{11}', \quad \text{and} \quad \mathbf{Q} = \mathbf{I} - \frac{1}{q}\mathbf{11}'$$

with

$$\mathbf{E}_{uu} = \sum_{k=1}^{\rho} \begin{Bmatrix} \lambda_{uk}\mathbf{c}_k\mathbf{c}'_k & \nu_{uk}\mathbf{c}_k\mathbf{d}'_k \\ \nu_{uk}\mathbf{d}_k\mathbf{c}'_k & \lambda_{uk}\mathbf{d}_k\mathbf{d}'_k \end{Bmatrix}$$

and

$$\mathbf{E}_{uv} = \begin{Bmatrix} \alpha_{uv}\mathbf{c}_v\mathbf{c}'_u & \beta_{uv}\mathbf{c}_v\mathbf{d}'_u \\ \beta_{uv}\mathbf{d}_v\mathbf{c}'_u & \alpha_{uv}\mathbf{d}_v\mathbf{d}'_u \end{Bmatrix}$$

for $u \neq v$, where

$$\lambda_{uu} = \frac{-3}{4C_u^2} \qquad \nu_{uu} = \frac{1}{4C_u^2}$$

$$\lambda_{uk} = \frac{1}{C_u C_k} \frac{3 - \rho_{uk}^4}{(\rho_{uk}^2 - \rho_{ku}^2)^2} \qquad \nu_{uk} = \frac{1}{C_u C_k} \frac{2\rho_{ku}^2}{(\rho_{uk}^2 - \rho_{ku}^2)^2} \quad \text{for } u \neq k$$

$$\alpha_{uv} = \frac{-1}{C_u C_v} \frac{\rho_{uv}^2 + \rho_{vu}^2}{(\rho_{uv}^2 - \rho_{vu}^2)^2} \qquad \beta_{uv} = \frac{-1}{C_u C_v} \frac{2}{(\rho_{uv}^2 - \rho_{vu}^2)^2}$$

$$\rho_{uv} = \left(\frac{C_u}{C_v}\right)^{1/2}$$

The parameters that depend on the model are

Model	r	$\mathbf{f}_u$	$\mathbf{F}_{uu}$	$\mathbf{F}_{uv}(u \neq v)$
(8.14(iv))	–	–	$\frac{1}{Cu}\begin{pmatrix}\mathbf{I} & \mathbf{0}\\ \mathbf{0} & \mathbf{I}\end{pmatrix}$	$\mathbf{0}$
(8.14(i))	$\frac{1}{pq}$	$\mathbf{0}$	$\frac{1}{Cu}\begin{pmatrix}\mathbf{P} & \mathbf{0}\\ \mathbf{0} & \mathbf{I}\end{pmatrix}$	$\mathbf{0}$
(8.14(ii))	$\frac{1}{pq}$	$\mathbf{0}$	$\frac{1}{Cu}\begin{pmatrix}\mathbf{I} & \mathbf{0}\\ \mathbf{0} & \mathbf{Q}\end{pmatrix}$	$\mathbf{0}$
(8.11)	$\frac{1}{pq}$	$\mathbf{0}$	$\frac{1}{Cu}\begin{pmatrix}\mathbf{P} & \mathbf{0}\\ \mathbf{0} & \mathbf{Q}\end{pmatrix}$	$\mathbf{0}$

The formulae for model (8.14 (iii)) are more complicated with

$$r^{-1} = \left(p - \sum_{k=1}^{\rho} \frac{(\mathbf{1}'\mathbf{c}_k)^2}{C_k}\right)\left(q - \sum_{k=1}^{\rho} \frac{(\mathbf{1}'\mathbf{d}_k)^2}{C_k}\right)$$

$$\mathbf{f}_u = \frac{1}{(C_u)^{1/2}} \begin{Bmatrix} \delta_u \mathbf{1} - \dfrac{1}{2C_u}\gamma_u \delta_u \mathbf{c}_u - \sum\limits_{k=1}^{\rho} \kappa_{uk}(\rho_{uk}\gamma_k\delta_u + \rho_{ku}\gamma_u\delta_k)\mathbf{c}_k/C_k \\ \gamma_u \mathbf{1} - \dfrac{1}{2C_u}\gamma_u \delta_u \mathbf{d}_u - \sum\limits_{k=1}^{\rho} \kappa_{uk}(\rho_{ku}\gamma_k\delta_u + \rho_{uk}\gamma_u\delta_k)\mathbf{c}_k/C_k \end{Bmatrix}$$

$$\mathbf{F}_{uu} = \frac{1}{C_u}\begin{pmatrix}\mathbf{I} & \mathbf{0}\\ \mathbf{0} & \mathbf{I}\end{pmatrix} + r\begin{pmatrix}\mathbf{u}_u\mathbf{u}_u' & \mathbf{u}_u\mathbf{v}_u'\\ \mathbf{v}_u\mathbf{u}_u' & \mathbf{v}_u\mathbf{v}_u'\end{pmatrix} \quad \text{and} \quad \mathbf{F}_{uv} = r\begin{pmatrix}\mathbf{u}_v\mathbf{u}_u' & \mathbf{u}_v\mathbf{v}_u'\\ \mathbf{v}_v\mathbf{u}_u' & \mathbf{v}_v\mathbf{v}_u'\end{pmatrix}$$

for $u \neq v$, where

$$\mathbf{1}'\mathbf{c}_u = \gamma_u, \quad \mathbf{1}'\mathbf{d}_u = \delta_u \quad \text{and} \quad \kappa_{uk} = \frac{\rho_{uk}}{\rho_{uk}^2 + \rho_{ku}^2}$$

The confidence ellipse for any vector of parameters $\mathbf{c}$ is then defined as

$$[\mathbf{c} - \mathrm{E}(\mathbf{c})]'\mathbf{V}(c)^{-1}[\mathbf{c} - \mathrm{E}(\mathbf{c})] = \chi$$

where χ is the threshold, at the desired probability level, of a χ^2 variable with ρ degrees of freedom, $\mathrm{E}(\mathbf{c})$ is the least-squares estimate of interest, and $\mathbf{V}(\mathbf{c})$ is the inverse of the covariance matrix given by the above formulae.

8.2 Diagnostic biplots

We now discuss the geometry which underpins biplots for biadditive models. The treatment here is when the models are exact; section 8.3 discusses practical issues when there is superimposed noise.

8.2.1 Exact rank two models

We begin by discussing the models (8.13) to which we add the simple additive model:

$$\mathbf{X} = m\mathbf{1}\mathbf{1}' + \mathbf{a}\mathbf{1}' + \mathbf{1}\mathbf{b}'. \tag{8.15}$$

All these models have rank two, at most. For example, (8.13 (i)) may be written in the form

$$\mathbf{X} = (m\mathbf{1} + \mathbf{a})\mathbf{1}' + \mathbf{c}\mathbf{d}'$$

which, being the sum of two unit rank matrices, has rank two, except in pathological cases where it has unit rank. Similarly, the other models under discussion have rank two, except for (8.13 (iv)), which always has unit rank. Note that this contrasts with data-matrices like that of Table 8.2 which, when put into the numerical form of an indicator matrix (Chapter 4), will normally have rank $p + q - 1$. If a matrix has rank two then it has an exact two-dimensional Euclidean representation. The representation of the general rank two matrix $\mathbf{X} = (\mathbf{u}_1, \mathbf{u}_2)(\mathbf{v}_1, \mathbf{v}_2)'$ is obtained by plotting p row-coordinates $(\mathbf{u}_1, \mathbf{u}_2)$ and q column-coordinates $(\mathbf{v}_1, \mathbf{v}_2)$ to give a two-dimensional biplot which is interpreted by evaluating inner products that reproduce the elements of $\mathbf{X}$. The previously discussed kind of linear biplot in which the second set of coordinates, representing variables, are shown as vectors marked with a scale, remains available but the symmetry of the classifications for the rows and columns of a two-way table, is more appropriately shown by using coordinates for both rows and columns. Another difference, is that biplots of data-matrices are associated with MDSs that show approximations to variously defined row-distances. Pythagorean distances between the rows (columns) of a two-way table may be of interest and can be represented in the biplot. Row-distances are found by ensuring that $(\mathbf{v}_1, \mathbf{v}_2)$ are orthonormal vectors and column-distances by ensuring that $(\mathbf{u}_1, \mathbf{u}_2)$ are orthonormal vectors. These scalings are easily achieved by using the SVD of **X**, but if both sets of vectors are so scaled then the scaling is wrong for reproducing the inner-product. Any two, but not all three, of the following may be represented in a biplot:

(i) row-distances
(ii) column-distances, and
(iii) inner-products.

Any rank-two matrix **X** may be represented in the form $(\mathbf{u}_1, \mathbf{u}_2)(\mathbf{v}_1, \mathbf{v}_2)'$ in an infinite number of ways, generally represented as $(\mathbf{u}_1, \mathbf{u}_2)\mathbf{T}\mathbf{T}^{-1}(\mathbf{v}_1, \mathbf{v}_2)'$ where **T** is any 2×2 non-singular matrix. One such representation of (8.15) is

$$\mathbf{X} = (m\mathbf{1} + \mathbf{a}, \mathbf{1})(\mathbf{1}, \mathbf{b})' \tag{8.16}$$

Figure 8.1 shows (8.16) as a biplot.

The main features of Figure 8.1 are:

(i) the points representing row-constants are collinear on a horizontal line;
(ii) the points representing column-constants are collinear on a vertical line;
(iii) the value x_{ij} is obtained as the inner product $OP_iOQ_j\cos\theta_{ij}$;
(iv) the distance between P_r and P_s is $a_r - a_s$;
(v) the distance between Q_r and Q_s is $b_r - b_s$; and
(vi) the row and column lines are orthogonal.

Of course, these properties depend, at least to some extent, on the particular decomposition (8.16). Recall that all decompositions

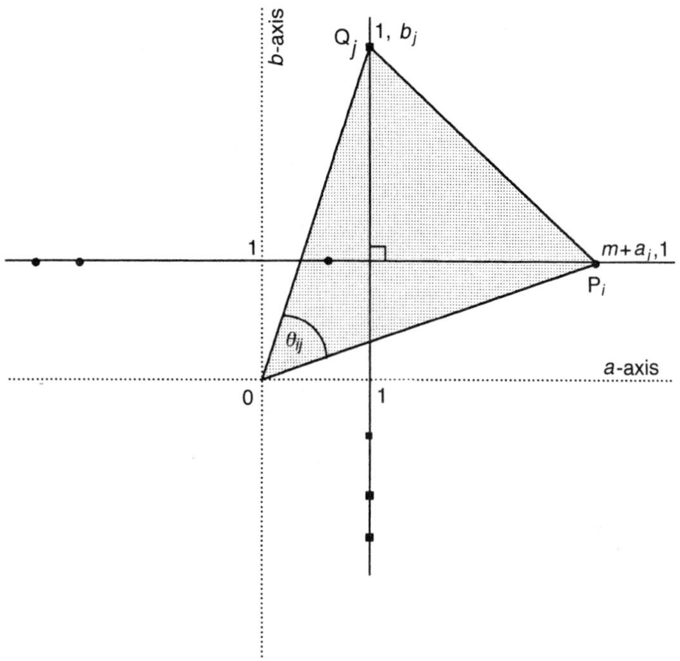

Fig. 8.1. *A biplot of the additive model (8.16). P_i is a typical row-point and Q_j is a typical column-point. Key: see Fig. 8.2.*

have the form

$$\mathbf{X} = (m\mathbf{1} + \mathbf{a},\mathbf{1})\mathbf{T}\mathbf{T}^{-1}(\mathbf{1},\mathbf{b})' \tag{8.17}$$

The inner-product is preserved by (8.17) and the linear transformation $\mathbf{T}$ will preserve all collinearities but the distances in (iv) and (v) now become proportional to $a_r - a_s$ and $b_r - b_s$, respectively. The interesting thing is that the orthogonality relationship (vi) remains valid, as can easily be verified by showing that the row-points $(m\mathbf{1} + \mathbf{a},\mathbf{1})\mathbf{T}$ lie on a line $(\mathbf{1},\mathbf{b})(\mathbf{T}^{-1})'$ with slope t_{11}/t_{12} and the column-points on a line with slope $-t_{12}/t_{11}$. Indeed, (vi) is a property only of biplots of additive models. If a biplot is found that gives collinear row- and column-points with the lines at right angles, then the model must be that represented by (8.15).

Thus the biplot may be used to diagnose additive models. This diagnostic use of biplots is due to Bradu and Gabriel (1978) who show that similar diagnostics can be found for the other models of (8.13). The Tukey model (8.13(iii)) may be written as

$$\mathbf{X} = (m\mathbf{1}, \mathbf{c})(\mathbf{1}, \mathbf{d})'$$

which again gives row and column collinearities, now parallel rather than orthogonal. However, the parallel property does not survive the transformation $\mathbf{TT}^{-1}$ and, in general, the biplot is of two intersecting lines. This is a simple demonstration of the fact that, except in the case of additive models, the particular decomposition chosen affects angles. When m is zero, we have the Fisher–Mackenzie model and $\mathbf{X}$ is of unit rank. Then the display is one-dimensional and the row and column collinearities coincide, though not the points on the collinearities. Thus, oblique intersecting lines diagnose the Tukey model and if these lines coincide we have the very special Fisher–Mackenzie model.

The row-regression model may be written as

$$\mathbf{X} = (m\mathbf{1} + \mathbf{a}, \mathbf{c})(\mathbf{1}, \mathbf{d})'$$

Now, the column-points are collinear on a vertical line but the row-points have a general two-dimensional scatter. In general, the transformation $\mathbf{TT}^{-1}$ preserves this configuration but the collinearity has a general direction. Similar remarks apply to column-regression models. With the model (8.3), both sets of points have general scatter in two dimensions but the model is then of rank three and two-dimensional representations are inadequate; the required generalizations are discussed below. Figure 8.2 illustrates the patterns associated with the different diagnostics.

8.2.2 Exact rank three and higher dimensional models

Model (8.3) is of rank three as can be seen by writing it in the form

$$\mathbf{X} = (m\mathbf{1} + \mathbf{a}, \mathbf{1}, \mathbf{c})(\mathbf{1}, \mathbf{b}, \mathbf{d})'. \tag{8.18}$$

Figure 8.3 shows (8.18) and is a three-dimensional counterpart to Fig. 8.1. In Fig. 8.3, P_i is a typical row-point and Q_j is a typical column-point. The row-points lie in one plane and the column-points lie in another plane; the two planes are orthogonal.

The main features of Fig. 8.3 are:

(i) the points representing row-constants are coplanar;
(ii) the points representing column-constants are coplanar;

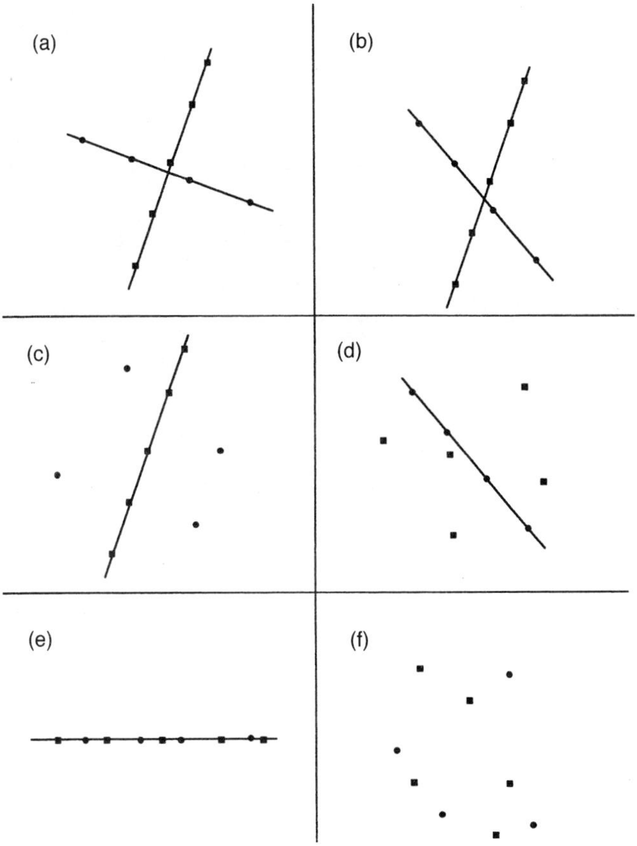

Fig. 8.2. *Diagnostic biplots for biadditive models: (a) additive; (b) Tukey (c) row-regression; (d) column-regression; (e) Fisher–Mackenzie; and (f) none of the previous models. Key:* • *row-points* ■ *column-points.*

(iii) the value x_{ij} is obtained as the inner product $\mathrm{OP}_i\mathrm{OQ}_j \cos \theta_{ij}$;

(iv) the distance between the projections of P_r and P_s onto the a-axis is $a_r - a_s$;

(v) the distance between the projections of Q_r and Q_s onto the b-axis is $b_r - b_s$;

(vi) the distance between the projections of P_r and P_s onto the c, d-axis is $c_r - c_s$;

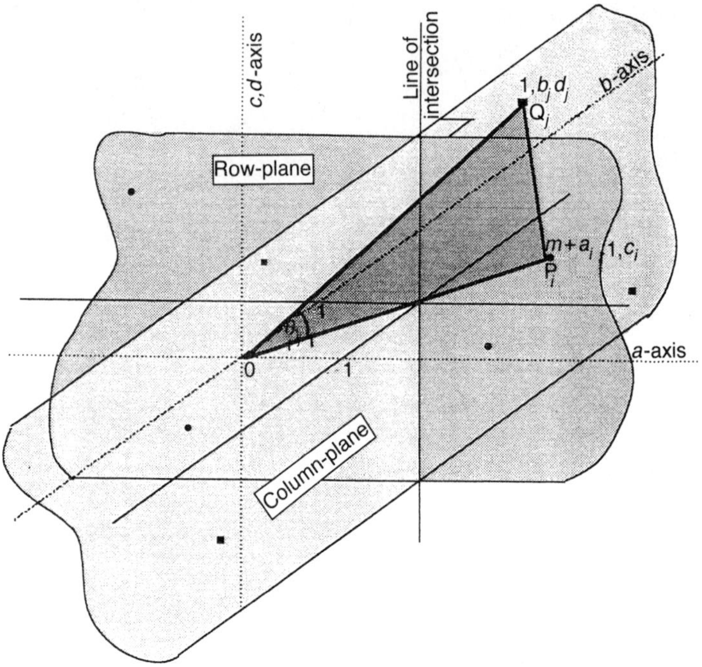

Fig. 8.3. *A three dimensional biplot of the biadditive model (8.18).*

(vii) the distance between the projections of Q_r and Q_s onto the c, d-axis $d_r - d_s$; and

(viii) the row and column planes are orthogonal.

Rather than use the c, d axis for projection, it is more convenient to use the parallel line given by the intersection of the row and column-planes. The mean of the row-points has coordinates $R(m, 1, 0)$ and the mean of the column-points has coordinates $C(1, 0, 0)$. Projecting OR onto the column-plane gives the b-axis, while projecting OC onto the row-plane gives a line through $(1, 1, 0)$ parallel to the a-axis; thus, these directions are those for determining the additive effects for rows and columns, respectively.

As with the additive model, we can ask how much of this structure survives transformations of the type $\mathbf{TT}^{-1}$ which now transforms (8.18) to

$$\mathbf{X} = (m\mathbf{1} + \mathbf{a}, \mathbf{1}, \mathbf{c})\mathbf{TT}^{-1}(\mathbf{1}, \mathbf{b}, \mathbf{d})' \tag{8.19}$$

These linear transformation preserve coplanarities but not angles. Thus the projections now only give proportionalities rather than absolute values of effects. As with the orthogonality of the lines derived from the additive model, the orthogonality of the planes survives transformation. For future reference, we supply a slightly more general proof of this result than that given for the equivalent result for the additive model.

Suppose $\mathbf{l}'\mathbf{x} = l$ and $\mathbf{m}'\mathbf{x} = m$ are two planes the directions of whose normals are $\mathbf{l}$ and $\mathbf{m}$, respectively. Then, before transformation, the angle θ between the planes is given by $\cos\theta = \mathbf{l}'\mathbf{m}$ and after transformation by

$$\cos^2\phi = \frac{(\mathbf{l}'\mathbf{m})^2}{(\mathbf{l}'\mathbf{A}\mathbf{l})(\mathbf{m}'\mathbf{A}^{-1}\mathbf{m})} \tag{8.20}$$

where $\mathbf{A} = \mathbf{T}\mathbf{T}'$. Here $\mathbf{l}$ and $\mathbf{m}$ are given as the directions of two planes such as the row and column planes derived from (8.18) where $\mathbf{l}' = (1, 0, 0)$ and $\mathbf{m}' = (0, 1, 0)$ and $\mathbf{A}$ is at choice. When $\mathbf{l}'\mathbf{m} = 0$ then for any choice of $\mathbf{A}$, $\cos^2\phi = 0$ and ϕ is always a right angle. We note that when $\cos\theta = \mathbf{l}'\mathbf{m} \neq 0$ then we may choose

$$\mathbf{A} = \lambda\mathbf{l}\mathbf{l}' + \mu\operatorname{cosec}^2\theta(\mathbf{m} - \cos\theta\mathbf{l})(\mathbf{m} - \cos\theta\mathbf{l})' + \mathbf{N}$$

where $\mathbf{N}$ is any matrix such that $\mathbf{l}'\mathbf{N} = \mathbf{m}'\mathbf{N} = 0$, $\mathbf{A}$ is of full rank and λ and μ are arbitrary positive numbers. Then

$$\mathbf{l}'\mathbf{A}\mathbf{l} = \lambda \quad \text{and} \quad \mathbf{m}'\mathbf{A}^{-1}\mathbf{m} = \lambda^{-1}\cos^2\theta + \mu^{-1}\sin^2\theta$$

and the denominator of (8.20) becomes $\cos^2\theta + \lambda\mu^{-1}\sin^2\theta$, which may be made as large as we like by choosing λ and μ appropriately. This means that $\mathbf{T}$ may be chosen in an infinity of ways to make the planes very nearly orthogonal, a point to which we shall return in section 8.3.

A rather more general framework is useful for establishing detailed results. We begin by returning to the notation

$$\begin{aligned}\mathbf{X} &= (m\mathbf{1} + \mathbf{a}, \mathbf{1}, \mathbf{c})\mathbf{T}\mathbf{T}^{-1}(\mathbf{1}, \mathbf{b}, \mathbf{d})' \\ &= (\mathbf{u}_2, \mathbf{u}_2, \mathbf{u}_3)(\mathbf{v}_2, \mathbf{v}_2, \mathbf{v}_3)' = \mathbf{U}\mathbf{V}'\end{aligned} \tag{8.21}$$

and define vectors

$$\mathbf{s} = \mathbf{a} - \frac{\mathbf{a}'\mathbf{c}}{\mathbf{c}'\mathbf{c}}\mathbf{c} \quad \text{and} \quad \mathbf{t} = \mathbf{b} - \frac{\mathbf{b}'\mathbf{d}}{\mathbf{d}'\mathbf{d}}\mathbf{d}$$

for which

$$\mathbf{s}'\mathbf{1} = \mathbf{s}'\mathbf{c} = \mathbf{t}'\mathbf{1} = \mathbf{t}'\mathbf{d} = 0$$

as well as

$$\mathbf{a}'\mathbf{1} = \mathbf{b}'\mathbf{1} = \mathbf{c}'\mathbf{1} = \mathbf{d}'\mathbf{1} = 0$$

From (8.21)

$$\mathbf{U}(\mathbf{V}'\mathbf{t}) = \mathbf{1}(\mathbf{b}'\mathbf{t}) \quad \text{and} \quad \mathbf{V}(\mathbf{U}'\mathbf{s}) = \mathbf{1}(\mathbf{a}'\mathbf{s}) \tag{8.22}$$

Because the right-hand sides of both equations in (8.22) are constant, it follows that the coordinates given by the rows of **U** and **V** lie in planes (the row- and column-planes, respectively) with normals given by

$$\mathbf{l} = \mathbf{V}'\mathbf{t} \quad \text{and} \quad \mathbf{m} = \mathbf{U}'\mathbf{s} \tag{8.23}$$

From (8.22) and the orthogonality relationships, we have $\mathbf{l}'\mathbf{m} = 0$, confirming the orthogonality of the row- and column-planes for any decomposition (8.21). From (8.22), we have

$$\mathbf{a}'\mathbf{U}(\mathbf{V}'\mathbf{t}) = \mathbf{c}'\mathbf{U}(\mathbf{V}'\mathbf{t}) = 0 \quad \text{and} \quad \mathbf{b}'\mathbf{V}(\mathbf{U}'\mathbf{s}) = \mathbf{d}'\mathbf{V}(\mathbf{U}'\mathbf{s}) = 0$$

It follows that ($\mathbf{U}'\mathbf{a}$, $\mathbf{U}'\mathbf{c}$) form a basis for all points in the row-plane and ($\mathbf{V}'\mathbf{b}$, $\mathbf{V}'\mathbf{d}$) form a basis for all points in the column-plane.

The coordinates of R and C, the centroids of the points in the row and column planes are $\mathbf{u} = \mathbf{1}'\mathbf{U}/p$ and $\mathbf{v} = \mathbf{1}'\mathbf{V}/q$ giving

$$\mathbf{U}\mathbf{v} = m\mathbf{1} + \mathbf{a} \quad \text{and} \quad \mathbf{V}\mathbf{u} = m\mathbf{1} + \mathbf{b} \tag{8.24}$$

showing that projection of the points in the row-plane onto the **v** direction gives the additive row effects and projection of the column-points onto the **u** direction gives the additive column effects. Because **v** does not lie entirely in the row-plane, projection of **U** onto **v** is the same as projection of **U** onto the component of **v** that lies in the row-plane. Because **l** is the normal to the row-plane, this component is

$$\boldsymbol{\alpha} = (\mathbf{I} - \mathbf{l}\mathbf{l}'/\mathbf{l}'\mathbf{l})\mathbf{v} \quad \text{and similarly} \quad \boldsymbol{\beta} = (\mathbf{I} - \mathbf{m}\mathbf{m}'/\mathbf{m}'\mathbf{m})\mathbf{u} \tag{8.25}$$

To confirm that $\boldsymbol{\alpha}$ and $\boldsymbol{\beta}$ give the correct projections, noting (8.23) and (8.24), we may evaluate

$$\mathbf{U}\boldsymbol{\alpha} = \mathbf{U}(\mathbf{I} - \mathbf{l}\mathbf{l}'/\mathbf{l}'\mathbf{l})\mathbf{v} = m\mathbf{1} + \mathbf{a} - \mathbf{1}(\mathbf{b}'\mathbf{t})(\mathbf{l}'\mathbf{v})/(\mathbf{l}'\mathbf{l})$$

which, apart from a constant, reproduces the vector **a**; similarly for **V**, **β** and **b**.

Above, we showed that **V′b**, **V′d** span the column-plane. Projection of **U** onto these directions gives

$$\mathbf{UV'b} = (\mathbf{b'b})\mathbf{1} + (\mathbf{b'd})\mathbf{c} \quad \text{and} \quad \mathbf{UV'd} = (\mathbf{b'd})\mathbf{1} + (\mathbf{d'd})\mathbf{c}$$

so that the projection of the row-points onto any direction in the column-plane gives the multiplicative effects **c** up to an additive constant and a factor of proportionality. Similarly, projection of the column-points onto any direction in the row-plane gives **d**. If we choose the intersection of the two planes then we have a direction **n**, say, which lies in both planes and hence may be used to give **c** and **d** simultaneously. The direction **n** is orthogonal to **l** and **m** and hence

$$\begin{aligned} n_1 &= l_2 m_3 - l_3 m_2 \\ n_2 &= l_3 m_1 - l_1 m_3 \\ n_3 &= l_1 m_2 - l_2 m_1 \end{aligned} \tag{8.26}$$

Of course, the elements of **l**, **m** and **n** will all require normalization if they are to be treated as direction cosines. Equation (8.26) suffices for computational purposes but more insight is gained by searching for explicit expressions for **Un** and **Vn** expressed in terms of the parameters of the biadditive model. To do this, we note that **n** may be obtained either as the projection of any vector in the row-plane onto the column-plane or the projection of any vector in the column-plane onto the row-plane, nothing that **U′a** and **V′b**, respectively, have already been identified as appropriate vectors. Thus, the projections

$$\gamma = \left(\mathbf{I} - \frac{\mathbf{ll'}}{\mathbf{l'l}}\right)\mathbf{V'b} \quad \text{and similarly} \quad \boldsymbol{\delta} = \left(\mathbf{I} - \frac{\mathbf{mm'}}{\mathbf{m'm}}\right)\mathbf{U'}\mathrm{a} \tag{8.27}$$

are both parallel to **n**. These give

$$\begin{aligned} \mathbf{U}\gamma &= (\mathbf{b'd})\mathbf{c} + \left[\mathbf{b'b} - \frac{(\mathbf{l'Vb})(\mathbf{b't})}{(\mathbf{l'l})}\right]\mathbf{1} \\ \mathbf{V}\boldsymbol{\delta} &= (\mathbf{a'c})\mathbf{d} + \left[\mathbf{a'a} - \frac{(\mathbf{m'Va})(\mathbf{a's})}{(\mathbf{m'm})}\right]\mathbf{1} \end{aligned} \tag{8.28}$$

showing that the values of **c** and **d** are recovered up to an additive constant and proportionality. The vectors γ and **δ** are not normalized but, being parallel, satisfy $(\boldsymbol{\gamma}'\boldsymbol{\gamma})(\boldsymbol{\delta}'\boldsymbol{\delta}) = (\boldsymbol{\gamma}'\boldsymbol{\delta})^2$. Substituting from (8.26) and noting that $\mathbf{a}'\mathbf{Ul} = \mathbf{b}'\mathbf{Vm} = 0$, gives

$$(\boldsymbol{\gamma}'\boldsymbol{\gamma})(\boldsymbol{\delta}'\boldsymbol{\delta}) = (\mathbf{a}'\mathbf{c})^2(\mathbf{b}'\mathbf{d})^2 \tag{8.29}$$

It follows that if $\boldsymbol{\gamma}$ and $\boldsymbol{\delta}$ are normalized, then the projections (8.28) are scaled so that the products $\mathbf{cd}'$ are determined exactly provided that origins representing the constant terms are identified on the line of intersection of the row- and column-planes.

The whole of the geometry discussed in this section is shown in Figure 8.4 which shows the orthogonal row and column-planes with the centroids R and C of the points contained in the planes. The directions $\boldsymbol{\alpha}$ and $\boldsymbol{\beta}$, which determine the additive effects, are obtained as the projections of OR and OC as shown. Multiplicative effects are obtained as projections onto $\mathbf{n}$, the line of intersection of the planes; M is an arbitrary point on this line. It is only the directions of $\boldsymbol{\alpha}$ and $\boldsymbol{\beta}$ in each plane which matter; furthermore, the points may be arbitrarily translated within each plane. Thus R and C both may be translated to a point M on the axis of intersection and $\boldsymbol{\alpha}$ and $\boldsymbol{\beta}$ rooted at M. Figure 8.5 shows this simplified geometry. Algebraic-

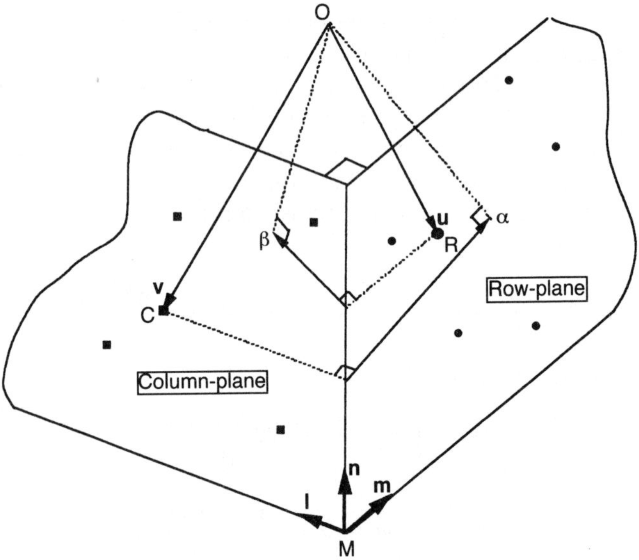

Fig. 8.4. *The geometry of the simple biadditive model (8.3). Key:* • *row-points;* ■ *column-points.*

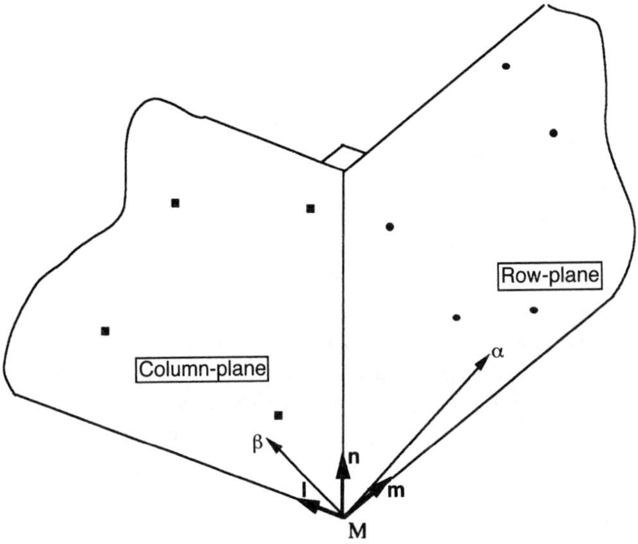

Fig. 8.5. *A simplification of Figure 8.4. See Fig. 8.4 for key.*

ally, this amounts to replacing

$$\mathbf{U} \quad \text{by} \quad \bar{\mathbf{U}} = \left(\mathbf{I} - \frac{\mathbf{11}'}{p}\right)\mathbf{U}$$

and

$$\mathbf{V} \quad \text{by} \quad \bar{\mathbf{V}} = \left(\mathbf{I} - \frac{\mathbf{11}'}{q}\right)\mathbf{V}$$

Then:

$$\begin{aligned} &\bar{\mathbf{U}}\mathbf{l} = 0 && \bar{\mathbf{V}}\mathbf{m} = 0 \\ &\bar{\mathbf{U}}\boldsymbol{\alpha} = \mathbf{a} && \bar{\mathbf{V}}\boldsymbol{\beta} = \mathbf{b} \\ &\bar{\mathbf{U}}\boldsymbol{\gamma} = (\mathbf{b}'\mathbf{d})\mathbf{c} && \bar{\mathbf{V}}\boldsymbol{\delta} = (\mathbf{a}'\mathbf{c})\mathbf{d} \end{aligned} \tag{8.30}$$

These results form the basis of the practical three-dimensional biplot. The inner-product is now

$$\bar{\mathbf{U}}\bar{\mathbf{V}}' = \left(\mathbf{I} - \frac{\mathbf{11}'}{p}\right)\mathbf{U}\mathbf{V}'\left(\mathbf{I} - \frac{\mathbf{11}'}{q}\right) = \left(\mathbf{I} - \frac{\mathbf{11}'}{p}\right)\mathbf{X}\left(\mathbf{I} - \frac{\mathbf{11}'}{q}\right) = \mathbf{cd}'$$

with typical element

$$c_i d_j = x_{ij} - x_{i.} - x_{.j} + x_{..}$$

so that it is the interaction which is recovered. Note that the additive effects remain available through the projections onto $\boldsymbol{\alpha}$ and $\boldsymbol{\beta}$.

The rank two models discussed in section 8.1 are recovered as two-dimensional cross-sections. The additive model (8.15) is obtained as the plane orthogonal to **n** so that the row- and column-planes degenerate into a pair of orthogonal lines. The Tukey model (8.13 (iii)) is obtained as a plane intersecting **n** in a non-orthogonal direction so that the resulting lines become oblique. The rows regression model (8.13 (i)) is merely the row-plane itself, where the column-plane collapses into a line which may be regarded as the line of intersection **n**; similarly for the columns regression model (8.13 (ii)). The Fisher–Mackenzie model (8.13 (iv)) is one-dimensional and collapses into a single line, **n**.

8.2.3 Interesting cross-sections of the three-dimensional space

On computer screens, it may be interesting to use three-dimensional graphics to show the space of Fig. 8.5, especially when interactive rotational facilities are available. However, there are occasions when two-dimensional cross-sections are informative and it is the purpose of this section to describe some useful two-dimensional plots. These will be described in terms of pairs of vectors as follows:

(i) the row-plane ($\boldsymbol{\alpha}$, **n**); also includes **l**
(ii) the column-plane ($\boldsymbol{\beta}$, **n**); also includes **m**
(iii) the additive-plane ($\boldsymbol{\alpha}$, $\boldsymbol{\beta}$)
(iv) the residual-plane (**l**, **m**).

The row- and column-points may be projected onto these planes and the axes in parenthesis may be furnished with scales in a similar manner to those described in previous chapters. The row-plane will contain its row-points and the projection of the column-points onto **n** may also be shown; this allows contributions of the multiplicative terms to be judged. Similar remarks pertain to the column-plane. Projections onto the directions $\boldsymbol{\alpha}$ and $\boldsymbol{\beta}$ of the additive plane give the additive effects.

8.2.4 *Other three-dimensional models*

The models listed in (8.14) have rank three when $\rho = 2$ as does the Fisher model with $\rho = 3$. Equation (8.14(i)) for generalized row-regression may be written as

$$\mathbf{X} = (m\mathbf{1} + \mathbf{a}, \mathbf{c}_1, \mathbf{c}_2)(\mathbf{1}, \mathbf{d}_1, \mathbf{d}_2)'$$

which represents a general three-dimensional scatter for the row-points **U** and coplanar scatter for the column-points **V**. Projection of the row-points onto **v**, the mean of the column-points, continues to give the additive effects **a**. Similar remarks apply to generalized column-regression. When **a** is null, we have the generalized Tukey model (8.14(iii)), and the decomposition represents two parallel planes but on imposing the usual arbitrary transformation **T**, the representation becomes two intersecting planes. Thus, the generalized Tukey model has a similar geometry to that of Fig. 8.4, but the planes are no longer orthogonal, analogous to the non-orthogonal lines of the two-dimensional case (Fig. 8.2(b)). Because in the Tukey models there are no natural constraints on the parameters, except for the relative scaling of the pairs of multiplicative terms, no additive parameters occur and there is no question of their graphical evaluation. However, with different parameterizations of the model, additive parameters that are linear combinations of the multiplicative parameters do occur; a discussion of these representations is given by Gower (1990a). We omit such a discussion here because, when used for diagnosis, it is natural to use the most general convenient form of model and search for special patterns like collinearities, coplanarities and orthogonalities which indicate that special sub-models might be applicable.

8.3 Practical three-dimensional diagnostic biplots

Before giving details of the practical three-dimensional diagnostic biplots, there are four issues that need discussion.

1. We note that the Eckart–Young theorem gives the best rank three approximation to **X** and, indeed, gives information on whether or not three dimensions are adequate for acceptable approximation. Even when three dimensions are insufficient, it may be worthwhile examining the best rank three

approximation to see if a biadditive model fits, while examining the higher dimensions separately. Only in exceptional circumstances will the three-dimensional scatter of points fall into coplanarities and collinearities; these are properties only of the exact models discussed above.
2. The biadditive models (8.11) and (8.14) with appropriate values of ρ generate special rank three matrices; that is why they are useful as diagnostics.
3. Suitable software for displaying three-dimensional images is needed; to be effective, the software must operate interactively.
4. For the diagnostics to be useful, the geometrical structure of row- and column-planes needs to be superimposed on the biplot as a template to which the scatter of points can be referred and this requires a decision as to how to assign the arbitrary scaling matrix **T**.

It is the last point that is the most difficult to resolve because, as we have already seen, although its choice does not affect collinearites and coplanarities, it does have a major effect on angles.

To determine **T**, Gower (1990a) suggested first fitting the biadditive model (8.3) by least-squares and using that parameterization to determine the associated pair of orthogonal row- and column-planes. We are then interested in seeing how the general three-dimensional scatter determined by the Eckart–Young theorem relates to these planes. Suppose that row coordinates **U** and column coordinates **V** represent any three-dimensional approximation to **X**. Suppose also that $(m\mathbf{1}+\mathbf{a}, \mathbf{1}, \mathbf{c})$ and $(\mathbf{1}, \mathbf{b}, \mathbf{d})$ give fitted values for the biadditive model (8.3). Then there exists a scaling matrix **T** such that

$$\mathbf{UT} \approx (m\mathbf{1}+\mathbf{a}, \mathbf{1}, \mathbf{c}) \quad \text{and} \quad \mathbf{V}(\mathbf{T}^{-1})' \approx (\mathbf{1}, \mathbf{b}, \mathbf{d})$$

When the model is exactly biadditive, a unique **T** will satisfy both of these equations simultaneously but, in other cases, an approximate solution must be sought. Gower (1990a) suggested that the multiple regression estimates should be found, to give

$$\mathbf{T}_r = (\mathbf{U}'\mathbf{U})^{-1}\mathbf{U}'(m\mathbf{1}+\mathbf{a}, \mathbf{1}, \mathbf{c}) \quad \text{and} \quad (\mathbf{T}_c^{-1})' \approx (\mathbf{V}'\mathbf{V})^{-1}\mathbf{V}'(\mathbf{1}, \mathbf{b}, \mathbf{d})$$

and pooled to define $\mathbf{T} = \frac{1}{2}(\mathbf{T}_r + \mathbf{T}_c)$. This is one way of determining a useful practical scaling and in the exact case recovers the representation shown on Fig. 8.3. Note that, even in the exact case, the coordinate system defined by the singular vectors of the

Eckart–Young theorem is not helpful for interpretation and we would prefer the orthogonal *a*-axis, *b*-axis and the *c*, *d*-axis of Fig. 8.3.

van Eeuwijk and Keizer (1995) consider another way of choosing $\mathbf{T}$. They look for approximately constant vectors in the SVD of $\mathbf{X} - x_{..}\mathbf{11}'$. Suppose $\mathbf{u}_1 \approx \mathbf{1}$ then it follows that $\mathbf{v}_1 \approx \mathbf{b}$; similarly when $\mathbf{v}_1 \approx \mathbf{1}$ then $\mathbf{u}_1 \approx \mathbf{a}$. Such pairs of vectors may be expected when the main effects dominate the solution, as often will happen. Having identified $\mathbf{a}$ and $\mathbf{b}$, the remaining singular vector pair will estimate $\mathbf{cd}'$ and their singular values may be distributed equally between the two vectors. van Eeuwijk and Keizer (1995) have considered how these scaling methods behave with varying degrees of noise in the biadditive model for the three choices:

(i) the Eckart–Young vectors with equal distribution of the singular values as in (8.10);
(ii) the Gower (1990a) scaling described above; and
(iii) the scaling as described above and more fully in van Eeuwijk and Keizer (1995).

Figure (8.6) shows some of their biplots. The software that produced these diagrams was written in Genstat 5; it incorporates interactive facilities for general three-dimensional rotations, for fitting and showing the planes, showing directions for projection and projections onto these directions (together with residuals), and for removing planes and lines. Figures 8.6(a–d) show three-dimensional biplots for the simple additive model (8.15). The biplot of Fig. 8.6(a) is for the model without noise so we have an exact two-dimensional biplot with the row and column points lying on orthogonal lines. In Fig. 8.6(b–d) there is 10 per cent noise with scaling for the best-fitting planes fitted by methods (i), (ii) and (iii), respectively. In all cases, the same Eckart–Young three-dimensional fit is shown but the different scalings give different emphases. In all cases, the projections giving the relative sizes of the additive main effects give an accurate picture of the known true situation but the effect of the noise is to increase the apparent rank of the model, manifesting itself in different ways with the different modes of scaling. In Fig. 8.6(b), the planes are oblique and there are some large residual displacements from the planes; in Fig. 8.6(c), the points are all close to coplanarity but there is major scatter within the planes; Figures 8.6(a) and (d) are similar so, in Fig. 8.6(d), the noise has been transferred to higher dimensions.

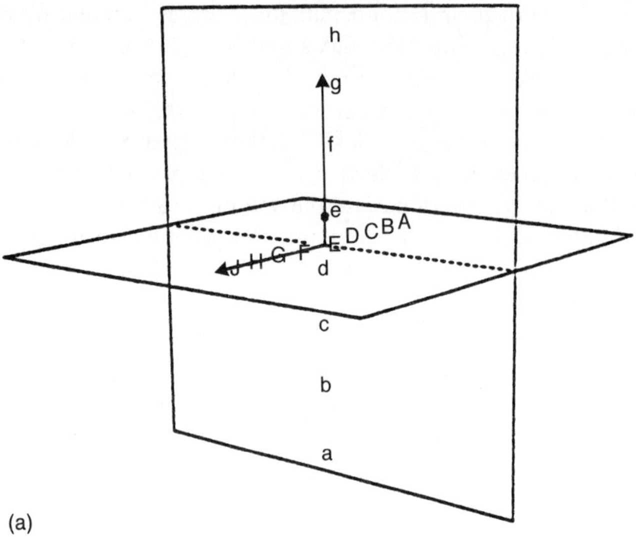

(a)

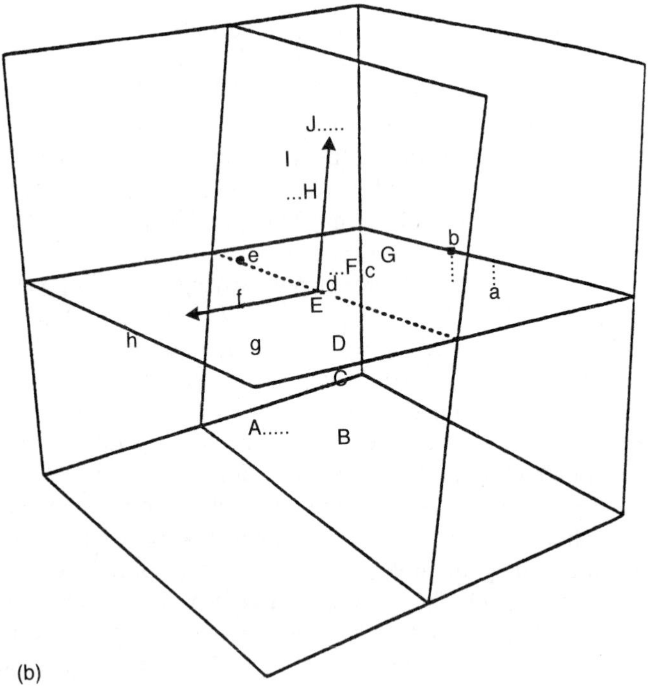

(b)

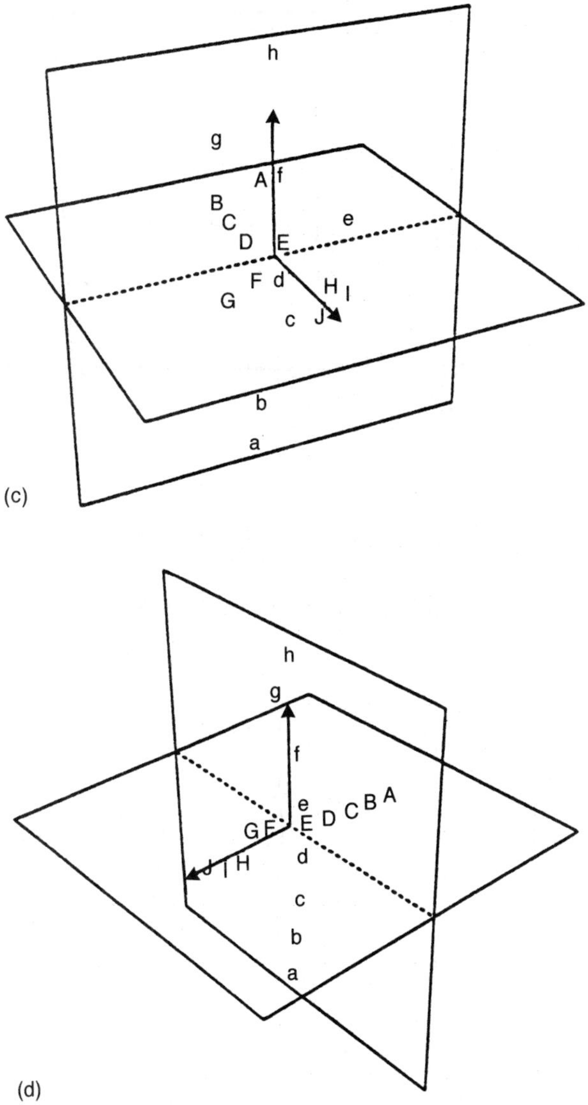

Fig. 8.6. *Biplots derived from a simple additive model: (a) is exact; (b),(c) and (d) 10% noise but different scaling methods (i), (ii) and (iii), as described in the text. In (e) and (f) the underlying model is (8.3): (e) with 35% interaction and no noise; and (f) with $17\frac{1}{2}$% interaction and $17\frac{1}{2}$% noise.*

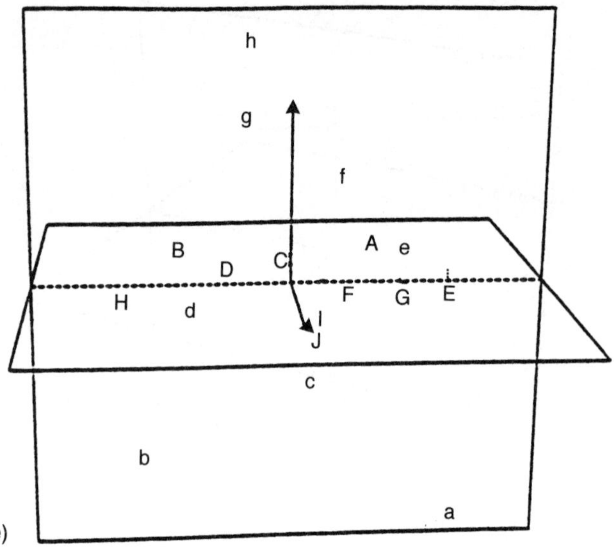

(e)

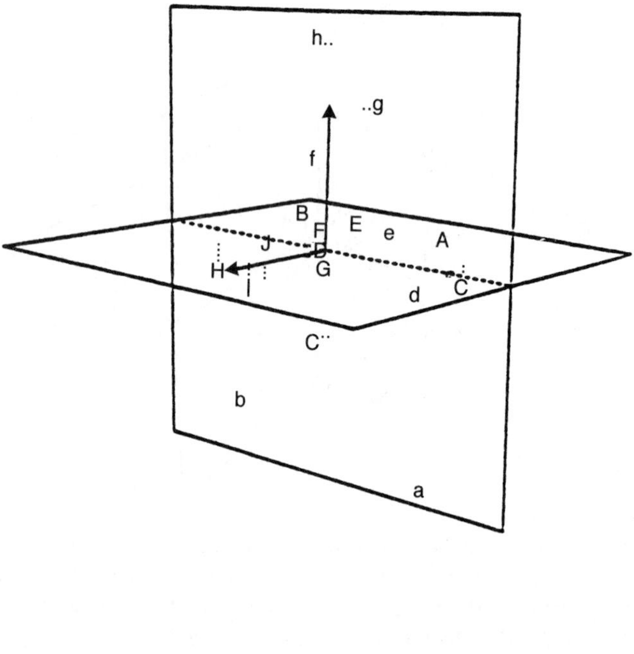

(f)

Figures 8.6(e) and (f) refer to the simple biadditive model (8.3), in (e) without noise and in (f) with the interaction and noise terms both contributing $17\frac{1}{2}$ per cent to total variation. Thus, in Fig. 8.6(e) the points fall exactly in the two planes while in Fig. 8.6(f) there is equal dispersion in the higher dimensions, not shown, and the points have the residuals shown in the biplot. The conclusion from this and similar studies is that, even with fairly modest levels of noise, diagnostic biplots can be misleading; the suggestion is that more complicated models than are necessary might be diagnosed. van Eeuwijk and Keizer (1995) write that it is possible to discriminate between (8.15) and (8.3), between (8.13 (iii)) and (8.14 (iii)) with $\rho = 2$, and between (8.13 (ii)) and (8.14 (ii)) with $\rho = 2$. The problem is that when the level of noise is substantial, in three dimensions, it is difficult to distinguish noise from high-order multiplicative interactions; and so we enter the realm of significance tests.

8.4 An example of a biplot for a two-way table

The artificial data for Fig. 8.6 will have to suffice for a three-dimensional biplot of a two-way table; a further example is given by Gower (1990a) which shows the several two-dimensional cross-sections mentioned in section 8.2.3. Figure 8.7 shows a biplot for two multiplicative interaction terms fitted to Table 8.1; these two terms account for 68.3 per cent of the total interaction variation. There is not much that needs saying. Unlike previous chapters that have dealt with data matrices where the columns refer to variables, there is no special reason to represent columns (or rows) as axes, so both are represented by sets of points. Both locations and genotypes are scaled by the square roots of the corresponding singular values as in (8.10). Interpretation is through inner products between location and genotype points. It follows that projections of the genotypes onto a line joining, say, location L4 to the origin gives a ranking of the genotypes at that location. Thus, genotypes G16, G17, G18, G19 and G20 have the largest, though rather modest, positive interactions at L4, whereas G1 has a large negative interaction. A scale could be applied to lines of projection that allowed numerical values of the interactions to be read off as for the biplots of Chapter 2 and elsewhere. The scale is omitted in the figure, but a dotted line through the origin at right angles to the L4 line, divides those genotypes with positive interactions from those with negative interactions. Genotypes close to the dotted line will have very small interactions at L4. The dotted line happens to pass close to L3 so the two lines partition the figure into four sets of genotypes, depending on whether the interaction is positive at both sites, or negative at both or positive at one and negative at the other. Thus we see that genotypes G2, G4, G6, G9, G10 and G12 all have positive, though in some cases small, interactions at both L3 and L4. Thus, G4 does quite well at L3 but has a poor interaction at L4, whereas G12 does the reverse; G9 does well at both sites. Of course, these remarks apply only to the interaction and a full appraisal would also include the location and genotype main effects. Unfortunately, the latter cannot be included in the figure because the interaction is two-dimensional so the full model has rank four. Corresponding to the row and column planes of the three-dimensional biplot one could examine three-dimensional cross-sections containing directions for the main effects. But this has not been done.

Confidence ellipses can be placed round each point; the formulae required for evaluating these ellipses are given in section 8.1.3 and

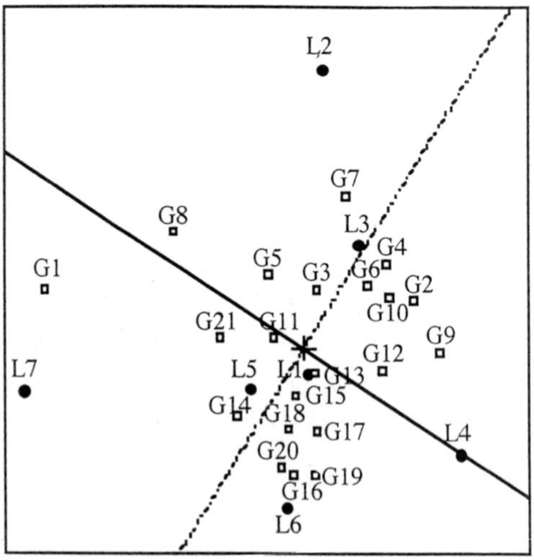

Fig. 8.7. *Biplot for the interaction term ($\rho = 2$) of model (8.10) fitted to the rye grass data of Table 8.1. The origin is marked by a cross.*

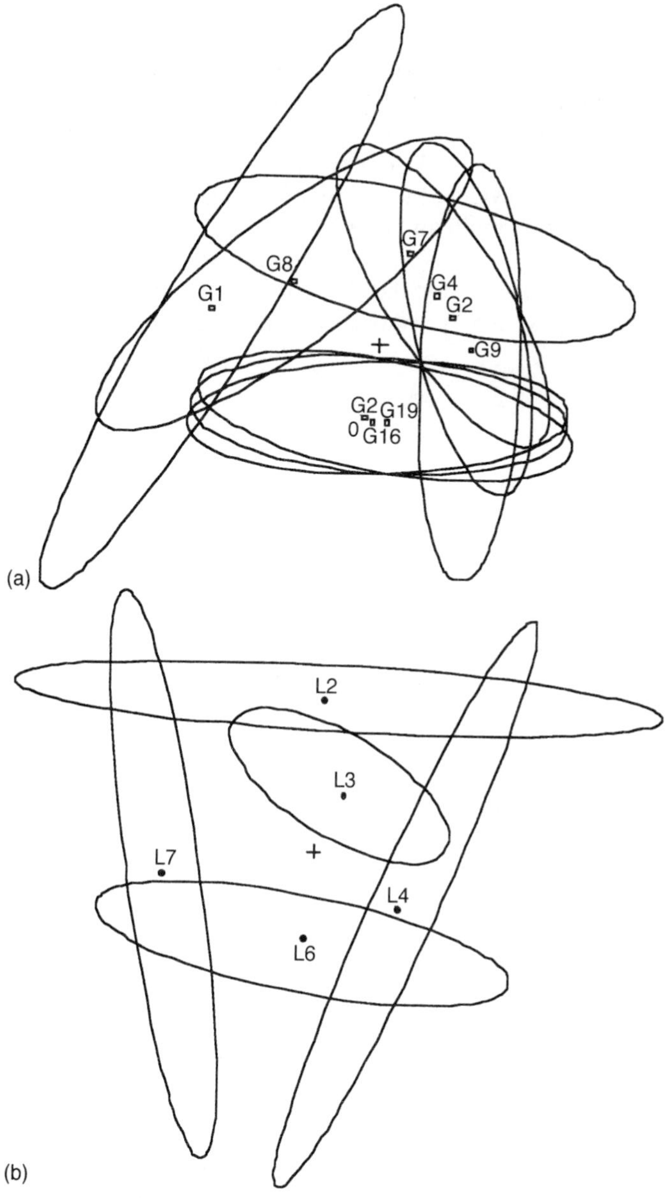

Fig. 8.8. *95% confidence ellipses for (a) the locations and (b) the genotypes. Only ellipses that do not contain the origin are shown.*

their derivation by Denis and Gower (1994). Figure 8.8 shows 95 per cent asymptotic confidence ellipses for locations and genotypes. These are shown in two separate diagrams to reduce the confusion of overlapping ellipses. If an ellipse contains the origin, it represents a significant interaction at the 95 per cent level; those locations and genotypes whose ellipses do not contain the origin are omitted. In this example, it is evident that many of the ellipses shown only just fail to include the origin and only locations L2, L4, L6 and L7 and genotypes G1, G7, G8, G16, G19 and G20 seem to contribute to interaction in a major way; the yields at other locations and for the other genotypes may be modelled adequately by main effects models.

CHAPTER 9

Correspondence analysis (CA)

9.1 Introduction

In much the same way that the biadditive models of Chapter 8 pertain to a quantitative variable classified by two categorical variables into a two-way table, correspondence analysis (CA) is also concerned with a two-way table, but now a contingency table. Thus the elements of the cells are counts and we are concerned with the two classifying categorical variables. The situation differs logically from the quantitative case because no third variable is involved. Thus, corresponding to Table 8.2, we would have only the first two columns, those giving the values of the categorical variables. We could proceed as with an MCA (Chapter 4) with $p = 2$ and indeed, as might be expected, there is a link between MCA and the method about to be discussed; this link is explored in section 10.2. Because the data of Table 8.1 arise from a balanced experiment, the first two columns of Table 8.2 lead to a contingency table all of whose elements are unity but observational data would normally show more interesting structure and it is this structure that CA is designed to expose in visual form. A curious feature of the practice of CA is the way that emphasis is usually placed on the display of the category levels rather than of the samples; this contrasts with PCA, where it is usual to concentrate on the display of the samples and to regard representation of the variables (as biplot axes) as of secondary interest. Although CA is concerned with contingency tables, its algebraic basis requires only that the row and column totals of the two-way table be positive. Not infrequently, the CA technique is used on positive quantitative tables when the methods of Chapter 8, perhaps with a logarithmic transformation of $\mathbf{X}$ would be more appropriate.

9.2 χ^2 distance

Although we are now concerned with a contingency table, for consistency with previous notation, we shall continue to refer to a table **X** as in Chapter 8 but shall now give it dimensions $r \times c$. In this chapter, we shall assume that $r \geqslant c$ so that normally we have that rank $(\mathbf{X}) = c$; when $r < c$ merely transpose **X**, so there is no loss of generality. The row and column totals of **X** will be written as diagonal matrices, **R** of size r and **C** of size c, respectively. CA uses χ^2 distance. Indeed it uses two χ^2 distances; one measuring ddistances between pairs of rows of **X** and the other between pairs of columns. The row ddistances are the Pythagorean ddistances between the points whose coordinates are given by the values in the rows of $\mathbf{Y}_r = \mathbf{R}^{-1}\mathbf{X}\mathbf{C}^{-1/2}$; similarly, the column ddistances are obtained from the columns of $\mathbf{Y}_c = \mathbf{R}^{-1/2}\mathbf{X}\mathbf{C}^{-1}$. We have already met χ^2 distance in the context of MCA where it has the simplified form (4.4) because there the row totals of the indicator matrix **G** all sum to p and the tabular values are confined to zero and unity.

The general definition of χ^2 ddistance between rows i and j of **X** is given by

$$d_{ij}^2 = \sum_{k=1}^{c} \frac{1}{c_k}\left(\frac{x_{ik}}{r_i} - \frac{x_{jk}}{r_j}\right)^2 \tag{9.1}$$

and similarly for χ^2 column ddistance. In terms of the elements of **X**, (9.1) has the form of a weighted Pythagorean distance in which the inverse nature of the diagonal matrices, which give the weights, shows that rare categories carry greater weight relative to common ones and hence tend to dominate. When this is not required, other methods and other distances should be used.

Definition (9.1) shows clearly that any two rows with the same *proportions* generate zero ddistance and hence coincident points and it follows that the other row distances are unaffected by amalgamating proportional rows. A less obvious advantage of χ^2 distance that is often stated in the CA literature (Lebart, Morineau and Warwick, 1984, p. 35) is the property of **distributional equivalence**. This states that if two rows (columns) of **X** are proportional, then their amalgamation will not affect the column (row) distances. If rows i and j are proportional, distributional equivalence is easily proved for χ^2 distance by considering the effect of amalgamation on the distance between columns 1 and 2. The only terms that change on amalgamation are those involving x_{i1}, x_{i2}, x_{j1}, and x_{j2}. Before amalgamation, the contribution of these terms to column ddistance is

$$\frac{1}{r_i}\left(\frac{x_{i1}}{c_1} - \frac{x_{i2}}{c_2}\right)^2 + \frac{1}{r_j}\left(\frac{x_{j1}}{c_1} - \frac{x_{j2}}{c_2}\right)^2$$

and after amalgamation

$$\frac{1}{r_i + r_j}\left(\frac{x_{i1} + x_{j1}}{c_1} - \frac{x_{j1} + x_{j2}}{c_2}\right)^2$$

On substituting $x_{j1} = \pi x_{i1}, x_{j2} = \pi x_{i2}$ and $r_2 = \pi r_1$, where π is a factor of proportionality, the two expressions are seen to be equal, thus establishing the property for columns 1 and 2. Clearly the choice of columns is arbitrary and the result is universally valid. Distributional equivalence is not valid for unweighted Pythagorean distance. It is a useful property because it shows that χ^2 distance, and hence CA, is not sensitive to variations of category-coding in which categories with similar properties are subdivided or amalgamated.

9.3 Analysis

There are several variants of CA, mostly differing in the scalings of the different dimensions in the associated graphical displays. In this section, we attempt to give an account of these variants but the reader is warned to expect confusion. Some variants differ because they have slightly different objectives and we shall try to expose these in the following.

9.3.1 *Principal components of χ^2 distance*

The matrices $\mathbf{Y}_r$ and $\mathbf{Y}_c$ can be analysed in the usual way by PCA, in which case all the results of Chapter 2 may be used for the graphical representation of points whose inter-distances approximate the χ^2 row (column) distances and directions, marked with scales, that allow the interpolation of new rows (columns) and predictions of individual values of $\mathbf{X}$. Thus, biplots of the kinds discussed in Chapter 2 are fully available. Indeed, this was the approach adopted for MCA in Chapter 4. The PCA approach involves regarding the rows (columns) as samples and the columns (rows) as variables and this asymmetric attitude, though appropriate to the data-matrix (indicator matrix) of MCA, contrasts with our treatment of quantitative values for a two-way table discussed in Chapter 8, and may seem unfortunate – but it is not fatal. It is not legitimate to superimpose the PCA plots of the points that generate approximations to the row and column distances because $\mathbf{Y}_r$ and $\mathbf{Y}_c$ have different singular values. It will be seen in the following that some variants of CA legitimize the simultaneous plots of row and column points but there are always caveats to their proper interpretation. It seems to us that PCA plots are straightforward and have much to offer but, so far as we are aware, they have not been used.

Links between the PCA approach to CA and canonical analysis

It has been pointed out that (9.1) is a weighted Pythagorean distance in the elements of **X**. Expressed in terms of the transformed coordinates $\mathbf{Y}_r$, χ^2 distance is unweighted, in precisely the same way that the canonical analysis of Chapter 5 leads to a transformed (canonical) space in which unweighted Pythagorean ddistances give Mahalanobis D^2. Indeed, (9.1) may be viewed as a simple form of D^2 in which the within-group matrix is **C** and hence diagonal, leading to the two-sided eigenvalue problem (Section A.2)

$$(\mathbf{X}'\mathbf{R}^{-2}\mathbf{X})\mathbf{W} = \mathbf{C}\mathbf{W}\mathbf{\Lambda}$$

with the same normalization of vectors $\mathbf{W}'\mathbf{C}\mathbf{W} = \mathbf{I}$ as in section A.2. From the same section (A.2) we have that $\mathbf{C}^{1/2}\mathbf{W}$ are the ortho-gonal eigenvectors of the symmetric matrix $\mathbf{C}^{-1/2}(\mathbf{X}'\mathbf{R}^{-2}\mathbf{X})\mathbf{C}^{-1/2}$. The coordinates $\mathbf{R}^{-1}\mathbf{X}$, which are row frequencies, or **profiles**, are transformed to $\mathbf{R}^{-1}\mathbf{X}\mathbf{W}$ in the canonical space, giving the innerproduct

$$\mathbf{R}^{-1}\mathbf{X}\mathbf{W}\mathbf{W}'\mathbf{X}'\mathbf{R}^{-1} = \mathbf{R}^{-1}\mathbf{X}\mathbf{C}^{-1}\mathbf{X}'\mathbf{R}^{-1} = \mathbf{Y}_r\mathbf{Y}'_r \qquad (9.2)$$

thus generating the χ^2 distances between the rows of **X**. Further, $\mathbf{W}'\mathbf{X}'\mathbf{R}^{-2}\mathbf{X}\mathbf{W} = \mathbf{\Lambda}$, which is diagonal, and hence the transformed coordinates are referred to principal axes in the canonical space. It is interesting to see the parallels with canonical analysis but, with the simplicity of a diagonal metric **C**, rather than tackle the two-sided eigenvalue problem, it is far more preferable to operate directly on the PCA of $\mathbf{Y}_r$. Similar remarks apply to the canonical analysis associated with $\mathbf{Y}_c$.

9.3.2 *Analysis of* χ^2

In CA, a compromise $\mathbf{Y}$ of $\mathbf{Y}_r$ and $\mathbf{Y}_c$ is defined as

$$\mathbf{Y} = \mathbf{R}^{-1/2}\mathbf{X}\mathbf{C}^{-1/2} \tag{9.3}$$

It can be seen tht (9.2) may be written $\mathbf{R}^{-1/2}\ \mathbf{Y}\mathbf{Y}'\mathbf{R}^{-1/2}$ and if the weights $\mathbf{R}^{-2}$ in the canonical analysis are replaced by $\mathbf{R}^{-1}$ then (9.2) may be written $\mathbf{Y}\mathbf{Y}'$. There are two further special reasons for interest in $\mathbf{Y}$ and section 9.3.4 gives yet another. First, let us consider the weighted least-squares problem of finding the rank ρ matrix $\hat{\mathbf{X}}$ which minimizes (section A.4)

$$\text{trace}\ [\mathbf{R}^{-1}(\mathbf{X}-\hat{\mathbf{X}})\mathbf{C}^{-1}(\mathbf{X}-\hat{\mathbf{X}})']$$

which may be written as the Euclidean norm

$$\|\mathbf{R}^{-1/2}(\mathbf{X}-\hat{\mathbf{X}})\mathbf{C}^{-1/2}\| = \|\mathbf{Y}-\mathbf{R}^{-1/2}\hat{\mathbf{X}}\mathbf{C}^{-1/2}\| \tag{9.4}$$

The solution is given by the Eckart–Young theorem using the singular value decomposition $\mathbf{Y}$ as

$$\mathbf{R}^{-1/2}\hat{\mathbf{X}}\mathbf{C}^{-1/2} = \mathbf{U}_\rho\boldsymbol{\Sigma}_\rho\mathbf{V}'_\rho$$

and hence

$$\hat{\mathbf{X}} = \mathbf{R}^{1/2}\mathbf{U}_\rho\boldsymbol{\Sigma}_\rho\mathbf{V}'_\rho\mathbf{C}^{1/2} \tag{9.5}$$

This shows that the *weighted* least-squares approximation to $\mathbf{X}$ is obtained from the singular value decomposition of $\mathbf{Y}$. Thus, with this weighted interpretation, it is $\mathbf{X}$ that is being approximated, rather than $\mathbf{Y}$, and the χ^2 interpretation is no longer fully valid.

The second reason for interest in $\mathbf{Y}$ is as follows. We have

$$\mathbf{Y}\mathbf{C}^{1/2}\mathbf{1} = \mathbf{R}^{-1/2}\mathbf{X}\mathbf{1} = \mathbf{R}^{1/2}\mathbf{1} \quad \text{and} \quad \mathbf{Y}'\mathbf{R}^{1/2}\mathbf{1} = \mathbf{C}^{-1/2}\mathbf{X}'\mathbf{1} = \mathbf{C}^{1/2}\mathbf{1}$$

which shows that $\mathbf{R}^{1/2}\mathbf{1}$ and $\mathbf{C}^{1/2}\mathbf{1}$ are a singular vector pair of $\mathbf{Y}$ corresponding to a unit singular value. These vectors and $\mathbf{Y}$ have all their elements non-negative and hence the Frobenius theorem of non-negative matrices (section A.11) shows that unity is the maximal singular value of $\mathbf{Y}$. Noting that $\mathbf{1}'\mathbf{R}\mathbf{1} = \mathbf{1}'\mathbf{C}\mathbf{1} = x_{..}$, the total of the elements of **X**, we may now write the singular value decomposition of **Y** as follows

$$\mathbf{Y} = \frac{\mathbf{R}^{1/2}\mathbf{1}\mathbf{1}'\mathbf{C}^{1/2}}{x_{..}} + \sum_{k=2}^{c} \sigma_k \mathbf{u}_k \mathbf{v}_k' \tag{9.6}$$

where the first term on the right-hand side has been normalized and $0 \leqslant \sigma_k \leqslant 1$, $k = 2, \ldots, c$. Equation (9.6) may be rearranged to give

$$\mathbf{R}^{-1/2}\mathbf{X}\mathbf{C}^{-1/2} - \frac{\mathbf{R}^{1/2}\mathbf{1}\mathbf{1}'\mathbf{C}^{1/2}}{x_{..}} = \sum_{k=2}^{c} \sigma_k \mathbf{u}_k \mathbf{v}_k' = \mathbf{U}\boldsymbol{\Sigma}\mathbf{V}' \tag{9.7}$$

showing that the right-hand side is the singular value decomposition of the left-hand side and hence that Eckart–Young approximation applies. Note that, in (9.7), **U** and **V** have been redefined from (9.5) to exclude the unit singular value and its associated singular vectors. An element of the matrix on the left-hand side of (9.7) is

$$\chi_{ij} = \frac{\left(x_{ij} - \dfrac{x_{i.}\,x_{.j}}{x_{..}}\right)}{\sqrt{(x_{i.}\,x_{.j})}} \tag{9.8}$$

whose square is proportional to the contribution of the (i,j)th cell to Pearson's χ^2 statistic for the independence of the row and column classifications. The factor of proportionality is $1/x_{..}$ and for this reason **X** is often normalized in CA to have unit total, so that its elements become proportions and (9.7) then requires no factor of proportionality. The equation (9.8) and its relationship to χ^2 reinforces the terminology of χ^2 distance and justifies biplots that preserve inner-product approximations to χ: $(\mathbf{U}\boldsymbol{\Sigma}, \mathbf{V}), (\mathbf{U}, \mathbf{V}\boldsymbol{\Sigma})$ or $(\mathbf{U}\boldsymbol{\Sigma}^{1/2}, \mathbf{V}\boldsymbol{\Sigma}^{1/2})$. Note that the distances generated by the coordinates $\mathbf{U}\boldsymbol{\Sigma}$, and $\mathbf{V}\boldsymbol{\Sigma}$ are functions of the Pythagorean distances between the rows and columns, respectively, of (9.7) and seem to have no intrinsic interest, except that the ddistances from the origin give the contributions of each row and column, respectively, to total χ^2.

9.3.3 Classical CA

To obtain approximations to the χ^2 distances, yet another plot may be used: plot $\mathbf{A} = \mathbf{R}^{-1/2}\mathbf{U}\boldsymbol{\Sigma}$ for the rows and $\mathbf{B} = \mathbf{C}^{-1/2}\mathbf{V}\boldsymbol{\Sigma}$ for the columns. From (9.7), it is evident that the first ρ columns of **A** and **B** give ρ-dimensional approximations to distances generated by the rows of $\mathbf{R}^{-1}\mathbf{X}\mathbf{C}^{-1/2}$ and the columns of $\mathbf{R}^{-1/2}\mathbf{X}\mathbf{C}^{-1}$, respectively; these are the χ^2 distances mentioned above in section 9.2. From (9.4) and (9.5), we have that the approximation minimizes

$$\|\mathbf{R}^{-1/2}(\mathbf{X}-\hat{\mathbf{X}})\mathbf{C}^{-1/2}\| = \|\mathbf{Y}-\mathbf{R}^{1/2}\mathbf{A}_\rho\Sigma_\rho^{-1}\mathbf{B}_\rho'\mathbf{C}^{1/2}\| \qquad (9.9)$$

Thus, the approximation no longer has a simple least-squares justification but the equality of singular values permits the superimposition of both row- and column-points. The warnings of interpreting row–column distances remain in force but the row and column coordinates are related in a way that has some use for interpretation. We have

$$\mathbf{A} = \mathbf{R}^{-1/2}\mathbf{U}\Sigma = \mathbf{R}^{-1/2}\mathbf{Y}\mathbf{V} = \mathbf{R}^{-1/2}\mathbf{Y}\mathbf{C}^{1/2}\mathbf{B}\Sigma^{-1} = \mathbf{R}^{-1}\mathbf{X}\mathbf{B}\Sigma^{-1}$$

Similarly,

$$\mathbf{B} = \mathbf{C}^{-1}\mathbf{X}'\mathbf{A}\Sigma^{-1}$$

Note that $\mathbf{R}^{-1}\mathbf{X}$ and $\mathbf{C}^{-1}\mathbf{X}'$ convert the absolute values of **X** to row and column frequencies as proportions of the relevant margins (**profiles**). We display these results for the coordinates as:

$$\left.\begin{aligned} \mathbf{A} &= \mathbf{R}^{-1}\mathbf{X}\mathbf{B}\Sigma^{-1} \\ \mathbf{B} &= \mathbf{C}^{-1}\mathbf{X}'\mathbf{A}\Sigma^{-1} \end{aligned}\right\} \qquad (9.10)$$

The biplot of **A** and **B** constitutes the classical form of CA encapsulated in much commercial software and, hence, often encountered in publications. Because $\mathbf{A}'\mathbf{R}\mathbf{A} = \Sigma^2, \mathbf{B}'\mathbf{C}\mathbf{B} = \Sigma^2$, with diagonal Σ^2, the coordinates given by **A** and **B** are referred to their principal axes relative to the metrics **R** and **C**, respectively. Therefore, Greenacre (1984) refers to **A** and **B** as the **principal coordinate** representation of the CA. Alternatively, the results may be written in terms of matrices

$$\mathbf{U}^* = \mathbf{R}^{-1/2}\mathbf{U} \quad \text{and} \quad \mathbf{V}^* = \mathbf{C}^{-1/2}\mathbf{V}$$

with the normalization given by

$$\mathbf{U}^{*\prime}\mathbf{R}\mathbf{U}^* = \mathbf{I} \quad \text{and} \quad \mathbf{V}^{*\prime}\mathbf{C}\mathbf{V}^* = \mathbf{I}$$

Then,

$$\mathbf{A} = \mathbf{U}^*\Sigma \text{ and } \mathbf{B} = \mathbf{V}^*\Sigma$$

Greenacre (1984) terms $(\mathbf{U}^*, \mathbf{V}^*)$ the **standard coordinates.**

Note that (9.7) gives

$$\mathbf{X} = \frac{\mathbf{R11'C}}{x_{..}} + \mathbf{R}\left(\sum_{k=2}^{c} \sigma_k \mathbf{u}_k^* \mathbf{v}_k^{*\prime}\right)\mathbf{C} = \mathbf{R}\left(\frac{\mathbf{11'}}{x_{..}} + \mathbf{U}^*\Sigma\mathbf{V}^{*\prime}\right)\mathbf{C} \quad (9.11)$$

which is usually referred to as the **reconstitution formula**, because it expresses the elements of the contingency table $\mathbf{X}$ in terms of its contributing dimensions in standard coordinates.

The equations (9.10) are known as the **transition formulae** of CA. They allow transition from one set of coordinates to the other. Their interest is marred by the presence of the factor Σ^{-1}. If it were absent, (9.10) would have the interesting interpretation that the ith row (column) coordinate would be at the centroid of the column (row) coordinates weighted by the frequencies in the ith row (column) of $\mathbf{X}$; these are the so-called **barycentric coordinates**. This implies that row-points are attracted towards column-points in which they have high frequencies in the relevant row and vice versa. Thus, the transition formulae give a distance-like interpretation to row–column contiguity. The factor Σ^{-1} reduces the force of this property but does not eliminate it.

An important use of the transition formulae, is to allow new rows or columns to be interpolated into a CA diagram. Corresponding to (9.10) we have

$$\begin{aligned} \mathbf{a} &= r^{-1}\mathbf{xB}\Sigma^{-1} \\ \mathbf{b} &= c^{-1}\mathbf{x}'\mathbf{A}\Sigma^{-1} \end{aligned} \quad (9.12)$$

where $\mathbf{x}$ is a new row with sum r and $\mathbf{x}'$ is a new column with sum c and $\mathbf{a}$ and $\mathbf{b}$ are their interpolated positions. Clearly, when $\mathbf{x}$ or $\mathbf{x}'$ coincide with the original rows or columns of $\mathbf{X}$, they interpolate into their correct positions. Rather than the term interpolation, the CA literature refers to **supplementary points.**

In PCA, the means of the variables are removed before computing eigenvectors. This procedure may also be used with any PCA of $\mathbf{Y}$ but it is unnecessary because $\mathbf{R}^{1/2}\mathbf{1}$ and $\mathbf{C}^{1/2}\mathbf{1}$ are the leading singular vector pair of $\mathbf{Y}$ corresponding to a unit singular value, and it follows that the first columns of $\mathbf{A}$ and $\mathbf{B}$ are vectors of units which would be removed on translating to the mean.

If one's priority is that the barycentric principle applies, then there would be no difficulty in taking the coordinates $\mathbf{A}$ that generate approximations to the row χ^2 distances or, better, the coordinates derived from the PCA of $\mathbf{Y}_r$ of section 9.3.1, and replacing $\mathbf{B}$ by $\mathbf{B}^* = \mathbf{C}^{-1}\mathbf{X}'\mathbf{A}$ which does not generate χ^2 distances but which does maintain the attraction properties; similary for $\mathbf{B}$ and $\mathbf{A}^* = \mathbf{R}^{-1}\mathbf{XB}$. A major cause of confusion in CA is that several different things might be approximated and not all can be represented in a single diagram. We can only show least-squares approximations simultaneously on one diagram for any two of:

(a) the row χ^2 distances,
(b) the column χ^2 distances, and
(c) the inner-product χ_{ij}.

Another cause of confusion is between interpretations based on the unweighted approximation to $\mathbf{Y}$ as in (a), (b) and (c), and the weighted approximation of $\mathbf{X}$ itself.

The main plots that are of interest are as follows:

(i) PCA of $\mathbf{Y}_r$ and $\mathbf{Y}_c$ to give points that approximate the row and column χ^2 distances. The dual set of coordinates may be chosen to satisfy the barycentric principal. The associated inner-product is available but not of great interest.
(ii) SVD of $\mathbf{Y}$ to give an inner-product approximation to the individual terms χ_{ij} of Pearson's χ^2 statistic (9.6).
(iii) Classical CA, where a SVD of $\mathbf{Y}$ is followed by the evaluation of $\mathbf{A}$ and $\mathbf{B}$ to give points that generate non-least-squares approximations to both row and column χ^2 distances but no inner product of special interest. The transition formulae provide some basis for interpreting row–column relationships in the symmetric plot; the asymmetric plot has similar properties to (i).
(iv) The weighted approximation to $\mathbf{X}$ as given by (9.5).

It is important that any published graphics make it clear just which of these, or other, representations is being represented.

9.3.4 *Canonical correlation*

The first paper on what was later to be known as CA seems to have been Hirschfeld (1935); Hirschfeld was to become better-known by his anglicized name H. O. Hartley. In the two-column data matrix form of **X**, we can ask what numerical values should be given to the categorical variables to maximize their correlation. This is a variant of the standard canonical correlation problem. Suppose the indicator matrices for the two variables are $\mathbf{G}_1$ and $\mathbf{G}_2$ as in Chapter 4 and that the scores for the row categories are given in a vector **a** and for the column categories in **b**. Initially, working in absolute values of the variables rather than in deviations from the mean, the correlation σ is given by

$$\sigma^2 = \frac{(\mathbf{a}'\mathbf{G}_1'\mathbf{G}_2\mathbf{b})^2}{(\mathbf{a}'\mathbf{G}_1'\mathbf{G}_1\mathbf{a})(\mathbf{b}'\mathbf{G}_2'\mathbf{G}_2\mathbf{b})} = \frac{(\mathbf{a}'\mathbf{X}\mathbf{b})^2}{(\mathbf{a}'\mathbf{R}\mathbf{a})(\mathbf{b}'\mathbf{C}\mathbf{b})}.$$

Differentiating with respect to **a** and **b** gives

$$\mathbf{X}\mathbf{b} = \frac{(\mathbf{a}'\mathbf{X}\mathbf{b})}{(\mathbf{a}'\mathbf{R}\mathbf{a})}\mathbf{R}\mathbf{a} \quad \mathbf{X}'\mathbf{a} = \frac{(\mathbf{a}'\mathbf{X}\mathbf{b})}{(\mathbf{b}'\mathbf{C}\mathbf{b})}\mathbf{C}\mathbf{b} \tag{9.13}$$

The scaling of the vectors **a** and **b** has no effect on σ so is arbitrary; hence we may choose the conventional constraints $\mathbf{a}'\mathbf{R}\mathbf{a} = \mathbf{b}'\mathbf{C}\mathbf{b} = 1$, so that $\mathbf{a}'\mathbf{X}\mathbf{b} = \sigma$. We also define

$$\mathbf{u} = \mathbf{R}^{1/2}\mathbf{a} \quad \text{and} \quad \mathbf{v} = \mathbf{C}^{1/2}\mathbf{b}$$

so that $\mathbf{u}'\mathbf{u} = \mathbf{v}'\mathbf{v} = 1$. With this notation (9.13) becomes

$$\left.\begin{aligned} \mathbf{Y}\mathbf{v} &= \sigma\mathbf{u} \\ \mathbf{Y}'\mathbf{u} &= \sigma\mathbf{v} \end{aligned}\right\} \tag{9.14}$$

where **Y** is given by (9.3). Equation (9.14) implies that **u** and **v** are a singular vector pair of **Y** and that σ is the corresponding singular value. The correlation is maximized by the maximum singular value which, as we have seen, is unity. This is because we have not removed the means when defining the correlation, but we have shown that the special first singular vectors effectively absorbs the mean. In detail, we have that for the first singular vectors

$$\frac{\mathbf{R}^{1/2}\mathbf{1}}{\sqrt{x_{..}}} = \mathbf{R}^{1/2}\mathbf{a}_1 \quad \text{and} \quad \frac{\mathbf{C}^{1/2}\mathbf{1}}{\sqrt{x_{..}}} = \mathbf{C}^{1/2}\mathbf{b}_1$$

so that

$$\mathbf{a}_1 = \frac{\mathbf{1}}{\sqrt{x_{..}}} \quad \text{and} \quad \mathbf{b}_1 = \frac{\mathbf{1}}{\sqrt{x_{..}}}$$

It follows that the $\mathbf{a}'\mathbf{R}\mathbf{a}$, when corrected for the mean, becomes

$$\mathbf{a}'\mathbf{G}_1'(\mathbf{I}-\mathbf{N})\mathbf{G}_1\mathbf{a}=\mathbf{a}'\mathbf{R}\mathbf{a}-\mathbf{a}'\mathbf{G}_1'\left(\frac{\mathbf{1}\mathbf{1}'}{n}\right)\mathbf{G}_1\mathbf{a}$$

Now $\mathbf{1}'\mathbf{G}_1\mathbf{a}=\mathbf{1}'\mathbf{R}\mathbf{a}=0$ because of the orthogonality relationships for all $\mathbf{a}\neq\mathbf{a}_1$. Thus $\mathbf{a}'\mathbf{G}_1'(\mathbf{I}-\mathbf{N})\mathbf{G}_1\mathbf{a}=\mathbf{a}'\mathbf{R}\mathbf{a}$ and similarly for the other terms. This shows that the solutions to (9.14) other than the leading term, pertain to correlations corrected for the mean and hence that the maximum sought occurs for the second singular vector. The solutions are identical to those of classical CA discussed in section 9.3.3. This correlational property gives another characterization of CA. Like the homogeneity/optimal-scores approach to MCA discussed in section 4.3.1, there is the difficulty that the solution sought is one-dimensional and that the inclusion of further singular vectors for $\sigma_2, \sigma_3, \ldots$ needs justification. A bonus is that we can now define coordinates for the sample units. There are $n=x_{..}$ of these but only rc different values, corresponding to the entries in the two-way table $\mathbf{X}$. The coordinates in the kth of ρ dimensions are

$$\mathbf{G}_1\mathbf{a}_k+\mathbf{G}_2\mathbf{b}_k$$

which becomes $a_{ik}+b_{jk}$ for the cell (i,j) or in terms of the singular vectors of $\mathbf{Y}$

$$\left(\frac{1}{\sqrt{r_i}}\right)u_{ik}+\left(\frac{1}{\sqrt{c_j}}\right)v_{jk}\quad\text{for } k=1,2,\ldots,\rho. \tag{9.15}$$

9.3.5 *Abundance matrices and canonical CA*

Abundance matrices occur in ecological applications and are two-way contingency tables classified by sites (here taken to refer to rows) and species (here taken to refer to columns). The (i,j)th cell gives the number of species j found at site i. Table 9.1 is an example of an abundance matrix. In ecology, it is often required to find scores for the row and column categories such that the weighted mean of the row-scores recovers, apart from a factor of proportionality, the column-scores and vice versa. This leads to scores $\mathbf{a}$ and $\mathbf{b}$ satisfying

$$\left.\begin{aligned}\mathbf{R}^{-1}\mathbf{X}\mathbf{b}&=\sigma\mathbf{a}\\ \mathbf{C}^{-1}\mathbf{X}'\mathbf{a}&=\sigma\mathbf{b}\end{aligned}\right\} \tag{9.16}$$

Table 9.1 Number of waders of each species at 15 sites

	1	*2*	*3*	*4*	*5*	*6*	*7*	*8*	*9*	*10*	*11*	*12*	*13*	*14*	*15*	*16*	*17*	*18*	*19*
A	12	2027	0	0	2070	39	219	153	0	15	51	8336	2031	14941	19	3566	0	5	0
B	99	2112	9	87	3481	470	2063	28	17	145	31	1515	1917	17321	3378	20164	177	1759	53
C	197	160	0	4	126	17	1	32	0	2	9	477	1	548	13	273	0	0	0
D	0	17	0	3	50	6	4	7	0	1	2	16	0	0	3	69	1	0	0
E	77	1948	0	19	310	1	1	64	0	22	81	2792	221	7422	10	4519	12	0	0
F	19	203	48	45	20	433	0	0	11	167	12	1	0	26	1790	2916	473	658	55
G	1023	2655	0	18	320	49	8	121	9	82	48	3411	14	9101	43	3230	587	10	5
H	87	745	1447	125	4330	789	228	529	289	904	34	1710	7869	2247	4558	40880	7166	1632	498
I	788	2174	0	19	224	178	1	423	0	195	162	2161	25	1784	3	1254	0	0	0
J	82	350	760	197	858	962	10	511	251	987	191	34	87	417	4496	15835	5327	1312	1020
K	474	930	0	10	316	161	0	90	0	39	48	1183	166	4626	65	127	4	0	0
L	77	249	160	136	999	645	15	851	101	723	266	495	83	1253	1864	4107	1939	623	527
M	22	144	0	4	1	1	0	10	0	2	9	125	5	411	0	3	0	0	0
N	0	791	0	0	4	38	1	56	1	30	54	95	0	1726	0	0	0	0	0
O	0	360	128	43	364	1628	63	287	328	641	850	83	67	48	6499	9094	5647	1333	582

which is precisely the one-dimensional form of (9.10). The ρ-dimensional solutions for **A** and **B** given by (9.10) are of interest and may be used to give biplots of the sites and species. Thus, once again, we establish an interest in classical CA. In plant ecology, it is sometimes reasonable to consider that site conditions change in a regular way, a **cline**, such as along the slope of a mountain from high to low altitudes or from dry to wetter land. Indeed, such trends may be indicated by the ordering given by the one-dimensional CA site scores. Associated with every site there may be a set of s environmental variables available in a matrix ${}_p\mathbf{Z}_s$, say, which are reasonably expected to influence trends. Then, we may constrain the CA scores to satisfy a linear regression $\mathbf{A} = \mathbf{Z}\boldsymbol{\beta}$, where the first column of **Z** may be assumed to be **1** to allow for a constant term in the regression. With this modification to **a** in (9.15), $\boldsymbol{\beta}$ satisfies the two-sided eigenvalue problem

$$\mathbf{Z}'\mathbf{X}\mathbf{C}^{-1}\mathbf{X}'\mathbf{Z}\boldsymbol{\beta} = \sigma^2\mathbf{Z}'\mathbf{R}\mathbf{Z}\boldsymbol{\beta} \tag{9.17}$$

with the usual normalization $\boldsymbol{\beta}'(\mathbf{Z}'\mathbf{R}\mathbf{Z})\boldsymbol{\beta} = \mathbf{I}$. From solutions to (9.17), we derive

$$\mathbf{A} = \mathbf{Z}\boldsymbol{\beta} \quad \text{and} \quad \mathbf{B} = \mathbf{C}^{-1}\mathbf{X}'\mathbf{Z}\boldsymbol{\beta}\boldsymbol{\Sigma}^{-1}$$

corresponding to the maximum effective eigenvalues. (Note that when the first column of **Z** is **1**, then $\mathbf{e}_1$ is always an eigenvector of (9.17)). The first ρ eigenvalues define a biplot in the usual way. Additionally, the eigenvectors $\boldsymbol{\beta}$ may be included as s points on the biplot to give information on the contributions to the trend of each environmental variable. When $s > p$, the equations $\mathbf{a} = \mathbf{Z}\boldsymbol{\beta}$ may be satisfied exactly for any **a**, so then the restriction is ineffective and we arrive back at the classical CA solution. This is the method of **canonical** CA, due to ter Braak (1986; 1994b) and available in CANOCO (ter Braak, 1990), ter Braak's suite of computer programs for CA and related methods. We think that the term canonical CA is somewhat unfortunate because all CA methods are based on algebraic canonical forms; we prefer ter Braak's other terminology of **restricted CA**, though this suffers from being unclear that it is a particular linear restriction that is being imposed.

The restricted CA problem may be formulated in a different way, by imposing the restriction directly onto (9.9), so we require the minimum of

$$\|\mathbf{R}^{-1/2}(\mathbf{X} - \hat{\mathbf{X}})\mathbf{C}^{-1/2}\| = \|\mathbf{Y} - \mathbf{R}^{1/2}\mathbf{Z}\boldsymbol{\beta}\boldsymbol{\Sigma}_\rho^{-1}\mathbf{B}_\rho'\mathbf{C}^{1/2}\| \tag{9.18}$$

The unknown matrix $\boldsymbol{\beta}\Sigma_\rho^{-1}\mathbf{B}'_\rho$ has rank ρ, so this minimization is a special case of the reduced rank least squares whose solution is given in (A.48). Specifically, in (A.48) set

$$\mathbf{Y}=\mathbf{R}^{-1/2}\mathbf{X}\mathbf{C}^{-1/2}, \quad \mathbf{A}=\mathbf{R}^{1/2}\mathbf{Z}, \quad \mathbf{B}=\mathbf{C}^{1/2}$$

and note that only the rank ρ matrix $\Gamma=\boldsymbol{\beta}\Sigma_\rho^{-1}\mathbf{B}'_\rho$ can be estimated and not its component parts. These settings imply that

$$\mathbf{P}=(\mathbf{Z}'\mathbf{R}\mathbf{Z})^{-1}, \ \mathbf{Q}=\mathbf{C}^{-1} \text{ and } \boldsymbol{\Gamma}_0=(\mathbf{Z}'\mathbf{R}\mathbf{Z})^{-1}\mathbf{Z}'\mathbf{R}^{1/2}\mathbf{Y}\mathbf{C}^{-1/2}$$

The ρ-dimensional approximation (A.40) to the generalized SVD $\boldsymbol{\Gamma}_0=\mathbf{S}\Sigma\mathbf{T}^{-1}$ with normalizations $\mathbf{S}'(\mathbf{Z}'\mathbf{R}\mathbf{Z})\mathbf{S}=\mathbf{I}$ and $\mathbf{T}'\mathbf{C}^{-1}\mathbf{T}=\mathbf{I}$ gives the solution which we shall now show agrees with (9.17). We have that

$$\begin{aligned}\boldsymbol{\Gamma}_0\mathbf{C}\boldsymbol{\Gamma}'_0&=(\mathbf{Z}'\mathbf{R}\mathbf{Z})^{-1}\mathbf{Z}'\mathbf{R}^{1/2}\mathbf{Y}\mathbf{Y}'\mathbf{Z}\mathbf{R}^{1/2}(\mathbf{Z}'\mathbf{R}\mathbf{Z})^{-1}\\&=\mathbf{S}\Sigma\mathbf{T}^{-1}\,\mathbf{C}(\mathbf{T}^{-1})'\Sigma'\mathbf{S}'\end{aligned}$$

which on substituting for $\mathbf{Y}$ and using the normalization constraints simplifies to

$$(\mathbf{Z}'\mathbf{R}\mathbf{Z})^{-1}\mathbf{Z}'\mathbf{X}\mathbf{C}^{-1}\mathbf{X}'(\mathbf{Z}'\mathbf{R}\mathbf{Z})^{-1}=\mathbf{S}\Sigma\Sigma'\mathbf{S}'$$

and on post-multiplying by $(\mathbf{Z}'\mathbf{R}\mathbf{Z})\mathbf{S}$ and pre-multiplying by $\mathbf{Z}'\mathbf{R}\mathbf{Z}$ this becomes

$$(\mathbf{Z}'\mathbf{X}\mathbf{C}^{-1}\mathbf{X}'\mathbf{Z})\mathbf{S}=(\mathbf{Z}'\mathbf{R}\mathbf{Z})\mathbf{S}\Sigma\Sigma'$$

which on comparison with (9.17) shows that $\mathbf{S}=\boldsymbol{\beta}$. From $\boldsymbol{\Gamma}_0=\mathbf{S}\Sigma\mathbf{T}^{-1}$, it is easy to show that $\mathbf{T}^{-1}=\mathbf{S}'\mathbf{Z}'\mathbf{X}\mathbf{C}^{-1}\Sigma^{-1}$. Thus, we have $\mathbf{A}=\mathbf{Z}\mathbf{S}$ and $\mathbf{B}=\mathbf{T}^{-1}$. (Note, here $\mathbf{A}$ and $\mathbf{B}$ refer to the plots for classical, and here restricted, CA and not to the matrices occurring in (A.48).)

Canonical CA is concerned with restricting the site scores of CA to be linearly related to environmental variables. We might also observe additional variables for the species giving, say, different chemical compositions for each of the species, to which we may require that the species scores be linearly related. Thus, we have two additional sets of variables, one given in a matrix $\mathbf{Z}_r$ and the other in a matrix $\mathbf{Z}_c$ of Equation (9.14), with $\mathbf{a}$ and $\mathbf{b}$ replaced by linear relationships $\mathbf{a}=\mathbf{Z}_r\boldsymbol{\beta}_r$ and $\mathbf{b}=\mathbf{Z}_c\boldsymbol{\beta}_c$, where $\boldsymbol{\beta}_r$ and $\boldsymbol{\beta}_c$ are regression coefficients, becomes

$$\left.\begin{aligned} \mathbf{R}^{-1}\mathbf{X}\mathbf{Z}_c\boldsymbol{\beta}_c &= \sigma\mathbf{Z}_r\boldsymbol{\beta}_r \\ \mathbf{C}^{-1}\mathbf{X}'\mathbf{Z}_r\boldsymbol{\beta}_r &= \sigma\mathbf{Z}_c\boldsymbol{\beta}_c \end{aligned}\right\} \tag{9.19}$$

We may consider multivariate multiple regression models $\mathbf{Z}_r\boldsymbol{\beta}_r$ and $\mathbf{Z}_c\boldsymbol{\beta}_c$ with $\boldsymbol{\beta}_r$ and $\boldsymbol{\beta}_c$ redefined as sets of regression coefficients assembled into matrices each with ρ columns, the dimensionality of the solution sought. Making this generalization, together with some reorganization, equation (9.19) may be rewritten as

$$\left.\begin{aligned} \mathbf{R}^{-1/2}\mathbf{X}\mathbf{C}^{-1/2}(\mathbf{C}^{1/2}\mathbf{Z}_c\boldsymbol{\beta}_c) &= \mathbf{R}^{1/2}\mathbf{Z}_r\boldsymbol{\beta}_r\Sigma \\ \mathbf{C}^{-1/2}\mathbf{X}'\mathbf{R}^{-1/2}(\mathbf{R}^{1/2}\mathbf{Z}_r\boldsymbol{\beta}_r) &= \mathbf{C}^{1/2}\mathbf{Z}_c\boldsymbol{\beta}_c\Sigma \end{aligned}\right\} \tag{9.20}$$

Equation (9.20) shows that in a full-dimensional solution

$$\mathbf{R}^{-1/2}\mathbf{X}\mathbf{C}^{-1/2} = \mathbf{R}^{1/2}\mathbf{Z}_r\boldsymbol{\beta}_r\Sigma\boldsymbol{\beta}_c'\Sigma_c'\mathbf{C}^{1/2}$$

with normalizations

$$\boldsymbol{\beta}_r'\mathbf{Z}_r'\mathbf{R}\mathbf{Z}_r\boldsymbol{\beta}_r = \mathbf{I} \quad \text{and} \quad \boldsymbol{\beta}_c'\mathbf{Z}_c'\mathbf{C}\mathbf{Z}_c\boldsymbol{\beta}_c = \mathbf{I}$$

A ρ-dimensional approximation minimizes $||\mathbf{Y} - \mathbf{Z}_1\boldsymbol{\Gamma}\mathbf{Z}_2'||$ with

$$\mathbf{Y} = \mathbf{R}^{-1/2}\mathbf{X}\mathbf{C}^{-1/2}$$

$$\mathbf{Z}_1 = \mathrm{R}^{1/2}\mathbf{Z}_r$$

$$\mathbf{Z}_2 = \mathbf{C}^{1/2}\mathbf{Z}_c$$

and

$$\boldsymbol{\Gamma} = \boldsymbol{\beta}_r\Sigma\boldsymbol{\beta}_c'$$

which has rank ρ. This is precisely the problem considered in (A.48) and has the solution based on the ρ leading vectors of the generalized singular value decomposition

$$\boldsymbol{\Gamma}_0 = (\mathbf{Z}_r'\mathbf{R}\mathbf{Z}_r)^{-1}\mathbf{Z}_r'\mathbf{X}\mathbf{Z}_c(\mathbf{Z}_c'\mathbf{C}\mathbf{Z}_c)^{-1} = \mathbf{S}\Sigma\mathbf{T}^{-1}$$

where

$$\mathbf{S}'(\mathbf{Z}_r'\mathbf{R}\mathbf{Z}_r)\mathbf{S} = \mathbf{I} \quad \text{and} \quad \mathbf{T}'(\mathbf{Z}_c'\mathbf{C}\mathbf{Z}_c)^{-1}\mathbf{T} = \mathbf{I}$$

Recall that $\mathbf{S}$ and $\mathbf{T}$ can be found easily from the two-sided eigenvalue expressions (A.24) and (A.26). When $\mathbf{Z}_c = \mathbf{I}$, we arrive back at the canonical CA solution and when also $\mathbf{Z}_r = \mathbf{I}$ we have the classical

CA solution. The solution gives $\hat{\boldsymbol{\Gamma}}$ and the separate components

$$\boldsymbol{\beta}_r = \mathbf{S} \quad \text{and} \quad \boldsymbol{\beta}_c = (\mathbf{T}')^{-1}$$

Having estimated $\boldsymbol{\Gamma}$ we may present it as a biplot with

$$\mathbf{A} = \mathbf{Z}_1\mathbf{S} \quad \text{and} \quad \mathbf{B} = \mathbf{Z}_2(\mathbf{T}')^{-1}$$

The methodology for restricted CA is closely related to that of reduced rank regression, discussed briefly in section 11.6; both depend on the approximation (A.43).

9.4 An example of CA

This example uses data given by Summers *et al.* (1987) and reported by Iloni (1991). Table 9.1 gives for 19 species of wader, the number of birds counted in the summer at 15 wetland and coastal

Table 9.2 List of species (numbered) and sites (lettered) with row and column totals of Table 9.1

SPECIES			*SITES*		
Code	*Name*	*C*	*Code*	*Name*	*R*
1	Oyster Catcher	2957	A	Namibia North-C	33484
2	White Fronted Plover	14865	B	Namibia North-W	54826
3	Kitt Litz's Plover	2552	C	Namibia South-C	1860
4	Three Banded Plover	710	D	Namibia South-W	179
5	Grey Plover	13473	E	Cape North-C	17499
6	Ringed Plover	5417	F	Cape North-W	6877
7	Bar Tailed Godwit	2614	G	Cape West-C	20734
8	Whimbrel	3162	H	Cape West-W	76067
9	Marsh Sandpiper	1007	I	Cape South-C	9391
10	Greenshank	3955	J	Cape South-W	33687
11	Common Sandpiper	1848	K	Cape East-C	8239
12	Turnstone	22434	L	Cape East-W	15113
13	Knot	12486	M	Transkei-C	737
14	Sanderling	61871	N	Natal-C	2796
15	Little Stint	22741	O	Natal-W	28045
16	Curlew Sandpiper	106037			
17	Ruff	21333			
18	Avocet	7332			
19	Black Winged Stilt	2740			
Total		309534	Total		309534

In a site name, W stands for wetlands and C for coast

sites in South Africa. The numbers are extremely variable, partly because of variation between the abundances of species and partly because not all sites are suited to all species. It is hoped that the analysis of proportions implicit in CA will adjust for the erratic nature of the raw data. Table 9.2 gives the names of the species and sites with their counts (i.e. the diagonal values of **R** and **C**) together with a number and letter code which is used to label the graphical display.

A two-dimensional fit accounts for 72.96 percent of the total squared singular values. Figure 9.1 shows the resulting plot. Three clusters of points for both sites and species are suggested and Table 9.3 shows Table 9.1 permuted to reflect this clustering. It will be seen that all sites in the first group have the postfix C in their name, which refers to coastal regions; the CA suggests that species

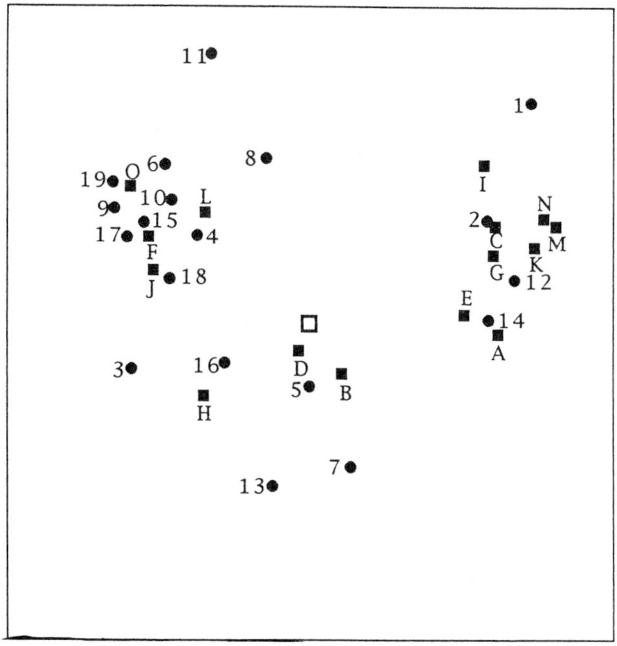

Fig. 9.1. *Correspondence analysis of 19 wader species from 15 wetlands in South Africa. Species are denoted by black circles and labelled to the left (or right) and wetland areas are denoted by black squares and labelled below (or above). The white open square denotes the origin.*

Table 9.3 Number of waders of each species at 15 sites permuted according to CA

	1	*2*	*12*	*14*	*3*	*5*	*7*	*13*	*16*	*4*	*6*	*8*	*9*	*10*	*11*	*15*	*17*	*18*	*19*
A	12	2027	8336	14941	0	2070	219	2031	3566	0	39	153	0	15	51	19	0	5	0
C	197	160	477	548	0	126	1	1	273	4	17	32	0	2	9	13	0	0	0
E	77	1948	2792	7422	0	310	1	221	4519	19	1	64	0	22	81	10	12	0	0
G	1023	2655	3411	9101	0	320	8	14	3230	18	49	121	9	82	48	43	587	10	5
I	788	2174	2161	1784	0	224	1	25	1254	19	178	423	0	195	162	3	0	0	0
K	474	930	1183	4626	0	316	0	166	127	10	161	90	0	39	48	65	4	0	0
M	22	144	125	411	0	1	0	5	3	4	1	10	0	2	9	0	0	0	0
N	0	791	95	1726	0	4	1	0	0	0	38	56	1	30	54	0	0	0	0
B	99	2112	1515	17321	9	3481	2063	1917	20164	87	470	28	17	145	31	3378	177	1759	53
D	0	17	16	0	0	50	4	0	69	3	6	7	0	1	2	3	1	0	0
H	87	745	1710	2247	1477	4330	228	7869	40880	125	789	529	289	904	34	4558	7166	1632	498
F	19	203	1	26	48	20	0	0	2916	45	433	0	11	167	12	1790	473	658	55
J	82	350	34	417	760	858	10	87	15835	197	962	511	251	987	191	4496	5327	1312	1020
L	77	249	495	1253	160	999	15	83	4107	136	645	851	101	723	266	1864	1939	623	527
O	0	360	83	48	128	364	63	67	9094	43	1628	287	328	641	850	6499	5647	1333	582

1, 2, 12 and 14 are especially associated with the coast. The second and third groups roughly divide the wetlands into a western region characterized by specis 3, 5, 7, 13, 16 and an eastern region characterized by the remaining species. However, the subdivision of the wetlands is not sharp and they share many species as is clear from the elements of the two submatrices of Table 9.3 concerned with the overlap. Closeness of pairs of row- (column-) points implies similar profiles independently of the species numbers and site densities. Recalling that the inner product interpretation of row-column associations approximates χ_{ij} (9.7), we see that rows and columns that subtend approximate right angles at the origin suggest independence. This is the case for the birds of the coastal region and the western wetlands and vice versa supporting the establishment of these clusters. It is also the case for the birds of the coastal regions with the eastern wetlands and vice versa, but we note the greater dispersion of species profiles in the eastern wetlands. The associations between the two wetland clusters subtend an obtuse angle at the origin, indicating negative contributions χ_{ij}; there are fewer numbers than expected, indicating different species dominations in the two areas.

CHAPTER 10

Relationships between CA and MCA

10.1 Introduction

We have seen that the PCA of quantitative data-matrix **X** can be handled either through the SVD of **X** or through the spectral decomposition of the variance-covariance matrix **X′X**. If one were interested in a least-squares fit to **X′X** itself then this would be given by *its* SVD, which happens to coincide with its spectral decomposition but the measures of fit to **X** and **X′X** differ. If one wished to fit **X′X** in *its* normalized form of a correlation matrix, then the unit diagonal might be excluded and only the off-diagonal correlational values fitted; indeed, this is the objective in most variants of factor analysis, but other approaches are discussed in Chapter 11. With categorical variables, the position is similar where each two-way contingency table of the Burt matrix corresponds to a single correlation coefficient. The desire for a simultaneous fit to all the 2×2 contingency tables provides a specially good reason for fitting the normalized Burt matrix directly. There is therefore some ambivalence as to whether one wishes to fit the original indicator matrix **G**, which can be done through the spectral decomposition of the Burt matrix $\mathbf{B} = \mathbf{G}'\mathbf{G}$ (see Chapter 4), or the contingency tables contained in **B**. The elements of the diagonal blocks of the Burt matrix are mostly zero and the others are units, so the decision on whether to include them in fitting processes is more critical than with correlation matrices. They should certainly be included if the objective is to fit the indicator matrix and its associated χ^2 distances, because then the spectral decomposition is merely a step in the computational process. However, when the objective is to fit the contingency tables, the diagonal blocks are irrelevant and should be excluded. When the diagonals are excluded we have what is termed **joint correspondence analysis** (JCA) (Greenacre, 1988; 1991), so-called because it gives a joint analysis of all the contingency tables. This chapter attempts to disentangle the relationships between CA, MCA and JCA paying special attention to whether one is primarily interested in approximating the indicator matrix or the Burt matrix.

10.2 CA as the MCA of an indicator matrix for two categorical variables

In Chapter 4, we developed MCA as a PCA of the indicator matrix

$$\frac{1}{\sqrt{p}}\mathbf{L}^{-\frac{1}{2}}\mathbf{G}$$

which is the method of section 9.2.1. Thus a strong link between CA and MCA has been established. Here, we begin by examining the converse of interpreting CA as a MCA of the indicator matrix for *two* categorical variables.

In the notation of Chapter 9, the Burt matrix (section 4.2.3) associated with the contingency table $\mathbf{X}$ is

$$\mathbf{B} = \begin{pmatrix} \mathbf{R} & \mathbf{X} \\ \mathbf{X}' & \mathbf{C} \end{pmatrix} \tag{10.1}$$

where $\mathbf{X}$ is the contingency table $\mathbf{G}_1'\mathbf{G}_2$ and $\mathbf{L}_1$ and $\mathbf{L}_2$ are replaced by $\mathbf{R}$ and $\mathbf{C}$, the row and column totals of $\mathbf{X}$ expressed in diagonal form. Setting $\mathbf{L} = \text{diag}(\mathbf{B})$, we recall (4.14) that MCA requires the solution to the eigenvalue problem

$$\tfrac{1}{2}\mathbf{BZ} = \mathbf{LZ\Lambda}$$

which may be rewritten

$$\tfrac{1}{2}[\mathbf{L}^{-\frac{1}{2}}\mathbf{B}\mathbf{L}^{-\frac{1}{2}}]\mathbf{L}^{\frac{1}{2}}\mathbf{Z} = \mathbf{L}^{\frac{1}{2}}\mathbf{Z\Lambda}$$

or, writing $\mathbf{Q} = \mathbf{L}^{\frac{1}{2}}\mathbf{Z}$ (note that here $\mathbf{Q}$ replaces $\mathbf{V}$ of Chapter 4 to distinguish it from the $\mathbf{V}$ of Chapter 9, both of which are needed in this chapter)

$$\mathbf{B}^*\mathbf{Q} = \frac{1}{2}\begin{pmatrix} \mathbf{I} & \mathbf{Y} \\ \mathbf{Y}' & \mathbf{I} \end{pmatrix}\mathbf{Q} = \mathbf{Q\Lambda} \tag{10.2}$$

where $\mathbf{Y}$ is given by (9.3). In the above, the factor $\frac{1}{2}$ is the $1/p$ of Chapter 4 with $p = 2$ categorical variables; this factor is superfluous here, but we retain it for consistency of notation and to ensure a maximum unit eigenvalue. We shall refer to the matrix $\mathbf{B}^*$ of (10.2), including the factor $\frac{1}{2}$, as the **normalized Burt matrix;** equivalent normalizations exist for all values of p. Replacing $\mathbf{Y}$ in (10.2) by its SVD, including the term with the unit singular value, exhibits the solution

$$\frac{1}{2}\begin{pmatrix} \mathbf{I}_r & {}_r\mathbf{U}_r\boldsymbol{\Sigma}_c\mathbf{V}'_c \\ \mathbf{V}\boldsymbol{\Sigma}'\mathbf{U}' & \mathbf{I}_c \end{pmatrix}\begin{pmatrix} \mathbf{U} & \mathbf{U}_c \\ (\mathbf{V},\mathbf{0}) & -\mathbf{V} \end{pmatrix} = \frac{1}{2}\begin{pmatrix} \mathbf{U} & \mathbf{U}_c \\ (\mathbf{V},\mathbf{0}) & -\mathbf{V} \end{pmatrix} \begin{pmatrix} \mathbf{I}+(\boldsymbol{\Sigma},\mathbf{0}) & \\ & \mathbf{I}-{}_c\boldsymbol{\Sigma} \end{pmatrix} \tag{10.3}$$

In (10.3) we have been more careful than usual to indicate the dimensions of the matrices involved and have shown them in a suffix notation. For convenience, and without loss of generality, we continue to assume that $r \geqslant c$. $\mathbf{U}_c$ indicates the first c columns of $\mathbf{U}$ and ${}_c\boldsymbol{\Sigma}$ indicates the first c rows of $\boldsymbol{\Sigma}$, and where necessary $\mathbf{V}$ and $\boldsymbol{\Sigma}$ are augmented by $r - c$ zero columns to make square matrices. Thus, the normalized Burt matrix reproduces the CA vectors that are columns of $\mathbf{U}$ and $\mathbf{V}$ with eigenvalues

$$\boldsymbol{\Lambda} = \frac{1}{2}\begin{pmatrix} \mathbf{I}+(\boldsymbol{\Sigma},\mathbf{0}) & \\ & \mathbf{I}-{}_c\boldsymbol{\Sigma} \end{pmatrix} \tag{10.4}$$

Thus, (10.4) gives eigenvalues

$$\lambda_k, \mu_k = \frac{1}{2}(1+\sigma_k), \frac{1}{2}(1-\sigma_k) \quad \text{for} \quad k = 1, 2, \ldots, c$$

and an eigenvalue of $\frac{1}{2}$ repeated $r - c$ times. The matrix of eigenvectors

$$\mathbf{Q} = \begin{pmatrix} \mathbf{U} & \mathbf{U}_c \\ (\mathbf{V},\mathbf{0}) & -\mathbf{V} \end{pmatrix}$$

of the MCA solution contains those of the CA solution but, so far as CA is concerned, there is redundant information because, apart from sign, the final c columns repeat the first c columns; the middle columns correspond to $\sigma = 0$ and are of no interest. Thus we confine attention to the first c columns which correspond to the biggest eigenvalues. $\lambda_1, \lambda_2, \ldots, \lambda_c$. These columns are in orthonormal form but every column has sums-of-squares equal to two; to recover the CA solution, care has to be taken to allow for this normalization. Also the first column corresponds to the uninteresting solution $\sigma = 1$ and is discarded. To recover the MCA solution, **Z**, **Q** must be premultiplied by $\mathbf{L}^{-\frac{1}{2}}$ which also gives the classical CA solutions (9.8). Thus, MCA computed from the spectral decomposition of $\mathbf{B}^*$, which in this instance is the same as its SVD, contains the vectors of the CA solution (9.8), apart from the trivial correction required for the normalization of the eigenvectors of (10.3); the CA scaling is recovered by setting $\boldsymbol{\Sigma} = 2\boldsymbol{\Lambda}_c - \mathbf{I}$.

There is an apparent anomaly in the measures of fit given by the two approaches. After eliminating the unit singular value, the Eckart–Young theorem gives the proportion of sum-of-squares fitted in a ρ-dimensional solution derived as the SVD of (9.3) as

$$\frac{\sigma_2^2 + \sigma_3^2 + \cdots + \sigma_\rho^2}{\sigma_2^2 + \sigma_3^2 + \cdots + \sigma_c^2} \tag{10.5}$$

whereas, recalling that the eigenvalues λ_k are the squares of the singular values of

$$\frac{1}{\sqrt{p}}\mathbf{G}\mathbf{L}^{-\frac{1}{2}}$$

with $p = 2$, the corresponding result for fitting to the latter is

$$\frac{\rho - 1 + \sigma_2 + \sigma_3 + \cdots + \sigma_\rho}{r + c - 2} \tag{10.6}$$

where the denominator comes from noting that, for the normalized Burt matrix, we have

$$\text{trace}\,(\tfrac{1}{2}\mathbf{L}^{-\frac{1}{2}}\,\mathbf{G}'\mathbf{G}\mathbf{L}^{-\frac{1}{2}}) = \tfrac{1}{2}(r + c)$$

which gives $\frac{1}{2}(r + c - 2)$ when subtracting the unit eigenvalue. The two measures of fit are not comparable because (10.5) is concerned with a weighted least-squares fit to ${}_r\mathbf{X}_c$, equivalent to an unweighted fit to **Y**, whereas (10.6) arises from a similar fit to the indicator matrix ${}_n\mathbf{G}_{(r+c)}$ in place of $\mathbf{X}$. Apart from the different sizes of these matrices and the different scalings, which are related to each other by (10.4), there is the major distinction that the rows and columns of $\mathbf{X}$ both become columns of $\mathbf{G}$, the rows of $\mathbf{G}$ pertaining to samples. We have that $n = x_{..}$ but there are only rc distinct values (i.e. rows of $\mathbf{G}$) corresponding to the rc cells of $\mathbf{X}$. Because of the distributional equivalence property (section 9.1) the rows of $\mathbf{G}$ may be amalgamated into the rc distinct category combinations without affecting the row or column χ^2 distances, although the row totals of $\mathbf{G}^*$, the modified form of $\mathbf{G}$, now become the frequencies in the cells of $\mathbf{X}$. Of course, these χ^2 distances are not the χ^2 distances between the rows or columns of $\mathbf{X}$. Although distributional equivalence guarantees the invariance of χ^2 distances, the modified row totals give different ρ-dimensional CA *approximations.* Thus the MCA of an indicator matrix for two variables is not the same thing as the corresponding CA of $\mathbf{X}$; the remarkable thing is that they are related and one can be derived from the other by the methods given earlier in this section.

10.3 Joint correspondence analysis (JCA)

In this section, we generalize the above analysis to more than two categorical variables. In doing this, one problem, discussed in section 10.3.1, continues to be how to choose the proper scaling to be associated with the vectors of $\mathbf{B}^*$. Additionally, when $p > 2$ the vectors of the normalized Burt matrix cannot be assumed optimal and section 10.3.2 discusses an iterative method for their improvement.

10.3.1 Non-iterative generalization

That the computations of MCA can conveniently be handled through the eigenstructure of the normalized Burt matrix, $\mathbf{B}^*$, leads to the possibility of additional misunderstanding but also to a useful variant of MCA. Being symmetric and positive semi-definite, the spectral decomposition of $\mathbf{B}^*$ is the same thing as its SVD. In the two-variable case, the ρ-dimensional Eckart–Young approximation to the normalized Burt matrix, including the unit diagonal blocks, accounts for a proportion of sums-of-squares fitted that is given by

$$\frac{\lambda_2^2+\lambda_3^2+\cdots+\lambda_\rho^2}{(\lambda_2^2+\lambda_3^2+\cdots+\lambda_c^2)+(\mu_2^2+\mu_3^2+\cdots+\mu_c^2)+\frac{1}{4}(r-c)}$$

$$=\frac{(1+\sigma_2)^2+(1+\sigma_3)^2+\cdots+(1+\sigma_\rho)^2}{r+c-2+2(\sigma_2^2+\sigma_3^2+\cdots+\sigma_c^2)} \qquad (10.7)$$

where the denominator derives from direct substitution from (10.4) or from noting that

$$(\mathbf{B}^*)^2=\tfrac{1}{4}\begin{pmatrix}\mathbf{I}+\mathbf{YY}' & 2\mathbf{Y}\\ 2\mathbf{Y}' & \mathbf{I}+\mathbf{Y}'\mathbf{Y}\end{pmatrix}$$

and hence

$$\text{trace}(\mathbf{B}^*)^2-1=\tfrac{1}{4}[r+c+2\text{trace}(\mathbf{Y}'\mathbf{Y})]-1$$

where

$$\text{trace}(\mathbf{Y}'\mathbf{Y})=1+\sigma_2^2+\sigma_3^2+\cdots+\sigma_c^2.$$

This fit to $\mathbf{B}^*$ given by (10.7) is to the matrix $\frac{1}{2}\mathbf{Y}$ of (10.2), rather than to $\mathbf{Y}$ itself, and includes the diagonal blocks $\frac{1}{2}\mathbf{I}$. Of more interest is to examine the actual ρ-dimensional fit to $\mathbf{Y}$, thus ignoring the diagonal blocks of $\mathbf{B}^*$. This is given by

$$\hat{\mathbf{Y}}=\frac{1}{2}\mathbf{U}(\mathbf{I}+\mathbf{\Sigma})_\rho\mathbf{V}'$$

where the proper normalization of $\mathbf{U}$ and $\mathbf{V}$ cancels the $\frac{1}{2}$ of $\frac{1}{2}\mathbf{Y}$ but the $\frac{1}{2}$ associated with the eigenvalues remains! Thus the residual matrix is

$$\mathbf{Y}-\hat{\mathbf{Y}}=\mathbf{U\Sigma V}'-\frac{1}{2}\mathbf{U}(\mathbf{I}+\mathbf{\Sigma})_\rho\mathbf{V}'$$

giving a fitted sum-of-squares

$$\tfrac{1}{4}[(1+\sigma_2)^2+(1+\sigma_3)^2+\cdots+(1+\sigma_\rho)^2]$$

and a residual sum-of-squares

$$\tfrac{1}{4}[(1-\sigma_2)^2+(1-\sigma_3)^2+\cdots+(1-\sigma_\rho)^2]+[\sigma_{\rho+1}^2+\sigma_{\rho+2}^2+\cdots+\sigma_c^2]$$

The residual and fitted sums-of-squares do not sum to the total because $\hat{\mathbf{Y}}$ is not a *least-squares* estimate of $\mathbf{Y}$. With this non-orthogonality, it is not legitimate to specify a proportion of fitted sum-of-squares as in (10.4–7).

All these difficulties arise from the wrong scaling for the eigenvectors of CA that is induced by the eigenvalues of $\mathbf{B}^*$. When $p = 2$, we know the correct scaling, so can easily recover a conventional CA from the decomposition (10.3) of $\mathbf{B}^*$. However, what happens when there are more than two categorical variables? For general values of p, we may still be interested in approximating the normalized Burt matrix to give some kind of average CA that simultaneously approximates all the two-way contingency tables. To achieve this we may continue to evaluate the spectral decomposition of $\mathbf{B}^*$. This will give Eckart–Young approximations for the whole of $\mathbf{B}^*$ and, in analogy with the case $p = 2$, we may accept its eigenvectors as an appropriate basis for plotting coordinates for the category levels but we know that, even when $p = 2$, the eigenvalues will give sup-optimal scaling when we wish to exclude approximation to the unit diagonal blocks. Hence, for general values of p, we are led to estimate the optimal scaling by excluding the diagonal blocks. Corresponding to (10.2) we have

$$\mathbf{B}^*\mathbf{Q} = \frac{1}{p}\begin{pmatrix} \mathbf{I} & \mathbf{Y}_{12} & \cdots & \mathbf{Y}_{1p} \\ \mathbf{Y}_{21} & \mathbf{I} & \cdots & \mathbf{Y}_{2p} \\ \vdots & \vdots & \cdots & \vdots \\ \mathbf{Y}_{p1} & \mathbf{Y}_{p2} & \cdots & \mathbf{I} \end{pmatrix}\begin{pmatrix} \mathbf{Q}_1 \\ \mathbf{Q}_2 \\ \vdots \\ \mathbf{Q}_p \end{pmatrix} = \begin{pmatrix} \mathbf{Q}_1 \\ \mathbf{Q}_2 \\ \vdots \\ \mathbf{Q}_p \end{pmatrix}\mathbf{\Lambda} \qquad (10.8)$$

where

$$\mathbf{Y}_{ij} = \mathbf{L}_1^{-\frac{1}{2}}\mathbf{G}_i'\mathbf{G}_j\mathbf{L}_j^{-\frac{1}{2}}$$

and we assume that only the first ρ column-vectors of $\mathbf{Q}$ are exhibited. The eigenvalues $\mathbf{\Lambda}$ give the wrong scaling and the problem is to determine a diagonal scaling matrix $\mathbf{\Gamma}$ (ρ columns) that minimizes, for given $\mathbf{Q}$

$$\sum_{i<j}^{p} \| \mathbf{Y}_{ij} - \mathbf{Q}_i\mathbf{\Gamma}\mathbf{Q}_j' \| \qquad (10.9)$$

the summation being over off-diagonal blocks only. Everything is known except $\mathbf{\Gamma}$, so the problem is linear and may be solved as a variant of multiple regression. When $p = 2$, there is only one term and from (10.3) we have that

$$\mathbf{Q}_1 = \frac{1}{\sqrt{2}}\mathbf{U} \quad \text{and} \quad \mathbf{Q}_2 = \frac{1}{\sqrt{2}}\mathbf{V}$$

so we immediately have

$$\mathbf{\Gamma} = 2\mathbf{\Sigma} \quad \text{and} \quad \mathbf{Q}_1\mathbf{\Gamma}\mathbf{Q}_2' = \mathbf{U\Sigma V}'$$

as required. For general values of p, (A.45) shows that the normal equations are

$$\left[\sum_{i<j}^{p}(\mathbf{Q}_i'\mathbf{Q}_i)^*(\mathbf{Q}_j'\mathbf{Q}_j)\right]\mathbf{\Gamma 1} = \sum_{i<j}^{p}\operatorname{diag}(\mathbf{Q}_i'\mathbf{Y}_{ij}\mathbf{Q}_j)\mathbf{1} \tag{10.10}$$

where the symbol * denotes a Hadamard product. Equation (10.10) gives the rescaling which minimizes (10.9). This process must improve the fit to the contingency tables. The vectors are uniquely determined from (10.8) but the settings of $\mathbf{\Gamma}$ for different values of ρ have to be recalculated from (10.10).

Because the columns of $\mathbf{Q}$ are orthonormal, we have

$$\sum_{i=j}^{p}\mathbf{Q}_i'\mathbf{Q}_i = \mathbf{I}$$

and (10.10) may be simplified as follows. Writing $\mathbf{S}_i = \mathbf{Q}_i'\mathbf{Q}_i$ we have

$$\sum_{i=1}^{p}\mathbf{S}_i = \mathbf{I}$$

and (10.10) becomes

$$\tfrac{1}{2}\left[\mathbf{S}_1^*\mathbf{I} + \mathbf{S}_2^*\mathbf{I} + \cdots + \mathbf{S}_p^*\mathbf{I} - \sum_{i=1}^{p}(\mathbf{S}_i^*\mathbf{S}_i)\right]\mathbf{\Gamma 1} = \sum_{i<j}^{p}\operatorname{diag}(\mathbf{Q}_i'\mathbf{Y}_{ij}\mathbf{Q}_j)\mathbf{1}$$

That is

$$\left[\mathbf{I} - \sum_{i=1}^{p}(\mathbf{Q}_i'\mathbf{Q}_i)^*(\mathbf{Q}_i'\mathbf{Q}_i)\right]\mathbf{\Gamma 1} = 2\sum_{i<j}^{p}\operatorname{diag}(\mathbf{Q}_i'\mathbf{Y}_{ij}\mathbf{Q}_j)\mathbf{1} \tag{10.11}$$

Equation (10.11) is the required simplification of (10.10) and may be solved for $\mathbf{\Gamma}$.

When $p = 2$, then $\sqrt{2}\mathbf{Q}_1 = \mathbf{U}_\rho$ and $\sqrt{2}\mathbf{Q}_2 = \mathbf{V}_\rho$ so that $2\mathbf{S}_1 = \mathbf{I}$ and $2\mathbf{S}_2 = \mathbf{I}$. Inserting these settings into (10.11) gives

$$[\mathbf{I} - \tfrac{1}{4}\mathbf{I} - \tfrac{1}{4}\mathbf{I}]\boldsymbol{\Gamma}\mathbf{1} = \text{diag}\,(\mathbf{U}'_\rho \mathbf{Y}_{12}\mathbf{V}_\rho)\mathbf{1} = \boldsymbol{\Sigma}_\rho \mathbf{1}$$

and hence $\boldsymbol{\Gamma} = 2\boldsymbol{\Sigma}_\rho$.

The fitted value is

$$\hat{\mathbf{Y}}_{12} = \mathbf{Q}_1 \boldsymbol{\Gamma} \mathbf{Q}'_2 = \mathbf{U}_\rho \boldsymbol{\Sigma}_\rho \mathbf{V}'_\rho$$

confirming that (10.10) and (10.11) give the correct result in the case of classical CA with two categorical variables.

Although we are not primarily interested in the diagonal blocks, it is interesting to note that these are estimated by $\mathbf{Q}_k \boldsymbol{\Gamma} \mathbf{Q}'_k$ for $(k = 1, 2, \ldots, p)$. Writing $\mathbf{Q}_k = (\mathbf{q}_{k1}, \mathbf{q}_{k2}, \ldots, \mathbf{q}_{k\rho})$ then

$$\mathbf{q}_{k1} = \frac{\mathbf{L}_k^{\frac{1}{2}}\mathbf{1}}{(\mathbf{1}'\mathbf{L}\mathbf{1})^{\frac{1}{2}}} \quad \text{and} \quad \mathbf{1}'\mathbf{L}_k^{\frac{1}{2}}\mathbf{q}_{kj} = 0$$

for $j \neq 1$ (section 4.3.3) and we have

$$\mathbf{Q}_k \boldsymbol{\Gamma} \mathbf{Q}'_k = \sum_{j=1}^{p} \gamma_j \mathbf{q}_{kj} \mathbf{q}'_{kj}$$

and inserting these values into the diagonal blocks of (10.8) we may verify that the vector

$$\mathbf{L}^{\frac{1}{2}}\mathbf{1} = (\mathbf{L}_1^{\frac{1}{2}}\mathbf{1}, \mathbf{L}_2^{\frac{1}{2}}\mathbf{1}, \ldots, \mathbf{L}_p^{\frac{1}{2}}\mathbf{1})'$$

remains an eigenvector of the modified matrix, now corresponding to the eigenvalue

$$\lambda = \frac{n\gamma_1}{(\mathbf{1}'\mathbf{L}\,\mathbf{1})^{1/2}} + (p-1)$$

rather than unity. It follows that if $\mathbf{p} = (\mathbf{p}_1, \mathbf{p}_2, \ldots, \mathbf{p}_p)$ is any other eigenvector of the modified matrix then $\mathbf{1}'\mathbf{L}^{1/2}\mathbf{p} = 0$. Furthermore, we shall now show that $\alpha_k = \mathbf{1}'\mathbf{L}_k^{1/2}\mathbf{p}_k = 0$ for $(k = 1,2,\ldots, p)$. The argument is similar to that given in section 4.3.2 for the corresponding result in classical MCA. Multiplying the kth row-block of the modified form of (10.8) by $\mathbf{p}$ and pre-multiplying the resulting vector by $\mathbf{1}'\mathbf{L}_k^{1/2}$ gives

$$\sum_{j \neq k}^{p} \alpha_j + \frac{n\gamma_1}{(\mathbf{1}'\mathbf{L}\mathbf{1})^{1/2}}\alpha_k = p\lambda\alpha_k$$

But

$$\sum_{j=1}^{p} \alpha_j = \mathbf{1}'\mathbf{L}^{1/2}\mathbf{p} = 0$$

and hence $\alpha_k = 0$. This result will be used in the next section. That $\mathbf{L}^{\frac{1}{2}}\mathbf{1}$ remains a vector, shows that, as previously, the matrix

$$\frac{\mathbf{L}^{1/2}\mathbf{1}\mathbf{1}'\mathbf{L}^{1/2}}{(\mathbf{1}'\mathbf{L}\mathbf{1})}$$

need not be removed from (10.8) before analysis, provided the first vector is ignored.

10.3.2 Iterative analysis

In section 10.3.1, a method was given for estimating $\mathbf{\Gamma}$ to improve the scaling of the eigenvectors of $\mathbf{B}^*$. When $p \neq 2$, we do not know that the eigenvectors of $\mathbf{B}^*$ are the best vectors to choose; can better be done by changing these vectors? The diagonal blocks fitted to (10.8) have the form $\mathbf{Q}_k\mathbf{\Gamma}\mathbf{Q}_k'$ for $(k = 1, 2, \ldots, p)$. If we now substitute these for the unit diagonal blocks, we may repeat the spectral decomposition of the modified matrix and arrive at revised vectors and thence a revised scaling $\mathbf{\Gamma}$. Proceeding iteratively, we minimize (10.9) not only over $\mathbf{\Gamma}$ but also over $\mathbf{Q}_1, \mathbf{Q}_2, \ldots, \mathbf{Q}_p$ to give a least-squares fit to all the contingency tables but excluding the diagonal blocks of $\mathbf{B}^*$. When the iterations have converged, $\mathbf{\Gamma}$ gives the ρ largest eigenvalues of the diagonal-modified version of $\mathbf{B}^*$. The rows of $\mathbf{L}_k^{-\frac{1}{2}}\mathbf{Q}_k$ $(k = 1, 2, \ldots, p)$ give the desired coordinates to be plotted for the levels of the kth variable. These may be assembled into the single matrix $\mathbf{Z} = \mathbf{L}^{-\frac{1}{2}}\mathbf{Q}$.

We showed at the end of section 10.3.1 that $\mathbf{L}^{\frac{1}{2}}\mathbf{1}$ remains an eigenvector after one iteration and that $\alpha_k = 0$ for $k = 1, 2, \ldots, p$. Precisely the same argument shows that this result is preserved on subsequent iterations and hence ignoring the first eigenvector in the converged solution gives an analysis of departures from the independence model and shows that the resulting ddistances may be viewed as χ^2 ddistances aggregated over all variables. The same results may be obtained by subtracting $\mathbf{L}^{1/2}\mathbf{1}\mathbf{1}'\mathbf{L}^{1/2}/\mathbf{1}'\mathbf{L}\mathbf{1}$ from $\mathbf{B}^*$ before starting the iterations but this is not necessary; the possibility of this adjustment applies equally to section 10.3.1. Greenacre

(1988) terms the resulting iterative method **joint correspondence analysis (JCA)**. When $p = 2$, JCA coincides exactly with CA. By adopting this method, we sacrifice the 'nice' nesting property of solutions given by the SVD, where to include extra dimensions it is necessary only to include the vectors associated with the next smaller singular value. JCA solutions in different number of dimensions, ρ, must each be computed *ab initio*, not only for $\mathbf{\Gamma}$ as in section 10.3.1 but also for $\mathbf{Q}_1, \mathbf{Q}_2, \ldots, \mathbf{Q}_p$. Further, by changing the diagonal blocks of the Burt matrix, the resulting matrix may not be p.s.d. It is not clear to us what affect this may have on interpretation but we note that if ρ is chosen too big, we may include negative eigenvalues in $\mathbf{\Gamma}$ and hence be unable to evaluate $\mathbf{\Gamma}^{\frac{1}{2}}$ to produce the inner product plots.

The rank of $\mathbf{B}^*$ will normally exceed ρ so the residual sum-of-squares (10.9) is

$$\gamma_{\rho+1}^2 + \gamma_{\rho+2}^2 + \cdots + \gamma_L^2$$

The solution is a proper least-squares fit to the modified $\mathbf{B}^*$. Specifically, writing $\mathbf{\Delta}$ for the contribution of the diagonal blocks to the sums-of-squares, we have the decomposition

$$\|\mathbf{B}^*\| = \mathbf{\Delta} + 2\sum_{i<j}^{p} \|\mathbf{Y}_{ij}\| = \mathbf{\Delta} + 2\sum_{i<j}^{p} \|\mathbf{Q}_i\mathbf{\Gamma}\mathbf{Q}_j'\|$$

$$+ 2\sum_{i<j}^{p} \|\mathbf{Y}_{ij} - \mathbf{Q}_i\mathbf{\Gamma}\mathbf{Q}_j'\| \qquad (10.12)$$

where

$$2\sum_{i<j}^{p} \|\mathbf{Y}_{ij} - \mathbf{Q}_i\mathbf{\Gamma}\mathbf{Q}_j'\| = \gamma_{\rho+1}^2 + \gamma_{\rho+2}^2 + \cdots + \gamma_L^2$$

is the residual sum-of-squares and

$$\mathbf{\Delta} + 2\sum_{i<j}^{p} \|\mathbf{Q}_i\mathbf{\Gamma}\mathbf{Q}_j'\| = \gamma_1^2 + \gamma_2^2 + \cdots + \gamma_\rho^2$$

is the fitted sum-of-squares.

The quantity $\mathbf{\Delta}$ is of little interest and cancels out on both sides of (10.12). The relevant total sum-of-squares is, after allowing for the eliminated 'independence' vector

$$2\sum_{i<j}^{p}\left\|\mathbf{Y}_{ij}-\frac{\mathbf{L}_i^{1/2}\mathbf{11}'\mathbf{L}_j^{1/2}}{\mathbf{1}'\mathbf{L1}}\right\| \tag{10.13}$$

which simplifies to

$$2\sum_{i<j}^{p}\|\mathbf{Y}_{ij}\|-\frac{(p-1)}{p} \tag{10.14}$$

and it is to (10.14) that the residual sum-of-squares should be compared. Of course, if one works without the divisor p in (10.8), so that $\mathbf{Y}_{ij}=\mathbf{L}_i^{-\frac{1}{2}}\mathbf{G}_i'\mathbf{G}_j\mathbf{L}_i^{-\frac{1}{2}}$, then (10.14) becomes

$$2\sum_{i<j}^{p}\|\mathbf{Y}_{ij}\|-p(p-1) \tag{10.15}$$

We also note the useful result that the sums of squares in the diagonal blocks after correction for the independence vector is given by

$$\sum_{i=1}^{p}\left\|\mathbf{I}-p\frac{\mathbf{L}_i^{1/2}\mathbf{11}'\mathbf{L}_i^{1/2}}{\mathbf{1}'\mathbf{L1}}\right\|=L-p \tag{10.16}$$

so the fitted and residual components of this quantity are subtracted from the ordinary MCA sums-of- squares for all methods that ignore the diagonal blocks. Not surprisingly, this adjustment brings about a dramatic improvement in fit that more truely represents the quality of the representation than does the uncorrected approach. Note that with the corrected MCA, the fitted and residual sums-of-square have to be re-evaluated to remove their components in the diagonal blocks. The resulting analysis of variance is non-orthogonal so care has to be taken with interpretation. However, the iterated JCA leads to an orthogonal analysis of variance.

With this analysis, the interpretation of distances between plotted category points raises some difficulty. Inner-product interpretations seem safer, approximating the actual elements of the contingency tables or the deviations from independence terms when ignoring the first vector.

To sum up, classical MCA is not a good method for the simultaneous approximation to all contingency tables but is legitimate for its MDS interpretation of approximating the weighted indicator matrix $\mathbf{GL}^{-1/2}$ discussed in Chapter 4, provided the special form of χ^2 distance (4.4) used there is deemed acceptable. A fairer assessment of the simultaneous fit is obtained by the regression method that excludes diagonal blocks described in section 10.3.1, whereas the iterative JCA of section 10.3.2, by definition, minimizes (10.9) and gives the optimal least-squares solution but has some interpretative problems. Greenacre (1991) reports a marked improvement; in his example, the precentage fitted by MCA is 46.6 percent which increases to 85.1 percent (diagonal excluded) when scaling is estimated by the regression method non-iteratively and to 98.5% when scaling is estimated iteratively by JCA. However, when non-orthogonality is taken into account or when the residual sum-of-squares is compared to a total sum-of-squares which contains extraneous diagonal terms, these results might need careful interpretation. We return to this point in the example discussed in section 10.4.

10.3.3 What do we plot?

Paralleling the different objectives of

(i) simultaneously approximating all the contingency tables and
(ii) approximating the χ^2 distances generated by the indicator matrix,

there are two kinds of biplot. We have derived $\mathbf{Q}_i\mathbf{\Gamma}\mathbf{Q}_j'$ as an approximation to $\mathbf{Y}_{ij}=\mathbf{L}_i^{-\frac{1}{2}}\mathbf{G}_i'\mathbf{G}_j\mathbf{L}_j^{-\frac{1}{2}}$. Thus by ploting $\mathbf{Q}_i\mathbf{\Gamma}^{\frac{1}{2}}$ for the ith variable and $\mathbf{Q}\mathbf{\Gamma}^{\frac{1}{2}}$ for all variables, we preserve the inner-product interpretation. This is the exact counterpart of one of the CA plots for rows and columns of a two-way contingency table described in section 9.3.2. The inner-product then approximates departures from the independence model as described for CA. This is valid for both the non-iterative and the iterative form of JCA. On the other hand, if the aim is to approximate the χ^2 distances and the indicator matrix, then it is best to use the methods of Chapter 4, but the vectors $\mathbf{Q}$ may be regarded as an approximation to $\mathbf{V}$ and the corresponding plots compared. Then the CLPs have coordinates

$$\mathbf{Z}=\frac{1}{\sqrt{p}}\mathbf{L}^{-\frac{1}{2}}\mathbf{Q}$$

and the samples have coordinates $\mathbf{YQ} = \mathbf{GZ}$. Thus, just as in MCA and the discussion of the extended matching coefficient in Chapter 4, we may represent the samples by $\mathbf{GZ}$ to give the usual vector-sum interpolation or, better, by $\mathbf{GZ}/p$ to give the samples at the centroids of the CLPs; the rescaling $\mathbf{\Gamma}$ is not relevant, except insofar as it determines the final form of $\mathbf{Q}$. In the non-iterative form of JCA, $\mathbf{Q}$ is the same as the $\mathbf{V}$ of MCA so there is no change but, if desired, the dispersions of the points can be rescaled to be given by $\mathbf{\Gamma}$ rather than by $\mathbf{\Sigma}$.

To construct prediction regions, requires that the approximation space $\mathscr{L}$ be embedded in a space $\mathscr{R}$ where distances are exact. With JCA, there is no obvious exact inter-sample distance, and so no space $\mathscr{R}$. One possibility would be to construct $\mathscr{R}$ as the space spanned by sample-coordinates $\mathbf{GZ}$ when $\rho = c$ and then proceeding with the Procrustean embedding described in section 3.3 and the construction of prediction-regions as described in section 4.4.5 and 7.3.2. An alternative approach, which we favour, is to match the CLPs $\mathbf{L}^{-1/2}\mathbf{Q}$ in $\mathscr{L}$ with $\mathbf{L}^{-1/2}$, the usual exact CLPs of MCA. All that has to be done is to replace the $\mathbf{V}$ of Chapter 4 by $\mathbf{Q}$ and then proceed to construct prediction regions as described in that chapter.

10.4 An example of JCA

The same data, Table 4.6, is used as for the MCA of Chapter 4. This allows JCA to be compared with MCA. The primary aim is to approximate all the two-way contingency tables. Table 10.1 shows the results given by the various methods discussed above. The total

Table 10.1 The quality of two-dimensional fits to all the two-way contingency tables of the Burt matrix

Method	*Total s.s*	*Fitted s.s*	*Residual s.s*
MCA	21.1699*	11.6902 (55.2%)	9.4798 (44.8%)
MCA less diagonal blocks	9.1699	8.0274 (87.5%)	3.4579 (37.7%)
JCA non-iterative	9.1699	5.7472 (62.7%)	3.1499 (34.4%)
JCA iterative form	9.1699	6.0680 (66.2%)	3.1020 (33.8%)

The independence vector is not included in any of the sums of squares (s.s)
*The MCA values may be derived from the squares of the eigenvalues accompanying Fig. 4.2, but the values given here are computed directly and more accurately

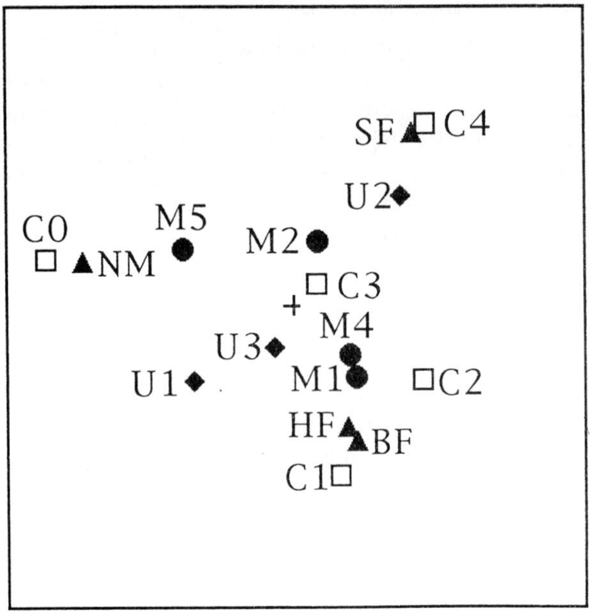

Fig. 10.1 *Inner-product biplot for the JCA of Table 4.6–iterative solution.*

sum-of-squares for MCA is

$$\sum_{ij}^{p} \|\mathbf{Y}_{ij}\| = \sum_{ij}^{p} \|\mathbf{Q}_i \boldsymbol{\Gamma} \mathbf{Q}_j'\| + \sum_{ij}^{p} \|\mathbf{Y}_{ij} - \mathbf{Q}_i \boldsymbol{\Gamma} \mathbf{Q}_j'\|$$

where the summation occurs over all p^2 blocks and $\boldsymbol{\Gamma} = \boldsymbol{\Sigma}$, whereas a similar orthogonal analysis of variance is valid for JCA, in its iterated form, except that summation occurs over all $p(p-1)$ non-diagonal blocks, each block occurring twice. Note that without diagonal blocks, the total sum-of-squares is reduced by $p(p-1) = 4 \times 3 = 12$, as given by (10.15). The middle two methods of Table 10.1 do not give an orthogonal analysis of variance and this can give some misunderstanding; the proportion of sum-of-squares fitted is not the complement of the residual sum-of-squares. The residual sums-of-squares of Table 10.1 steadily decrease, as they must, but the improvement over the MCA solution, ignoring the diagonal blocks, is slight.

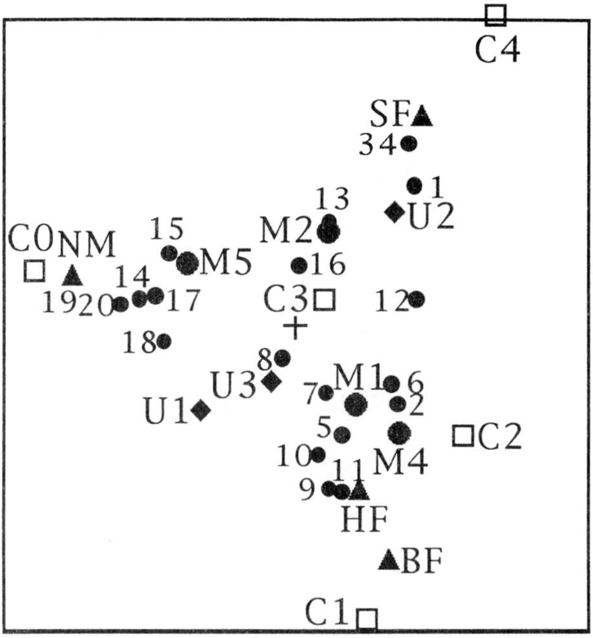

Fig. 10.2 *JCA of the data in Table 4.6. Representation of chi square distance iterative solution.*

Figure 10.1 shows the inner-product display of the iterated JCA as described in section 10.3.3. The configuration is similar to that of MCA (Fig. 4.2) but there are some significant changes. For example, now SF and C4 are more strongly associated and M4 has moved towards the centre, indicating less association with C2. Because MCA is approximating the χ^2 distances implied by the indicator matrix, Fig. 10.1 should give the better representation of the associations. To derive a representation of JCA that gives more weight to the χ^2 distances, we show in Fig. 10.2 the plot discussed in section 10.3.3 for this purpose. The scaling of the category coordinates differs from that of Fig. 10.1 but, even though the sample points are added a trifle heuristically, the resulting plot is surprisingly close to that of Fig. 4.2. For a graphical method, there is little to choose between Figs. 4.4 and 10.2. The improvements in fit suggested by Table 10.1 are not reflected by parallel improvements in the plots.

CHAPTER 11

Other topics

11.1 Introduction

In the previous chapters, we have set up some general principles for the construction of biplots. The fundamental underpinning is the concept of a generalized coordinate system, termed a **reference system**, which associates the values of both continuous and categorical variables with every sample position. To a MDS, which approximates proximities between samples, is added an approximation to the reference system. The approximation of the reference system depends on whether it is to be used for interpolation or prediction. Because a coordinate system is such a fundamental concept, it is clear that these principles may be exploited in more general contexts than those which we have so far discussed. In this chapter, we briefly review some selected extensions. We hope that the reader will feel equipped to make further extensions of a similar kind.

11.2 Biplots in non-metric MDS

There are two essentially different approaches each of which has, or potentially has, many variants. We outline the essentials of both approaches and their biplot implications but do not go into detail.

11.2.1 Distance transformation methods

The fundamental difference between metric and non-metric scaling is the occurrence of the transformation function

$$\delta_{ij} = \tau(d_{ij}) + \varepsilon_{ij}$$

in (3.1) where ε_{ij} is a measure of discrepancy. In general, $\tau(.)$ is unknown and requires estimation but when it is specified in advance the discrepancies vanish and we arrive at a metric scaling of the known proximies $\tau(d_{ij})$; examples are the identity transformation (Chapter 3) and, say, the square-root transform. When $\tau(.)$ is unknown, there are two fitted distances corresponding to d_{ij}:

(i) the values δ_{ij} obtained by the MDS, and
(ii) the values $\hat{\delta}_{ij} = \tau(d_{ij})$ lying on the transformation curve.

The fit of the transformation $\tau(d_{ij})$ to δ_{ij} is then measured in terms of the agreement between δ_{ij} and $\hat{\delta}_{ij}$. Figure 11.1 shows two typical forms of transformation. The goodness of fit is invariant to monotonic transformations of the d_{ij} because residuals are measured parallel to the horizontal axis of Fig. 11.1; stretching the vertical axis has no effect on $\hat{\delta}_{ij}$ and δ_{ij}. Thus for given τ, coordinates $\mathbf{Z}$ generate δ_{ij} by minimizing a function of the form (3.1) thus giving pairs (d_{ij}, δ_{ij}), to which a new transformation function τ is fitted. The process is then repeated until convergence. The MDS solution is that $\mathbf{Z}$ which gives a non-trivial local and, ideally, a global minimum of (3.1). Many functions have the general form of (3.1) and there are several measures of the goodness-of-fit of $\tau(d_{ij})$ to

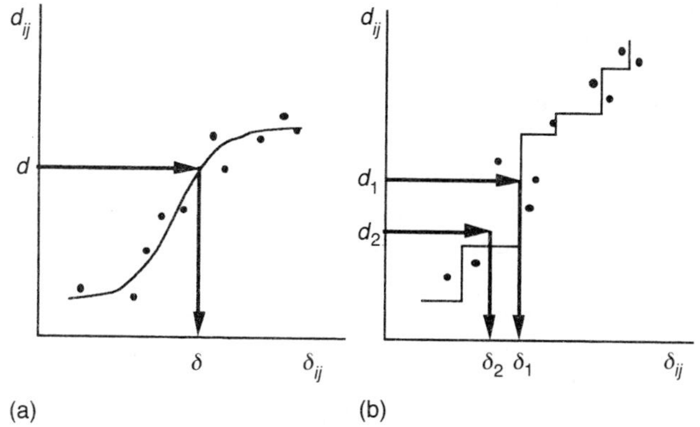

Fig. 11.1. *Non-metric MDS. The points show the relationship between observed distances d_{ij} and the values δ_{ij} fitted by MDS. The fitted transformation $\delta_{ij} = \tau(d_{ij})$ between observed and fitted proximities is shown for: (a) a smooth monotonic and (b) a monotonic step function. The bold lines indicate interpolations that are discussed in the text.*

δ_{ij}, some of which are confusingly similar to (3.1). How the transformations are specified and the algorithmic details are the core of MDS research and are beyond the scope of this book. (See Cox and Cox (1994) and the references cited therein for an introduction to MDS. Note that their notation, and much of that of the MDS literature, differs from ours, especially in the reversal of the roles of d_{ij}, for us the observed proximities, and δ_{ij}, for us the fitted distances.)

When the MDS algorithm has converged to give the sample coordinates **Z**, the transformation function $\tau(.)$ is known, at least empirically. Then biplot axes may be constructed as follows. Construct pseudosamples as described in Chapters 6 and 7. These may be interpolated into $\mathscr{L}$, the space containing **Z**, by the transformation

$$\hat{\delta}_{in+1} = \tau(d_{in+1})$$

and using these new values in (A.64) and (A.65) to find the interpolated position of the pseudosample. For continuous variables with pseudosamples defined by a single parameter, trajectories may be drawn as described in Chapter 6; for categorical variables, CLPs may be found by taking the centroids of sets of pseudosamples as described in Chapter 7. Thus, the only new feature is the transformation step. If the functional form of $\tau(.)$ is known, as happens, for example, when the transformation is defined as a polynomial or a B-spline, the procedure is trivial but when the transformation is known only empirically, special considerations apply. Figure 11.1(a) shows graphical interpolation of a distance d in a smooth transformation to give a value δ. Interpolating into the step function of Fig. 11.1(b) raises some problems. For d_1 there is no difficulty because the three points which uniquely determine the vertical section of the monotone regression bracket d_1. Neither the two lower points nor the three upper points bracket d_2. This leads to ambiguity of interpolation because it is irrelevant where the step is drawn between these two sets of points. One solution is to interpolate mid-way between the two relevant vertical sections, as indicated in the figure.

The above considerations permit the construction of a reference system in $\mathscr{L}$ but how this reference system is to be interpreted needs consideration. In earlier chapters, there was an exact reference system in $\mathscr{R}$ but this is not so in $\mathscr{L}$; even the positions of the original samples will not usually correspond to their coordinates as given by the approximate reference system. It can be expected that the approximation will be better the more dimensions m are given to $\mathbf{Z}$. One possibility is to embed the reference system in a space of several dimensions (e.g. choose $m = 10$) and then repeat the MDS in $\mathscr{L}$ of $\rho = 2$, say, dimensions. The reference system may then be approximated in $\mathscr{L}$ using the methods of Chapters 3, 7 and 8. This would give a predictive biplot. Interpolative biplots were shown to be problematic, even with linear metric scaling other than principal coordinates, and the position is surely worse in the non-metric case.

11.2.2 *Variable transformation methods*

A second form of non-metric scaling has been developed by Meulman (1986; 1992) which applies transformations to the original variables rather than to observed or derived proximities. Thus, the kth variable is transformed to $\mathbf{y}_k = \tau(\mathbf{x}_k)$ for $(k = 1, 2, \ldots, p)$ which we shall write as $\mathbf{Y} = \tau(\mathbf{X})$, a notation which includes more general transformations than those actually used. Pythagorean distances $\mathbf{D}$ are then defined on $\mathbf{Y}$ in $\mathscr{R}$ and are approximated by $\mathbf{\Delta}$ generated by coordinates $\mathbf{Z}$ in $\mathscr{L}$, found by least-squares scaling, although other criteria could be used. Again, iterative methods are essential for combining the transformation and MDS steps of an algorithm. Because this approach works directly in terms of the variables, it is better suited to biplot interpretations than the MDS methods discussed in section 11.2.

The transformations $\mathbf{y}_k = \tau_k(\mathbf{x}_k)$ directly allow pseudosamples to be incorporated into the display of $\mathbf{Z}$ in $\mathscr{L}$. For a quantitative variable, a pseudosample $\mu\mathbf{e}_k$ maps into a point with coordinates $\tau(\mu\mathbf{e}_k) = (\tau_1(0), \tau_2(0), \ldots \tau_k(\mu), \ldots, \tau_p(0))$ in $\mathscr{R}$ and the Pythagorean distances of this point from the mappings of the original points may be found and the pseudosample then mapped into $\mathscr{L}$ by the algebraic interpolation methods for least-squares-scaling discussed in section 3.4.1. The locus of these mappings as μ varies give a non-linear biplot for the kth variable. Meulman and Heiser (1993) discuss this approach and give examples.

An alternative approach with some attractive features follows from noting that in $\tau(\mu\mathbf{e}_k), \tau_k(\mu)$ is the only coordinate involving μ and hence as μ varies the locus generates a linear coordinate axis in $\mathscr{R}$; further, the p axes are orthogonal, assuming all variables to be quantitative (see below for categorical variables). Of course,

$\tau_k(\mu) - \tau_k(\mu + \text{constant})$ is normally a function of μ and hence the markers for equal intervals of the kth variable are irregularly spaced in $\mathscr{R}$. Thus, once $\tau(\mathbf{X})$ has been determined, the reference system in $\mathscr{R}$ is determined exactly with linear axes marked with irregular scales. The mechanism of earlier chapters is applicable with very little modification. In particular with metric scaling methods, such as least-squares scaling, the orthogonal Procrustean embedding method of section 3.3.1 is applicable. The normal planes to the kth axis in $\mathscr{R}$ remain parallel but because the markers are irregularly spaced they will remain irregular in their back-projections onto $\mathscr{L}$. This gives a linear predictive biplot where predictions are made as before by the marker at the point of orthogonal projection. We would expect this to improve on the regression method (section 3.3.2) which does not cope with irregular scales Just as with metric least-squares scaling, interpolation is more difficult but the minimal error projection Procrustes method of section 3.4.2 remains available.

Methods other than least-squares scaling can be used for the MDS while retaining the same procedures for transformation of variables. If PCA is used for the scaling then something like the classical linear biplot axes will be obtained. These will be linear and have the same directions for prediction and interpolation but will be equipped with irregular scales. The two scales will continue to be inversely related (section 2.3).

When the kth variable is categorical $\mathbf{y}_k = \tau_k(\mathbf{x}_k)$ defines a transformation of each category-level to a numerical value, or values, in one or more dimension. A one-dimensional transformation defines a linear arrangement of CLPs. Ordered categorical variables may be handled by imposing order constraints on the transformed values of the CLPs, which may lead to equal values (quantifications) for two, or more, category levels. When $m > 1$ dimension is allowed, the resulting m-dimensional configuration of CLPs is a flattened simplex in $\mathscr{R}$. In all cases, the reference system is exact and the construction of prediction regions is available as described in sections 4.5.2, 7.4.2 and A.13. The more general forms of back-projection (section A.9) from m dimensions onto ρ dimensions will be relevant.

The methods described in this section remain to be implemented and validated, but we believe them to be potentially very useful additions to the armoury of useful practical biplots.

11.3 Biplots in generalized bilinear models

A generalized linear model (GLM) is one where the expectation of a variable is a function of a linear model. The expection is taken relative to any probability distribution but the theory is most fully developed for the exponential family (McCullagh and Nelder, 1989). Thus

$$E(\mathbf{y}) = \mathbf{f}(\mathbf{x}'\boldsymbol{\beta})$$

The parameters $\boldsymbol{\beta}$ may be estimated by maximum likelihood using an iterative weighted least-squares algorithm. The inverse of $f(\cdot)$ is known as the **link function**; it is worth noting the concept of transforming the scale of observations that also occurs in non-metric scaling (section 11.2). The GLMs form a powerful and flexible family of models whose most important members are, perhaps, the classical linear model, the log-linear models for contingency tables and logistic regression. Apart from this flexibility, an advantage of GLMs is that when the model is structured, then the well-understood mechanism of interpretation in terms of nested and crossed factors is immediately available; in particular, interactions of more than two factors are easily handled. Corresponding to the analysis of variance of a linear model there is an analysis of deviance (log-likelihood ratio). Several authors have pointed out relationships between a log-linear analysis of a two-way contingency table and the correspondence analysis of the same table (e.g. van der Heijden and de Leeuw, 1985). Depending on one's point of view, either one may be regarded as giving an approximation to the other.

In Chapter 8, we saw how biadditive models could be used to model interactions multiplicatively to give special forms of biplot. Recently, there has been interest in generalized biadditive models (GBMs) for which, in general

$$E(\mathbf{y}) = f(E(\mathbf{y})) = f(\boldsymbol{\alpha}'\mathbf{X}\boldsymbol{\beta}).$$

It is to be understood that $\boldsymbol{\alpha}$ and/or $\boldsymbol{\beta}$ may contain units, thus admitting additive and constant terms into the model. The most popular member of this family is the log-biadditive model; for the details of fitting such models see de Falguerolles and Francis (1994) and van Eeuwijk (1995). Clearly, once a GBM has been fitted, it may be exhibited as a biplot using precisely the methods of Chapter 8. Thus, in this line of research, the emphasis is on extending the

GLMs to include biadditive and bilinear terms rather than attempting to show that GLMs and CA have features in common.

11.4 Biplots in three-way analyses

In recent years, there have been several developments concerned with the analysis of multiway data; a useful general reference is Coppi and Bolasco (1991). These methods are generally referred to as multiway or multimode methods. There is some imprecision in the usage of the terms multiway and multimode but this need not concern us here because our remarks pertain mainly to the analysis of sets of multivariate data matrices or to sets of distance matrices, so the three modes referred to here are sample units, variables and replicates, the first two of which may generate distance matrices. Biplots for different types of a single data matrix or a single distance matrix have been the concern of previous chapters. Certain types of analysis combine the individual analyses into a group average and represent the individuals as deviations from this average. The question then arises as to if and how information on variables might be incorporated into the display of group averages.

One such analysis is **generalized Procrustes analysis (GPA)** (Gower, 1975). In GPA, we have M matrices $\mathbf{Y}_1, \mathbf{Y}_2, \ldots, \mathbf{Y}_M$. Rather than think of these as data-matrices, it is better to regard them as giving coordinates of M configurations which may have been derived by any MDS method from distance matrices $\mathbf{D}_1, \mathbf{D}_2, \ldots, \mathbf{D}_M$ which, in turn, may themselves have been derived from data-matrices $\mathbf{X}_1, \mathbf{X}_2, \ldots, \mathbf{X}_M$. In certain special cases, the $\mathbf{Y}_i$ and the $\mathbf{X}_i$ may coincide, and in others the $\mathbf{Y}_i$ may be derived by MDS; see Dijksterhuis and Gower (1991/2) for a non-technical discussion of the various forms of pre-scaling that may be used to derive commensurable matrices $\mathbf{Y}_m (m = 1, 2, \ldots, M)$ from incommensurable data matrices $\mathbf{X}_m (m = 1, 2, \ldots, M)$. After pre-scaling, the first step of GPA is to centre each matrix $\mathbf{Y}_m$ by subtracting the column means and, usually, to normalize so that the centred matrices satisfy $\| \mathbf{Y}_m \| = 1$ for $m = 1, 2, \ldots, M$. Then, orthogonal matrices $\mathbf{Q}_m$ are found that minimize

$$\sum_{m=1}^{M} \| \mathbf{Y}_m \mathbf{Q}_m - \mathbf{M} \| \tag{11.1}$$

where

$$\mathbf{M} = \frac{1}{M}\sum_{m=1}^{M} \mathbf{Y}_m \mathbf{Q}_m \tag{11.2}$$

is the group average. The computational details of this and related problems are reviewed by Gower (1994) but it should be clear that at the minimum of (11.1) each $\mathbf{Q}_m$ is given by the orthogonal Procrustes fit of $\mathbf{Y}_m$ to $\mathbf{M}$ (A.9). The orientation of $\mathbf{M}$ is arbitrary, but it is convenient to fix it by referring the group average configuration to its principal axes, and we assume that this has been done.

The MDS of the individual $\mathbf{Y}_m$ may have been equipped with biplot axes by the methods previously discussed. Indeed, these axes may be regarded as being rigidly embedded in spaces $\mathscr{R}_1, \mathscr{R}_2, \ldots, \mathscr{R}_M$. The orientations $\mathbf{Q}_m$ have embedded the M configurations in a common space $\mathscr{R}$ which also holds the group average $\mathbf{M}$. The biplot axes for $\mathscr{R}$ may be defined as the average of the corresponding axes in $\mathscr{R}_1, \mathscr{R}_2, \ldots, \mathscr{R}_M$. For quantitative variables, linear biplot axes average into linear axes, non-linear axes average into non-linear axes and CLPs average into a CLP at their centroid.

Gower and Dijksterhuis (1994) show how this process works in practice by discussing data on nine brands of coffee, on each of which the same six quantitative and five categorical variables were recorded by seven assessors. Thus, $n = 9$, $p = 11$ and $M = 7$. The results for each assessor were represented in two dimensions using the generalized biplot methodology of Chapter 7 (section 7.2.1.). The most simple form was used, in which contributions to overall ddistance in (6.6) were treated as Pythagorean for quantitative variables and as the extended matching dissimilarity for categorical variables. As explained in section 7.2.1 this leads to particularly simple computation with linear biplots for the quantitative variables and CLPs for the categorical variables with the properties discussed in section 4.5.1. The resulting interpolative biplots are shown in Fig. 11.2, together with the GPA group average configuration; all configurations are displayed relative to the principal axes of the group average. The biplot axes and CLPs of the group average are obtained as the average of the corresponding elements for the seven assessors. There is clear heterogeneity among the assessors, so the group average itself and its biplot should be interpreted with caution. With this kind of data, a points of view analysis might be helpful with a different biplot representing each point of view, but with only seven assessors this is not worth attempting.

This sketch illustrates how biplots may be incorporated into three-way methods. Clearly, similar methods are available for all methods when the reference system may be regarded as being rigidly embedded in individual spaces $\mathscr{R}_m (m = 1, 2, \ldots, M)$ which are then combined. Similar methods may also be used with other methods. For example, with **individual difference scaling**, Carroll and Chang (1970), individual configurations $\mathbf{Y}_m$ are related to a group average $\mathbf{Y}$ by

$$\mathbf{Y}_m = \mathbf{Y}\mathbf{W}_m$$

where the matrices $\mathbf{W}_m$ are diagonal and represent different weights given to the axes of the group average. There is some unidentifiabil-

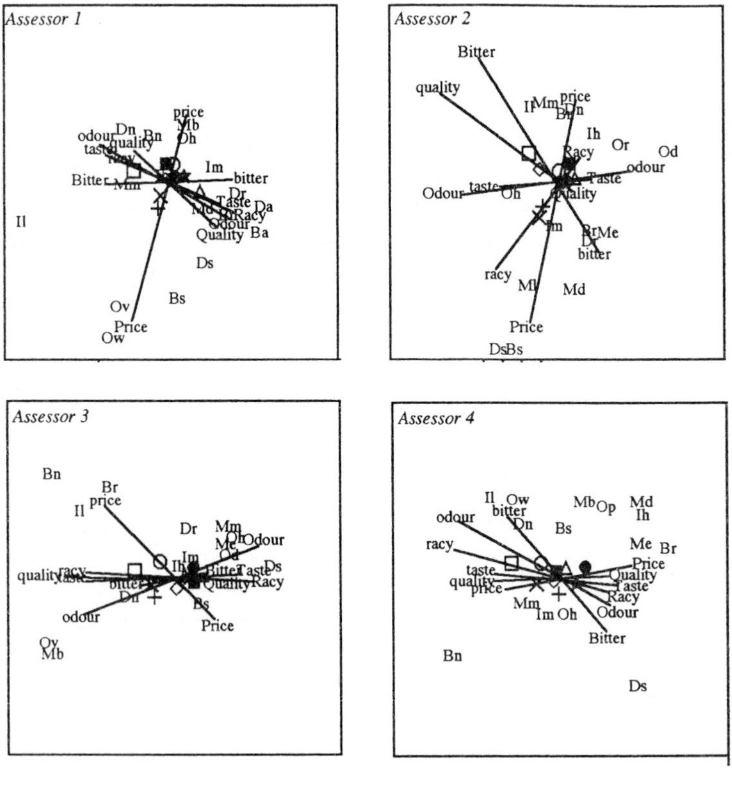

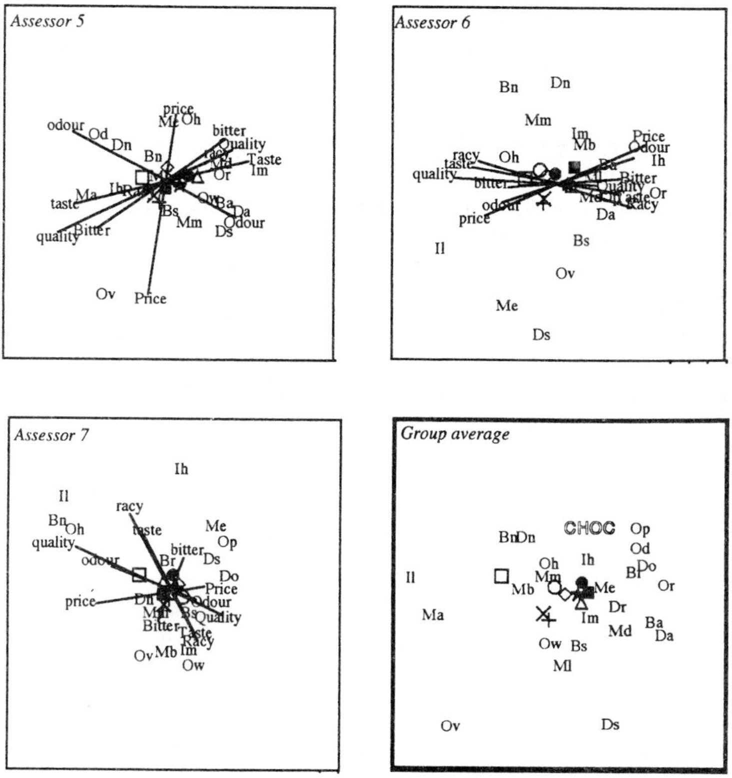

Fig. 11.2. *A generalized Procrustes analysis of coffee images for seven assessors, together with their group average. (Modified from Fig. 3 of Gower and Dijksterhuis, 1994.)*

ity in the parameterization which may be remedied either by choosing

$$\text{(i)} \quad \frac{1}{M}\sum_{m=1}^{M} \mathbf{W}_m = \mathbf{I}$$

or

$$\text{(ii)} \quad \sum_{m=1}^{M} \mathbf{W}_m^2 = \mathbf{I}$$

With (i) the group average coincides with the average of the individual configurations and with (ii) the sum of the individual

Pythagorean ddistances sum to the ddistances within the group average configuration. In both cases, biplot axes for the group average may be obtained as the average of the biplots for the individual configurations.

Carlier and Kroonenberg (1995) discuss biplots in the context of extensions of CA to analyse three-way contingency tables, using conventional CA biplots for various fitted two-way arrays or two-way slices of fitted three-way arrays. Principal components common to M data-matrices (Krzanowski, 1979; Flury, 1984; 1987) offers another area ripe for incorporating biplot information.

11.5 Biplots for special classes of matrices

In this section, we give an overview of several biplot displays associated with special matrices. Most, but not all, are based on the SVD (section A.3) and its Eckart – Young approximation (section A.4).

11.5.1 The covariance, correlation and coefficient of variation biplots

In the following, we shall continue to assume that $\mathbf{X}$ is expressed in deviations from its (column) means. In Chapter 2, we presented the classical biplot of $\mathbf{X}$ through its SVD

$$\mathbf{X} = (\mathbf{U\Sigma})\mathbf{V}'$$

where the bracketing indicates the plotting of the rows of $\mathbf{U\Sigma}$ to give positions for the samples and the rows of $\mathbf{V}$ to give unit points for interpolative biplots. We could equally well have bracketed the decomposition to give

$$\mathbf{X} = \mathbf{U}(\mathbf{\Sigma V}')$$

for which the inner-product interpretations remain unaltered but the row- and column-points have new interpretations. The rows of $\mathbf{V\Sigma}$ give distances derived from $\mathbf{S} = \mathbf{X}'\mathbf{X}$ by (A.51). Thus the ddistance of the kth column point from the origin is s_{kk}, the sum-of-squares for the kth variable, and the ddistance between the hth and kth points is $s_{hh} + s_{kk} - 2s_{hk}$. It follows that the cosine of the angle subtended at the origin by these two points is r_{hk}, the correlation between the hth and kth variables.

Thus the column-points have useful interpretations. How about the row-points? We have

$$\begin{aligned}\mathbf{B} = \mathbf{X}(\mathbf{X}\mathbf{X})^{-1}\mathbf{X}' &= \mathbf{U}\boldsymbol{\Sigma}\mathbf{V}'(\mathbf{V}\boldsymbol{\Sigma}'\boldsymbol{\Sigma}\mathbf{V}')^{-1}\mathbf{V}\boldsymbol{\Sigma}'\mathbf{U}' \\ &= \mathbf{U}\boldsymbol{\Sigma}(\boldsymbol{\Sigma}'\boldsymbol{\Sigma}')^{-1}\boldsymbol{\Sigma}'\mathbf{U}' = \mathbf{U}\mathbf{U}'\end{aligned}$$

where $\mathbf{U}$ is an orthonormal matrix with p columns. Thus, the ddistances between the row-points may be said to have Mahalanobis form. However, the metric concerned is $\mathbf{X}'\mathbf{X}$ rather than being independent of $\mathbf{X}$ (this is the sphering transformation of exploratory data analysis). The result is that $\mathbf{B}$ is idempotent, which implies that sums-of-squares of projections of the n points onto any direction is unity. (This is obvious from noting that $\mathbf{U}$ is orthonormal so that sums-of-squares onto a unit vector $\mathbf{l}$ all give $\mathbf{l}'\mathbf{U}'\mathbf{U}\mathbf{l} = \mathbf{l}'\mathbf{l} = 1$.) Thus the p-dimensional scatter of the n points is essentially spherical and is uninteresting. If the p-dimensional representation lacks interest, then there seems little point in paying any attention to its approximation in fewer dimensions. Although distances between points with coordinates given by the n rows of $\mathbf{U}_\rho$ seem to have little interesting interpretive value, the vectors so defined may be equipped with predictive scales, using minor variants of the methods already described, and used to give graphical evaluations of the Eckart–Young ρ-dimensional approximation to $\mathbf{X}$.

This form of biplot admits several variants. If $\mathbf{X}$ is first divided by $\sqrt{(n-1)}$ then sums-of-squares and products become sample variances. If $\mathbf{X}$ is normalized to have unit column sums-of-squares then the ddistance between the hth and kth points is $2(1 - r_{hk})$. Thus, the greater the correlation the nearer are the two points, at least in the exact representation but not necessarily in approximations. In exact representations, all p points should lie on a unit circle. The unit circle is often shown on correlation biplots to give visual appreciation of the degree of approximation in ρ dimensions for each of the variables. We return to the representations of correlations in section 11.5.3.

Another variant is the coefficient of variation biplot, in which the uncentred $\mathbf{X}$ is divided by $\sqrt{(n-1)}$ and by the means $\mathbf{m} =$ diag $(\mathbf{1}'\mathbf{X}/n)$ i.e. the biplot is based on the SVD of $\mathbf{X}\mathbf{m}^{-1/2}/\sqrt{(n-1)}$. Then, after centering, the distance of the kth point from the origin is the coefficient of variation of that variable. In the exact representation, the cosines of angles subtended at the origin remain correlations but their approximations differ in fewer dimensions. The coefficient of variation biplot is due to Underhill (1990), a paper which gives examples contrasting it with covariance and correlation biplots; further examples are given by Iloni (1991).

Yet another variant has been developed by ter Braak (1983), specifically for the display of measures of ecological diversity based on abundance matrices (section 9.3.6). Now $\mathbf{X}$ is a table where x_{ij} gives the number (i.e. abundance) of species j at site i as in Table 9.1. The table is first transformed into relative abundances $x_{ij}/x_{i.}$ or site profiles. With this transformation the sum-of-squares of the rows is known as α-diversity and the ddistance between rows is known as β-diversity, measures that are familiar to ecologists. Biplots of the form $(\mathbf{U\Sigma})\mathbf{V}'$ are produced, without first removing the site means; α-diversity and β-diversity are approximated by distances from the origin and distances between pairs of row-points, respectively; the entries of the transformed table are approximated by inner-products between row- and column-points, the calculation of whose numerical values can be facilitated by providing the row-points with scales. The site means may be removed in the same way that the means of variables are removed in PCA and, in this case, α-diversity is approximated relative to the projection of the true zero point onto the approximation space. A disadvantage is that very abundant species can dominate the analysis and therefore log transformations may be advantageous but then the diversity measures need reappraisal. As well as ter Braak (1983), examples are given by Iloni (1991).

Before closing this section, we mention again the decomposition

$$\mathbf{X} = (\mathbf{U\Sigma})(\mathbf{\Sigma}'\mathbf{\Sigma})^{-1/2}(\mathbf{\Sigma}'\mathbf{V}')$$

which is essentially the form used in CA. We may plot $\mathbf{U\Sigma}_\rho$ and $\mathbf{V\Sigma}_\rho$ to approximate row and column distances but lose the inner-product interpretation, unless one accepts the mental gymnastics implied by weighting by $(\mathbf{\Sigma}'\mathbf{\Sigma})^{-\frac{1}{2}}$. The effects of these decompositions may be summed up by saying that there are three things that might be approximated: (i) the row distances, (ii) the column distances, and (iii) the inner-product. Any two of these may be shown simultaneously in a biplot, but not all three.

11.5.2 Special matrices

Although not strict biplot methods, it is useful to consider the Eckart–Young approximation to symmetric and to skew-symmetric matrices. For completeness, we outline the properties of biplots of symmetric matrices, but regard them of no great importance; biplots of skew-symmetric matrices are much more interesting.

Symmetric matrices

The SVD of a symmetric matrix **X** depends on whether it is p.s.d or not. Bailey and Gower (1990) noted that the Eckart–Young theorem, when applied to symmetric matrices implies that the diagonal elements are given half the weight of non-diagonal elements in the least-squares approximation. Different weightings lead to different properties of the fitted matrix and to different biplots. When the diagonal is null and fitted by null values, as with the MDS of distance matrices or with skew-symmetric matrices, the problem is avoided. In other cases, algorithms that give equal weight could be developed, but we do not discuss this possibility here. In this section, **X** is not necessarily centred. If **X** is symmetric we have

$$\text{(i)}\quad \mathbf{X} = \mathbf{U\Sigma U}' \quad \text{for } \mathbf{X} \text{ p.s.d.}$$

and

$$\text{(ii)}\quad \mathbf{X} = \mathbf{U\Sigma KU}' \quad \text{for } \mathbf{X} \text{ not p.s.d,}$$

where **K** is diagonal with $k_{ii} = 1$ when σ_i is an eigenvalue of **X** and $k_{ii} = -1$ when $-\sigma_i$ is an eigenvalue of **X**; when $\boldsymbol{\sigma}_i = 0$, the corresponding value is arbitrary, but we shall set $k_{ii} = 1$.

These results are essentially the spectral decompositions of section A.1 but modified because the singular values **Σ**, unlike eigenvalues, are always non-negative. In case (i), a biplot may be displayed with

$$\mathbf{X} = (\mathbf{U\Sigma})\mathbf{U}'$$

the rows of which give points generating approximations to the Pythagorean ddistances between the rows of **X** and the columns giving directions equipped with scales for interpolation or prediction. These scales are computed similarly to those of Chapter 2 but require some modification because **X** is not centred at the origin. However, because the row and column vectors of **X** are the same, this involves plotting redundant information which can be avoided by plotting solely the rows of $\mathbf{U\Sigma}^{\frac{1}{2}}$. This preserves the inner-product interpretation of $\mathbf{X}$ and its least-squares approximation, but some tedious modifications, not given here, are needed if scales are required for biplot axes. In the full dimensional representation, the ddistance between the ith and jth points is

$$x_{ii} + x_{jj} - 2x_{ij}$$

and this is approximated in ρ dimensions. However, the approximation to ddistance may be improved by using MDS displays – see the discussion in section 11.5.3 for the special case when x_{ij} is a correlation.

When $\mathbf{X}$ has negative eigenvalues, case (ii), the form

$$\mathbf{X} = (\mathbf{U\Sigma K})\mathbf{U}'$$

remains available without modification but because $(\mathbf{\Sigma K})^{\frac{1}{2}}$ is not real, the second form is not available. If we plot $\mathbf{U\Sigma}^{\frac{1}{2}}$ the ddistances remain as above, but the evaluation of inner-products has to be done with a knowledge of $\mathbf{K}$ and this is awkward.

Whether or not $\mathbf{X}$ is p.s.d, all these plots have some diagnostic value in the manner of Chapter 8. In two-dimensional biplots, collinearities with the origin indicate a unit rank matrix, and collinearities not through the origin indicate a model of the form $m\mathbf{11} + \mathbf{aa}'$. Similarly, in three dimensions, coplanarities indicate a model $m\mathbf{11} + \mathbf{aa}' + \mathbf{bb}'$ where $m = 0$ if the plane includes the origin.

Skew-symmetric matrices

Skew-symmetric matrices rarely occur naturally as data but square non-symmetric matrices are common. Examples are: trade data for imports and exports, social mobility data on, for example, professions of fathers and sons, and psychological confusion data which depend on the order in which stimulii are presented. One method of analysis is to partition the data into symmetric and skew-symmetric components, which are analysed separately. This is a natural thing to do because different mechanisms often govern symmetry and departures from symmetry. Note that the zero diagonal values of skew-symmetric matrices imply that there are no problems of differential weighting of diagonal and non-diagonal values in ordinary least-squares approximations, as there are with symmetric matrices. The SVD of a skew-symmetric matrix $\mathbf{X}$ is

$$\mathbf{X} = \mathbf{U}\boldsymbol{\Sigma}\mathbf{J}\mathbf{U}'$$

where

$$\mathbf{J} = \mathrm{diag}\begin{pmatrix} 0 & 1 \\ -1 & 0 \end{pmatrix} \quad \text{and} \quad \boldsymbol{\Sigma} = \mathrm{diag}(\sigma_1, \sigma_1, \sigma_2, \sigma_2, \sigma_3, \sigma_3, \ldots)$$

with final diagonal values, respectively, of $+1$ and 0 when $\mathbf{X}$ is of odd order. Thus, we may write a ρ-dimensional approximation as

$$\mathbf{X}_\rho = \sum_1^\rho \sigma_i(\mathbf{u}_{2i-1}\mathbf{u}'_{2i} - \mathbf{u}_{2i}\mathbf{u}'_{2i-1})$$

Note that by ρ-dimensional, we now really mean 2ρ-dimensional, because the singular values occur in equal pairs, so there is no way of choosing any best single direction in the two-dimensional space spanned by $\mathbf{u}_{2i}$ and $\mathbf{u}_{2i-1}$. Thus, graphical displays are in sets of two-dimensions which have been termed **bimensions**; usually, we shall be concerned with a single bimension. Just as for symmetric matrices, we may present both row and column information separately but, noting the occurrence of $\mathbf{U}$ on both sides of the decomposition, seek ways of avoiding this duplication of information by plotting just the rows of $\mathbf{U}$ as coordinates. There is a novel way of interpreting such diagrams which we shall describe for the first bimension. To simplify the notation, we write $\mathbf{u}_1 = \mathbf{u}$ and $\mathbf{u}_2 = \mathbf{v}$. Then, the two-dimensional approximation to x_{ij} has the form

$$x_{ij} = \sigma(u_i v_j - u_j v_i)$$

which is proportional to the area of a triangle with coordinates $\mathrm{P}_i(u_i, v_i)$, $\mathrm{P}_j(u_j, v_j)$ and the origin. The geometry is shown in Fig. 11.3. The area of the triangle $\mathrm{OP}_i\mathrm{P}_j$ approximates x_{ij} and areas determined in a clockwise sense are positive. The dotted line gives the locus of all points which determine the same area with P_i.

Unlike most other diagrams in this book, Fig. 11.3 shows the orthogonal axes which support the coordinate representations. In practice, these will be ignored as usual. Indeed, with skew-symmetry, there is even less reason than usual to show these axes, because the singular values are equal in pairs and hence the rotation of the axes in each bimension is entirely arbitrary. The geometry of Fig. 11.3 is non-Euclidean, with association being given by the area

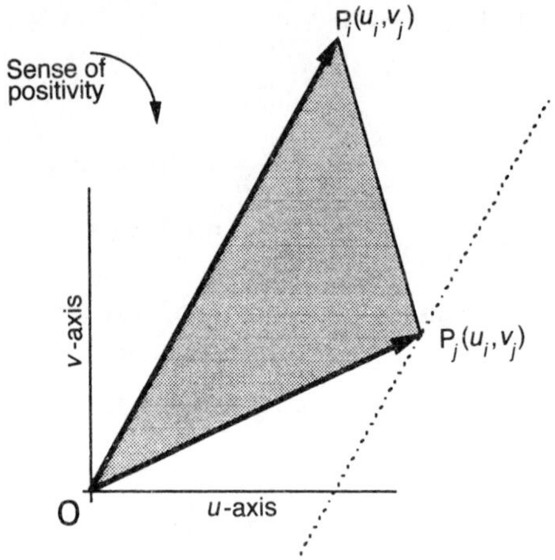

Fig. 11.3. *Geometry of representing a skew-symmetric matrix X.*

subtended at the origin by two points. It follows that all points that give the same approximation, as determined by area, with P_i as does P_j must lie on the line parallel to OP_i through P_j; this is equivalent to the equal distance property of Euclidean geometry given by a circle through P_j centred at P_i. Skewness is modelled by noting that the areas OP_iP_j and OP_jP_i are in opposite senses and hence have opposite signs. The area of any triangle $P_iP_jP_k$ is proportional to $x_{ij} + x_{jk} + x_{ki}$ and, depending on whether or not the triangle contains the origin, we have (inconsistent triad) $i \prec j \prec k$, $k \prec i$ or (consistent triad) $i \prec j \prec k$, $i \prec k$, where $\prec$ is to be read as 'is worse than'. As a diagnostic plot, collinearity indicates that $\mathbf{X}$ has the special skewlinear form, $\mathbf{X} = \mathbf{u1}' - \mathbf{1u}'$. Other properties of this geometry and some examples are given by Gower (1977) and Constantine and Gower (1978).

11.5.3 Other correlational biplots

The column-points given by the correlational biplot described in section 11.5.1, based on the SVD of the normalized and centred version of $\mathbf{X}$, may be regarded as giving the Eckart–Young approximation to the symmetric correlation matrix $\mathbf{X}'\mathbf{X}$; the unit diagonal values are included. It is unnecessary to approximate these known values except insofar as they permit the use of the unit circles described in section 11.5.1. Hills (1969) suggested that it might be better to use MDS methods to approximate either the matrix with elements (i) $\{1 - r_{hk}\}$ or (ii) $\{1 - r_{hk}^2\}$. The first form takes account of the signs of the correlations, while the second does not. In both cases, unit correlation leads to coincident points. The furthest apart pair of points in case (i) occurs for the most negative correlation and, in case (ii), for the correlation whose absolute value is nearest zero. In both cases, only unit correlation gives coincident points. Hills used PCO for his MDS, in which case one is not obliged to transform the correlations to be complements of unity, but any method of metric or non-metric scaling could be used. The advantage of this type of approach is that, because we are using distance interpretations, the observed and fitted values of the diagonal terms are both automatically zero and do not enter into approximations. It follows that Hills' method must give better fits than the methods of section 11.5.2 when using the same number, ρ, of approximating dimensions. Another advantage is that the differential weighting of the diagonal and non-diagonal terms of ordinary least-squares fits to symmetric matrices is avoided. The disadvantage is that information on the sample units is not included so, in its existing from, the method is not a biplot method, though we believe that this deficiency can be remedied.

Various methods of factor analysis also give representations of correlations. This is a vast subject and we can give only the most basic introduction here; see Lawley and Maxwell (1971) for further information. Like PCA, factor loadings $\mathbf{V}_\rho$ are sought which approximate

$$\mathbf{X}'\mathbf{X} = \mathbf{V}_\rho'\mathbf{V}_\rho + \mathbf{E}$$

where $\mathbf{E}$ is a non-negative diagonal matrix. Exact solutions of this type will not normally exist. If one can assume multinormality, this model may be fitted by maximum likelihood, as described by Lawley and Maxwell (1971), giving solutions that are independent of the possibly incommensurable measurement scales of the variables. In other cases, we have the same kinds of problems of commensurability as with PCA (section 2.5) and then it is usual to normalize the data so that $\mathbf{X}'\mathbf{X}$ becomes a correlation matrix. A least-squares fit may then be defined which minimizes

$$\|\mathbf{X}'\mathbf{X} - \mathbf{V}_\rho'\mathbf{V}_\rho - \mathbf{E}\|$$

The resulting version of factor analysis is termed **principal factor analysis** because it is equivalent to a PCA of $\mathbf{X}'\mathbf{X}-\mathbf{E}$ which differs from $\mathbf{X}'\mathbf{X}$ only in its diagonal values. Of course, $\mathbf{E}$ is unknown and iterative algorithms have to be used, usually of alternating least-squares form – see especially the well-known MINRES algorithm. Unlike PCA, the solutions for increasing values of ρ are not nested and have to be computed *ab initio*. The rows of $\mathbf{V}_\rho$ may be plotted in the same way as the eigenvectors of PCA. A full biplot may be obtained by including sample points, either by evaluating **factor scores** (Lawley and Maxwell, 1971), or by embedding in the space $\mathscr{R}$ using the methods of Chapter 3. JCA (Chapter 10) uses very similar methods but with categorical variables.

11.6 Reduced rank regression

Multivariate multiple regression is concerned with observations on n samples for p dependent variables, given in a matrix $\mathbf{Y}$, and q independent variables, given in a matrix $\mathbf{X}$. It is assumed that these are related by linear regressions whose least-squares estimate $\mathbf{B}=\mathbf{B}_0$ minimizes $\|\mathbf{Y}-\mathbf{X}\mathbf{B}\|$ where $\mathbf{B}_0$ is a $p \times q$ matrix of regression coefficients given by

$$\mathbf{B}_0=(\mathbf{X}'\mathbf{X})^{-1}\mathbf{X}'\mathbf{Y}$$

We may represent $\mathbf{B}_0$ as a biplot, using the methods of Chapters 2 and 8, based on the SVD $\mathbf{B}_0=(\mathbf{U}\mathbf{\Sigma})\mathbf{V}'$ and may also plot the rows of $\mathbf{X}\mathbf{U}\mathbf{\Sigma}$ which form inner-products with $\mathbf{V}$ that give the fitted values $\hat{\mathbf{Y}}=\mathbf{X}\mathbf{B}_0$. We note that $\mathbf{B}_0$ does not represent a samples $\times$ variables data-matrix, so the methods of Chapter 8 are indicated. However, it makes little sense to remove either row or column means of regression coefficients; we are interested in approximating the whole matrix and therefore take the decomposition of its raw form. A similar representation is used in the following.

Reduced rank regression, or redundancy analysis, is a variant of multivariate multiple regression which minimizes $\|\mathbf{Y}-\mathbf{X}\mathbf{B}\|$ where $\mathbf{B}$ is constrained to be of specified rank ρ. This is the problem of (A.48) with $\mathbf{A}=\mathbf{X}, \mathbf{B}=\mathbf{I}$ and $\Gamma_0=(\mathbf{X}'\mathbf{X})^{-1}\mathbf{X}'\mathbf{Y}=\mathbf{B}_0$. Accordingly, we have to minimize $\|\mathbf{X}(\mathbf{B}_0-\mathbf{B})\|$ and define the generalized SVD (A.22):

$$\mathbf{B}_0=\mathbf{S}\mathbf{\Sigma}\mathbf{T}^{-1} \tag{11.3}$$

where $\mathbf{T}$ is orthogonal and $\mathbf{S}$ normalized to satisfy $\mathbf{S}'(\mathbf{X}'\mathbf{X})\mathbf{S}=\mathbf{I}$. It follows that $\mathbf{S}$ and $\mathbf{T}$ satisfy

$$(\mathbf{X}'\mathbf{Y}\mathbf{Y}'\mathbf{X})\mathbf{S}=(\mathbf{X}'\mathbf{X})\mathbf{S}\boldsymbol{\Sigma}\boldsymbol{\Sigma}'$$

$$(\mathbf{Y}'\mathbf{X}(\mathbf{X}'\mathbf{X})\mathbf{X}'\mathbf{Y})\mathbf{T}=\mathbf{T}\boldsymbol{\Sigma}'\boldsymbol{\Sigma} \tag{11.4}$$

where $\mathbf{T}$ is orthogonal and $\mathbf{S}$ normalized to satisfy $\mathbf{S}'(\mathbf{X}'\mathbf{X})\mathbf{S}=\mathbf{I}$. The required estimate may be written in either of the equivalent forms:

$$\hat{\mathbf{B}}=\mathbf{B}_0\mathbf{T}_\rho\mathbf{T}^\rho \quad \text{(i)}$$

$$\hat{\mathbf{B}}=\mathbf{S}\boldsymbol{\Sigma}_\rho\mathbf{T}^\rho \quad \text{(ii)} \tag{11.5}$$

As is evident from (11.4), a problem with this approach is that the estimates are not linearly invariant to the choice of scale of $\mathbf{Y}$. To remove this effect, $\mathbf{Y}$ may be scaled by some suitable matrix $\boldsymbol{\Omega}$, so that we operate on $\mathbf{Y}\boldsymbol{\Omega}$. As will be evident shortly, we have to consider only $\boldsymbol{\Omega}'\boldsymbol{\Omega}$. ter Braak (1990), first considered the problem of biplots in reduced rank regression and in ter Braak (1994a) considers three choices of $\boldsymbol{\Omega}'\boldsymbol{\Omega}$:

(a) normalize $\mathbf{Y}$ as in PCA and then choose $\boldsymbol{\Omega}'\boldsymbol{\Omega}=\mathbf{I}$
(b) choose $\boldsymbol{\Omega}'\boldsymbol{\Omega}=(\mathbf{R}'\mathbf{R})^{-1}$ where $\mathbf{R}=(\mathbf{Y}-\mathbf{X}\mathbf{B}_0)$
(c) chose $\boldsymbol{\Omega}'\boldsymbol{\Omega}=(\text{diag}\,\mathbf{R}'\mathbf{R})^{-1}$

and favours (c).

Now, in place of $\|\mathbf{X}(\mathbf{B}_0-\mathbf{B})\|$, we require $\mathbf{B}$ which minimizes

$$\begin{aligned}\|\mathbf{X}(\mathbf{B}_0-\mathbf{B})\boldsymbol{\Omega}\| &= \text{trace}\,[\boldsymbol{\Omega}'\boldsymbol{\Omega}(\mathbf{B}_0-\mathbf{B})'\mathbf{X}'\mathbf{X}(\mathbf{B}_0-\mathbf{B})]\\ &= \|(\boldsymbol{\Omega}'\boldsymbol{\Omega})^{\frac{1}{2}}(\mathbf{B}_0-\mathbf{B})'(\mathbf{X}'\mathbf{X})^{\frac{1}{2}}\|\end{aligned} \tag{11.6}$$

which is the Eckart–Young theorem in two metrics (A.42) with solutions (A.40) and (A.41). Thus, from the SVD

$$(\boldsymbol{\Omega}'\boldsymbol{\Omega})^{\frac{1}{2}}\mathbf{B}_0(\mathbf{X}'\mathbf{X})^{\frac{1}{2}}=\mathbf{U}\boldsymbol{\Sigma}\mathbf{V}' \tag{11.7}$$

we obtain, as in (A.22) and (A.23), the generalized decomposition

$$\mathbf{B}_0=\mathbf{S}\boldsymbol{\Sigma}\mathbf{T}^{-1}$$

with normalizations $\mathbf{S}'(\boldsymbol{\Omega}'\boldsymbol{\Omega})\mathbf{S}=\mathbf{I}$ and $\mathbf{T}'(\mathbf{X}'\mathbf{X})^{-1}\mathbf{T}=\mathbf{I}$. The ρ-dimensional approximations are given by (A.42) and (A.43).

Thus, the biplot for the regression coefficients is to plot $\mathbf{S}_\rho\boldsymbol{\Sigma}_\rho=\mathbf{B}_0\mathbf{T}_\rho$ for the rows and $(\mathbf{T}^\rho)'$ for the columns and use the usual inner-product interpretation for b_{ij}. In addition, we note that the fitted values $\mathbf{X\hat{B}}$ may be written $(\mathbf{XS}_\rho\mathbf{\Sigma}_\rho)\mathbf{T}^\rho$. Hence, by plotting the n points with coordinates given by the rows of $\mathbf{XS}_\rho\mathbf{\Sigma}_\rho$ we obtain approximations to the fitted values by forming inner-products with the already existing column points $(\mathbf{T}^\rho)'$. Hence we take the rows of $(\mathbf{T}^\rho)'$ to give directions of biplot axes which may be marked with scales to aid prediction. Predictions are given for the fitted values and for the columns of the matrix of regression coefficients. Note that when $\rho = 2$, the biplot representation is exact but when $\rho > 2$, a two-dimensional biplot gives a weighted least-squares approximation.

Ter Braak (1994a) also shows how the biplot can accommodate information that allows t-ratios to be estimated. In fact, the t-coefficients are given by the elements of the following modification of the matrix (11.7):

$$[\text{diag}(\mathbf{\Omega}'\mathbf{\Omega})]^{\frac{1}{2}}\mathbf{B}[\text{diag}(\mathbf{X}'\mathbf{X})]^{\frac{1}{2}} \tag{11.8}$$

It is clear that (11.8) may be formed approximately as the inner-product

$$\{[\text{diag}(\mathbf{\Omega}'\mathbf{\Omega})]^{\frac{1}{2}}\mathbf{S}_\rho\mathbf{\Sigma}_\rho\}\{\mathbf{T}^\rho[\text{diag}(\mathbf{X}'\mathbf{X})]^{\frac{1}{2}}\}$$

which involves only a change of scale, not direction, of the previous biplot. ter Braak (1994a) shows how lines and circles may be added to the biplot to help to assess significance.

We may also consider the minimization of $\|\mathbf{Y} - \mathbf{Z}_1\mathbf{BZ}_2'\|$ over $\mathbf{B}$, restricted to have given rank ρ whose solution is given in (A.48). As a generalization of restricted multivariate multiple regression, this seems to have little interest but when $\mathbf{Y}$ derives from a two-way contingency table, this model may offer a useful generalization of canonical CA as discussed in section 9.3.6.

11.7 Missing values

When some values are missing in $\mathbf{X}$, most biplot methods are not directly applicable but can easily be made so by using estimated values in place of those that are not recorded or which, for some reason, are regarded as better replaced. Examples of the latter have already been met in JCA (Chapter 10) where the diagonal blocks of the Burt matrix have to be iteratively estimated and, in section

11.5.3, where in the factor analysis of a correlation matrix, diagonal values are replaced by estimates. With less regular patterns of missing values, similar estimation methods may be used. With quantitative variables, well-known methods for inputing missing values are the algorithms of Beale and Little (1975), Titterington and Jiang (1983) and Titterington (1984), but these imply that the multinormal assumption is realistic.

In the context of linear biplots, Gabriel and Zamir (1979) adopt a more direct approach. They discuss alternating least-squares algorithms that minimize the weighted least-squares criterion

$$\| \mathbf{W}*(\mathbf{X} - \hat{\mathbf{X}}) \|$$

where * denotes the Hadamard product, $\hat{\mathbf{X}}$ is of specified rank ρ, and $\mathbf{W}$ is a matrix of weights applicable to the elements of $\mathbf{X}$. To handle missing values, we are concerned only with the special case where $w_{ij} = 0$ when x_{ij} is missing, else $w_{ij} = 1$. At convergence, we have a least-squares estimate $\hat{\mathbf{X}}$ of rank ρ, which may be regarded as the Eckart–Young estimate when values of $\mathbf{X}$ are missing; indeed, when no values are missing $\hat{\mathbf{X}}$ should coincide with the Eckart–Young result. With the algorithms used, which work by fitting a succession of unit rank matrices, $\hat{\mathbf{X}}$ will not be in SVD form, but this may easily be found after convergence as a separate step; then the methods of Chapter 2 may be used without further modification. Although this approach does not overtly 'fill in' the missing values of $\mathbf{X}$, the result is the same as if every missing value x_{ij} is replaced by $\hat{x}_{ij}$ and this is done by the algorithms. Note that the solutions for different ranks are not nested, as they are when no values are missing, so if different values of ρ are to be explored, then the iterations must be repeated *ab initio*. The alternating least-squares algorithms can converge on a local optimum and Gabriel and Zamir (1979) give reccommendations on the best algorithm and choice of initialization.

The standard way of filling in missing values in CA is based on the reconstitution formula (9.9). For the (i, j)th entry of $\mathbf{X}$ the reconstitution is

$$x_{ij} = r_i c_j \left(\frac{1}{x_{..}} + \sum_{k=2}^{c} \sigma_k \mathbf{u}^*_{jk} \mathbf{v}^*_{jk} \right) \tag{11.9}$$

The ρ-dimensional approximation to (11.9) may be put into an iterative form

$$x_{ij}^{m+1} = x_{i.}^{m} x_{.j}^{m} \left(\frac{1}{x_{..}^{m}} + \sum_{k=2}^{\rho} \sigma_k^m \mathbf{u}_{jk}^{*m} \mathbf{v}_{jk}^{*m} \right) \tag{11.10}$$

deriving values at the $(m+1)$th step from those at the mth step. Starting with initial estimates for the missing cells of **X**, (11.10) may be applied to all the missing cells, leaving the other cells untouched, and the process iterated until convergence. At convergence, the values currently estimating the missing cells will complete **X** which may then be treated by the methods of Chapter 9; the current SVD will be the one required. Classical CA is an analysis of the residuals from fitting the independence model ((9.6) and (9.7)). Several authors have pointed out that residuals from more elaborate models might be similarly analysed (van der Heijden, de Falguerolles and de Leuuw, 1989). In particular, de Leeuw and van der Heijden (1988) show that the above method based on $\rho = 0$ is equivalent to doing a CA based on residuals from the **conditional independence model** $\alpha_i \beta_j$ and where missing values are replaced by their expected values as given by maximum likelihood estimates.

Metric and non-metric MDS of proximities is unaffected by missing values; the criterion (3.1) is merely evaluated over all non-missing pairs (i, j). Thus, there is no problem in obtaining a graphical display of the samples. However, missing values in **X** need imputation before any of the methods discussed earlier for constructing the reference system might be applied. Existing imputation methods seem unsuitable and further work is needed.

APPENDIX A

Algebraic results

This appendix provides some basic algebraic results needed in the main text. We are not especially concerned with formal proofs, but give proofs when these offer additional insight and can be done conveniently. Some of the results are very well known, others less so and some we believe to be new. The more important results are highlighted in italics.

A.1 Spectral decomposition

Any symmetric matrix $\mathbf{A}$ *of degree n may be written* $\mathbf{A} = \mathbf{V}\boldsymbol{\Lambda}\mathbf{V}'$ *where* $\mathbf{V}$ *is orthogonal and* $\boldsymbol{\Lambda}$ *is diagonal with its diagonal elements (its eigenvalues) written in non-increasing order. When* $\mathbf{A}$ *is positive definite, then the eigenvalues are all positive; when* $\mathbf{A}$ *is positive semi-definite, then some eigenvalues are zero and the remainder are positive.*

We are concerned with solutions to the simultaneous equations

$$\mathbf{A}\mathbf{v} = \lambda\mathbf{v} \tag{A.1}$$

Solutions exist only when λ satisfies the determinantal equation det $(\mathbf{A} - \lambda\mathbf{I}) = 0$, known as the **characteristic polynomial**. This polynomial is of degree n and so has n roots, which may not be unique. To each root λ_i, there is a corresponding solution $\mathbf{v}_i$, known as an **eigenvector**. Note that the scale of the eigenvectors is arbitrary, so we may assume that they are normalized to give $\mathbf{v}'\mathbf{v} = 1$. When λ_i and λ_j are distinct eigenvalues we have

$$\left.\begin{aligned}\mathbf{A}\mathbf{v}_i &= \lambda_i\mathbf{v}_i\\ \mathbf{A}\mathbf{v}_j &= \lambda_j\mathbf{v}_j\end{aligned}\right\} \tag{A.2}$$

and hence

$$\left.\begin{aligned} \mathbf{v}_j'\mathbf{A}\mathbf{v}_i = \lambda_i \mathbf{v}_j'\mathbf{v}_i \\ \mathbf{v}_i'\mathbf{A}\mathbf{v}_j = \lambda_j \mathbf{v}_i'\mathbf{v}_j \end{aligned}\right\} \tag{A.3}$$

The left-hand sides of (A.3) are equal and hence subtraction gives

$$\lambda_i \mathbf{v}_i'\mathbf{v}_j = \lambda_j \mathbf{v}_i'\mathbf{v}_j$$

Thus if $\lambda_i \neq \lambda_j$ then $\mathbf{v}_i'\mathbf{v}_j = 0$. This establishes the orthogonality of the columns of $\mathbf{V}$ when the eigenvalues are all distinct. Then, all solutions to (A.1) can be presented simultaneously as

$$\mathbf{AV} = \mathbf{V\Lambda} \tag{A.4}$$

The ordering of the columns of $\mathbf{V}$, correspondingly $\mathbf{\Lambda}$, is arbitrary so we may arrange them so that the diagonal values of $\mathbf{\Lambda}$ do not increase. Because of the orthogonality of $\mathbf{V}$, (A.4) gives the spectral decomposition of $\mathbf{A}$ as

$$\mathbf{A} = \mathbf{V\Lambda V}' \tag{A.5}$$

and the reduction to diagonal form

$$\mathbf{\Lambda} = \mathbf{V}'\mathbf{AV} \tag{A.6}$$

Also, for any vector $\mathbf{v}_i$, we have that $\mathbf{v}_i'\mathbf{A}\mathbf{v}_i = \lambda_i$. Hence, if $\mathbf{A}$ is p.d. then $\lambda_i > 0$ and if $\mathbf{A}$ is p.s.d. then $\lambda_i \geqslant 0$.

When an eigenvalue is repeated, it has two independent vectors which may be taken to be orthogonal. We do not prove this result, but it is strongly suggested by the informal argument that, however close may be λ_i and λ_j, they are associated with orthogonal vectors, as shown above; in the limit $\lambda_i = \lambda_j = \lambda$, say, and the vectors remain orthogonal. (Note that this argument fails for asymmetric matrices, where the vectors are not orthogonal; indeed, for such matrices, equal eigenvalues are not necessarily associated with independent eigenvectors.) With symmetric matrices, if $\mathbf{v}_i$ and $\mathbf{v}_j$ are an orthogonal pair of vectors for a repeated eigenvalue λ, then any linear combination $\alpha\mathbf{v}_i + \beta\mathbf{v}_j$ satisfies $\mathbf{A}(\alpha\mathbf{v}_i + \beta\mathbf{v}_j) = \lambda(\alpha\mathbf{v}_i + \beta\mathbf{v}_j)$ and hence is also an eigenvector associate with λ. Thus, eigenvectors associated with a repeated eigenvalue are not unique but they occupy the unique subspace spanned by $\mathbf{v}_i$ and $\mathbf{v}_j$. Clearly, the argument extends to any number of equal eigenvalues and the results (A.4), (A.5) and (A.6) remain valid, though not unique, when some eigenvalues are repeated.

A.2 The two-sided eigenvalue problem

Given two symmetric matrices, $\boldsymbol{A}$ (general) and $\boldsymbol{B}$ (p.d), then there exists a non-singular transformation matrix $\mathbf{W}$ such that $\boldsymbol{W'BW}=\boldsymbol{I}$ and $\boldsymbol{W'AW}=\boldsymbol{\Lambda}$, a diagonal matrix with its diagonal elements (its eigenvalues) written in non-increasing order.

Now, we are concerned with the two-sided eigenvalue equation

$$\mathbf{Aw}=\lambda\mathbf{Bw} \tag{A.7}$$

whose solutions occur for values of λ satisfying the characteristic equation det $(\mathbf{A}-\lambda\mathbf{B})=0$. When λ_i and λ_i are distinct eigenvalues we have

$$\begin{aligned}\mathbf{Aw}_i&=\lambda_i\mathbf{Bw}_i\\ \mathbf{Aw}_j&=\lambda_j\mathbf{Bw}_j\end{aligned} \tag{A.8}$$

and hence

$$\begin{aligned}\mathbf{w}_j'\mathbf{Aw}_i&=\lambda_i\mathbf{w}_j'\mathbf{Bw}_i\\ \mathbf{w}_i'\mathbf{Aw}_j&=\lambda_j\mathbf{w}_i'\mathbf{Bw}_j\end{aligned} \tag{A.9}$$

The left-hand sides of (A.9) are equal and hence subtraction gives

$$\lambda_i\mathbf{w}_i'\mathbf{Bw}_j=\lambda_j\mathbf{w}_i'\mathbf{Bw}_j$$

Thus, if $\lambda_i \neq \lambda_j$, then $\mathbf{w}_i'\mathbf{Bw}_j=0$ showing that $\mathbf{w}_i$ and $\mathbf{w}_j$ are orthogonal relative to $\mathbf{B}$. When $i=j$, we may normalize $\mathbf{w}_i$ so that $\mathbf{w}_i'\mathbf{Bw}_i=1$. The set of all such solutions satisfy

$$\mathbf{W'BW}=\mathbf{I} \tag{A.10}$$

and

$$\mathbf{AW}=\mathbf{BW\Lambda} \tag{A.11}$$

where $\mathbf{\Lambda}$ is diagonal. Pre-multiplying (A.11) by $\mathbf{W}'$ and using (A.10) immediately gives a transformation to diagonal form

$$\mathbf{W'AW}=\mathbf{\Lambda} \tag{A.12}$$

The normalization (A.10) may be written

$$\mathbf{WW'}=\mathbf{B}^{-1} \tag{A.13}$$

When some eigenvalues are equal, the results remain valid and can be justified by the arguments given at the end of section A1. A detailed discussion of the case when $\mathbf{B}$ is not positive definite is given by de Leeuw (1982).

From (A.5) we see that a p.s.d symmetric matrix $\mathbf{B}$ has a unique symmetric square root matrix

$$\mathbf{B}^{\frac{1}{2}} = \mathbf{V}'\mathbf{\Lambda}^{\frac{1}{2}}\mathbf{V} \tag{A.14}$$

Because $\mathbf{B}$ is assumed to be p.s.d, it follows that, after a little rearrangement, (A.11) may be written

$$(\mathbf{B}^{-\frac{1}{2}}\mathbf{A}\mathbf{B}^{-\frac{1}{2}})\mathbf{B}^{\frac{1}{2}}\mathbf{W} = (\mathbf{B}^{\frac{1}{2}}\mathbf{W})\mathbf{\Lambda} \tag{A.15}$$

Comparing (A.15) with (A.4) we see that $\mathbf{U} = \mathbf{B}^{\frac{1}{2}}\mathbf{W}$ and $\mathbf{\Lambda}$ are eigenvectors and eigenvalues of the symmetric matrix $\mathbf{B}^{-\frac{1}{2}}\mathbf{A}\mathbf{B}^{-\frac{1}{2}}$. Using the normalization $\mathbf{U}'\mathbf{U} = \mathbf{I}$ of the eigenvectors of Section A.1 gives $\mathbf{W}'\mathbf{B}\mathbf{W} = \mathbf{I}$ in agreement with (A.10). The vectors of the two-sided eigenvalue problem are given by $\mathbf{W} = \mathbf{B}^{-\frac{1}{2}}\mathbf{U}$. Thus, for computational purposes, the two-sided eigenvalue problem may be solved as a simple variant of the one-sided problem.

A.3 Singular value decomposition (SVD)

A.3.1 Generalized SVD in the metric $\mathbf{B}$

Suppose $\mathbf{A} = \mathbf{X}'\mathbf{X}$ in (A.11), then $\mathbf{A}$ is p.s.d. with s, say, non-zero (positive) eigenvalues and let Σ_s^2 be the $s \times s$ diagonal matrix of the positive eigenvalues and $\mathbf{W}_s$ represent the first s columns of the corresponding eigenvectors $\mathbf{W}$. Then (A.11) may be rewritten

$$\mathbf{A}\mathbf{W}_s = \mathbf{B}\mathbf{W}_s\Sigma_s^2 \tag{A.16}$$

and hence

$$\mathbf{W}_s'\mathbf{A}\mathbf{W}_s = \Sigma_s^2$$

which may be rewritten

$$\Sigma_s^{-1}\mathbf{W}_s'\mathbf{X}'\mathbf{X}\mathbf{W}_s\Sigma_s^{-1} = \mathbf{I}$$

It follows that $\mathbf{U}_s = \mathbf{X}\mathbf{W}_s\Sigma_s^{-1}$ is an orthonormal matrix with s columns. Augment $\mathbf{U}_s$ by $n - s$ columns, orthogonal to those of $\mathbf{U}_s$ and to each other, to give an orthogonal matrix $\mathbf{U}$. Similarly, augment Σ_s by corresponding zeros to give Σ of order n. Then

$$\mathbf{X}\mathbf{W}\Sigma^{-1} = \mathbf{U}$$

Thus

$$\mathbf{X} = \mathbf{U}\boldsymbol{\Sigma}\mathbf{W}^{-1} \tag{A.17}$$

(A.17) is the generalized SVD of $\boldsymbol{X}$, which is expressed as the product of an orthogonal matrix $\boldsymbol{U}$ a diagonal matrix $\boldsymbol{\Sigma}$ whose non-zero values are known as singular values, and the inverse of the eigenvectors of (A.11).

From (A.13), we may rewrite (A.17) in the form

$$\mathbf{X} = \mathbf{U}\boldsymbol{\Sigma}\mathbf{W}'\mathbf{B} \tag{A.18}$$

which is a more usual representation and gives rise to the expression that (A.18) is the SVD of $\mathbf{X}$ in the metric $\mathbf{B}$. When $\mathbf{A}$ is defined as a weighted sum-of-squares-and-products matrix weighted by a p.d matrix $\mathbf{N}$ to give, $\mathbf{A} = \mathbf{X}'\mathbf{N}\mathbf{X}$ then, in the above derivation, $\mathbf{X}$ is replaced by $\mathbf{N}^{\frac{1}{2}}\mathbf{X}$ replacing (A.18) by

$$\mathbf{X} = \mathbf{N}^{-1/2}\mathbf{U}\boldsymbol{\Sigma}\mathbf{W}'\mathbf{B} \tag{A.19}$$

This form is most often used when $\mathbf{N}$ is a diagonal matrix of weights.

When $\mathbf{B} = \mathbf{I}, \mathbf{W}$ becomes the orthogonal matrix $\mathbf{V}$ and (A.17) becomes

$$\mathbf{X} = \mathbf{U}\boldsymbol{\Sigma}\mathbf{V}' \tag{A.20}$$

which is the classical SVD of $\mathbf{X}$, with weighted form

$$\mathbf{X} = \mathbf{N}^{-1/2}\mathbf{U}\boldsymbol{\Sigma}\mathbf{V}' \tag{A.21}$$

A.3.2 Generalized SVD in the metrics $\boldsymbol{P}$ and $\boldsymbol{Q}$

The expression (A.17) suggests the existence of the most general form of SVD:

Given any rectangular matrix $\boldsymbol{X}$ and any conformable symmetric positive definite matrices $\boldsymbol{P}$ and $\boldsymbol{Q}$ then we may write

$${}_p\mathbf{X}_q = {}_p\mathbf{S}_p\boldsymbol{\Sigma}_q\mathbf{T}_q^{-1} \tag{A.22}$$

where $\boldsymbol{S}$ and $\boldsymbol{T}$ satisfy the normalizations:

$$(i)\,\boldsymbol{S}'\boldsymbol{P}^{-1}\boldsymbol{S} = \boldsymbol{I} \quad \text{and} \quad (ii)\,\boldsymbol{T}'\boldsymbol{Q}\boldsymbol{T} = \boldsymbol{I} \tag{A.23}$$

Note the inverse of $\mathbf{P}$ in the normalization shown in (A.23). Clearly, this could be avoided if we were to replace $\mathbf{P}$ by its inverse, but this would have undesirable consequences on the representations (A.24) and (A.25), below. We think that the convention we adopt here is the best compromise, but readers are warned that it may differ from representations, which are algebraically equivalent, given by other authors.

The above derivation of (A.17) was constructive. To prove the validity of the decomposition (A.22), we now give the more usual deductive type of proof. Without loss of generality, we assume that $p \leqslant q$; if not, transpose $\mathbf{X}$. This assumption guarantees the existence of all the inverses used in the following, except when $\mathbf{X}$ is of deficient row-rank (i.e. when rank $\mathbf{X} < p$) in which case the result remains valid but the proof requires modification not given here.

Consider the two-sided eigenvalue problem

$$(\mathbf{X}'\mathbf{P}^{-1}\mathbf{X})\mathbf{T} = \mathbf{Q}\mathbf{T}(\mathbf{\Sigma}'\mathbf{\Sigma}) \tag{A.24}$$

which is (A.11), with $\mathbf{A} = \mathbf{X}'\mathbf{P}^{-1}\mathbf{X}, \mathbf{B} = \mathbf{Q}$ and $\mathbf{\Lambda} = \mathbf{\Sigma}'\mathbf{\Sigma}$, whose eigenvectors $\mathbf{T}$ may be normalized to satisfy part (ii) of (A.23). Pre-multiplying by $\mathbf{X}\mathbf{Q}^{-1}$ and post-multiplying by $\mathbf{\Sigma}'(\mathbf{\Sigma}\mathbf{\Sigma}')^{-1}$ gives

$$\mathbf{X}\mathbf{Q}^{-1}\mathbf{X}'(\mathbf{P}^{-1}\mathbf{X}\mathbf{T}\mathbf{\Sigma}'(\mathbf{\Sigma}\mathbf{\Sigma}')^{-1}) = \mathbf{P}(\mathbf{P}^{-1}\mathbf{X}\mathbf{T}\mathbf{\Sigma}'(\mathbf{\Sigma}\mathbf{\Sigma}')^{-1})\mathbf{\Sigma}\mathbf{\Sigma}'$$

Hence, setting

$$(\mathbf{S}^{-1})' = \mathbf{P}^{-1}\mathbf{X}\mathbf{T}\mathbf{\Sigma}'(\mathbf{\Sigma}\mathbf{\Sigma}')^{-1} \tag{A.25}$$

we have a two-sided eigenvalue problem for $(\mathbf{S}^{-1})'$

$$(\mathbf{X}\mathbf{Q}^{-1}\mathbf{X}')(\mathbf{S}^{-1})' = \mathbf{P}(\mathbf{S}^{-1})'(\mathbf{\Sigma}\mathbf{\Sigma}') \tag{A.26}$$

where the normalization of the vectors satisfies

$$\begin{aligned}(\mathbf{S}^{-1})\mathbf{P}(\mathbf{S}^{-1})' &= (\mathbf{\Sigma}\mathbf{\Sigma}')^{-1}\mathbf{\Sigma}\mathbf{T}'\mathbf{X}'\mathbf{P}^{-1}\mathbf{X}\mathbf{T}\mathbf{\Sigma}'(\mathbf{\Sigma}\mathbf{\Sigma}')^{-1}\\ &= (\mathbf{\Sigma}\mathbf{\Sigma}')^{-1}\mathbf{\Sigma}\mathbf{T}'\mathbf{Q}\mathbf{T}(\mathbf{\Sigma}'\mathbf{\Sigma})\mathbf{\Sigma}'(\mathbf{\Sigma}\mathbf{\Sigma}')^{-1}\\ &= (\mathbf{\Sigma}\mathbf{\Sigma}')^{-1}\mathbf{\Sigma}(\mathbf{\Sigma}'\mathbf{\Sigma})\mathbf{\Sigma}'(\mathbf{\Sigma}\mathbf{\Sigma}')^{-1} = \mathbf{I}\end{aligned}$$

Thus, the normalization of $\mathbf{S}$ satisfies part (i) of (A.23).

Post-multiplying (A.21) by $\mathbf{\Sigma}\mathbf{\Sigma}'$ and pre-multiplying by $\mathbf{P}$ and, as a consequence of the normalization, substituting $\mathbf{S}$ for $\mathbf{P}(\mathbf{S}^{-1})'$ gives

$$\mathbf{S}\mathbf{\Sigma}\mathbf{\Sigma}' = \mathbf{X}\mathbf{T}\mathbf{\Sigma}'$$

which, because we are assuming that $p \leqslant q$, guarantees the equality of the first p columns of $\mathbf{S}\mathbf{\Sigma}$ and $\mathbf{X}\mathbf{T}$. For the $q - p$ null vectors $\mathbf{T}_0$

of (A.24), we have that $(\mathbf{X}'\mathbf{P}^{-1}\mathbf{X})\mathbf{T}_0 = \mathbf{0}$, which on pre-multiplying by $(\mathbf{X}\mathbf{X}')^{-1}\mathbf{X}\mathbf{P}$ gives $\mathbf{X}\mathbf{T}_0 = \mathbf{0}$. Thus, $\mathbf{S}\boldsymbol{\Sigma}$ and $\mathbf{X}\mathbf{T}$ are both zero in their final $q-p$ columns and we have that $\mathbf{S}\boldsymbol{\Sigma} = \mathbf{X}\mathbf{T}$ for all q columns; post-multiplying by $\mathbf{T}^{-1}$ gives (A.22), as required.

Setting $\mathbf{P} = \mathbf{I}$ and $\mathbf{Q} = \mathbf{I}$ gives orthogonal vectors as solutions to (A.24) and (A.26) and then (A.22) gives the classical result (A.20).

A.4 Approximation

A.4.1 Principal components analysis (PCA)

$\mathscr{L}$, a ρ-dimensional subspace of $\mathscr{R}$, is spanned by ρ linearly independent vectors which are the columns of the $p \times \rho$ matrix $\mathbf{L}$; without loss of generality, we can assume that the columns of $\mathbf{L}$ are orthonormal, so that $\mathbf{L}'\mathbf{L} = \mathbf{I}$. The projection of $\mathbf{X}$ onto $\mathscr{L}$ is given by $\mathbf{X}\mathbf{L}\mathbf{L}'$ (section A.8) with residuals $\mathbf{X}(\mathbf{I} - \mathbf{L}\mathbf{L}')$. Principal components are the settings of $\mathbf{L}$ that minimize $\|\mathbf{X}(\mathbf{I} - \mathbf{L}\mathbf{L}')\|$ which is the same as maximizing trace $(\mathbf{X}\mathbf{L}\mathbf{L}'\mathbf{X}')$. We have

$$\begin{aligned}\operatorname{trace}(\mathbf{X}\mathbf{L}\mathbf{L}'\mathbf{X}') &= \operatorname{trace}(\mathbf{L}'\mathbf{X}'\mathbf{X}\mathbf{L}) \\ &= \operatorname{trace}(\mathbf{L}'\mathbf{V}\boldsymbol{\Lambda}\mathbf{V}'\mathbf{L}) = \operatorname{trace}(\mathbf{P}'\boldsymbol{\Lambda}\mathbf{P}) \qquad \text{(A.27)}\end{aligned}$$

where $\mathbf{X}'\mathbf{X} = \mathbf{V}\boldsymbol{\Lambda}\mathbf{V}'$ is the spectral decomposition (A.5) and $\mathbf{P} = \mathbf{V}'\mathbf{L}$ is orthonormal. Note that if $\mathbf{P}$ is a matrix that maximizes (A.27) then $\mathbf{P}\mathbf{Q}$ is another solution for arbitrary $\rho \times \rho$ orthogonal matrices $\mathbf{Q}$. Assuming that $\mathbf{P}$ is one of these solutions, (A.27) may be written

$$\operatorname{trace}(\mathbf{X}\mathbf{L}\mathbf{L}'\mathbf{X}') = \operatorname{trace}(\mathbf{P}'\boldsymbol{\Lambda}\mathbf{P}) = \operatorname{trace}\left(\sum_{i=1}^{n} \lambda_i \mathbf{p}_i \mathbf{p}_i'\right) \qquad \text{(A.28)}$$

where $\mathbf{p}_i$ represents the ith column of $\mathbf{P}'$. Writing $\mathbf{P} = \{p_{ij}\}$ and $\mathbf{F} = \{p_{ij}^2\}$ then (A.28) may be written

$$\operatorname{trace}(\mathbf{X}\mathbf{L}\mathbf{L}'\mathbf{X}') = \sum_{i=1}^{n} \lambda_i \left(\sum_{j=1}^{\rho} f_{ij}\right) \qquad \text{(A.29)}$$

The non-negativity of $\mathbf{F}$ and the orthonormality of $\mathbf{P}$ shows that

(i) $0 \leqslant f_{ij} \leqslant 1$ for $i = 1, 2, \ldots, n$ and $j = 1, 2, \ldots, \rho$

(ii) $\sum_{i=1}^{n} f_{ij} = 1$ for $j = 1, 2, \ldots, \rho$

(iii) $0 \leqslant \sum_{j=1}^{\rho} f_{ij} \leqslant 1$ for $i = 1, 2, \ldots, n$

Thus (A.29) is a linear function of the $n\rho$ elements f_{ij} subject to the linear constraints (i), (ii), and (iii). It follows that the maximization of trace $(\mathbf{XLL'X'})$ is a linear programming problem whose solution is at a vertex of the feasible region defined by the constraints. The constraints (i) show that the feasible region is bounded by the $n\rho$-dimensional unit cube. The other two sets of constraints show that not all vertices of the cube are admissible. For example, suppose we make $f_{ij} = 1$, then (ii) and (iii) show that all other elements in the ith row and jth column of $\mathbf{F}$ must be zero. Thus, at the maximum $\mathbf{F}$ must have a unit in every column in such a way that the units occur in ρ mutually exclusive rows, the remaining $n - \rho$ rows of $\mathbf{F}$ being zero. Such vertices of the unit cube define the boundaries of the feasible region. Thus

$$\sum_{j=1}^{\rho} f_{ij} = 1$$

for some set of ρ rows of $\mathbf{F}$ and

$$\sum_{j=1}^{\rho} f_{ij} = 0$$

for the remaining $n - \rho$ rows, showing that (A.29) is the sum of ρ of the eigenvalues and clearly this sum is maximum when the non-zero rows of $\mathbf{F}$ refer to the ρ largest eigenvalues. As in section A.1, we assume that the eigenvalues are given in non-increasing order so that $\mathbf{F}$ is some permutation of the columns of

$$\begin{pmatrix} \mathbf{I}_\rho \\ \mathbf{0} \end{pmatrix}$$

This implies that

$$\mathbf{P} = \begin{pmatrix} \mathbf{K}_\rho \\ \mathbf{0} \end{pmatrix}$$

where $\mathbf{K}_\rho$ differs from a permutation matrix only in that its non-zero values may be -1 as well as $+1$. Note the orthogonality of the columns of $\mathbf{P}$, even though this constraint has not been imposed formally. Thus

$$\mathbf{V'L} = \begin{pmatrix} \mathbf{K}_\rho \\ \mathbf{0} \end{pmatrix} \mathbf{Q} \quad \text{and} \quad \mathbf{L} = \mathbf{V} \begin{pmatrix} \mathbf{K}_\rho \\ \mathbf{0} \end{pmatrix} \mathbf{Q} \tag{A.30}$$

where the arbitrary orthogonal matrix $\mathbf{Q}$ is reintroduced. Thus $\mathbf{L}$ is a linear combination of the columns of $\mathbf{V}_\rho$, the first ρ columns of $\mathbf{V}$. When $\rho = 1, \mathbf{L} = \mathbf{v}_1$ and when $\rho = 2$, $\mathscr{L}$ is spanned by $\mathbf{v}_1$ and $\mathbf{v}_2$ and so on. In this way, the columns of $\mathbf{V}$ itself provide a nested set of solutions in successively higher numbers of dimensions. This nesting of solutions is an important property of PCA. When $\lambda_\rho = \lambda_{\rho+1}$, the ρ-dimensional solution is not unique because we may set the final column of $\mathbf{V}_\rho$ to be any unit vector in the space spanned by $\mathbf{v}_\rho$ and $\mathbf{v}_{\rho+1}$.

In the above, $\mathbf{X}$ may be any matrix. For reasons connected with Huygen's principle (2.1), PCA is more usually concerned with the centred matrix $(\mathbf{I} - \mathbf{N})\mathbf{X}$. Then $\mathbf{X}'\mathbf{X}$ becomes the variance-covariance matrix $\mathbf{X}'(\mathbf{I} - \mathbf{N})\mathbf{X}$ and when the columns of $(\mathbf{I} - \mathbf{N})\mathbf{X}$ are normalized, this variance-covariance matrix becomes a correlation matrix. The basic results of PCA may be summed up as follows.

> *Given a centred data-matrix $(\boldsymbol{I} - \boldsymbol{N})\boldsymbol{X}$ then its pricipal components are the eigenvectors, given by the columns of $\boldsymbol{V}$, of the covariance (correlation) matrix $\boldsymbol{X}'(\boldsymbol{I} - \boldsymbol{N})\boldsymbol{X}$. $\boldsymbol{V}_\rho$ the first ρ eigenvectors, span a ρ-dimensional space $\mathscr{L}$ which has the property that $\|(\boldsymbol{I} - \boldsymbol{N})\boldsymbol{X}(\boldsymbol{I} - \boldsymbol{L}_\rho\boldsymbol{L}'_\rho)\|$ is minimized when $\boldsymbol{L}_\rho = \boldsymbol{V}_\rho$.*

A.4.2 The Eckart–Young theorem

The choice of orthogonal projection in PCA is not arbitrary. Any rank ρ matrix, of the same size as ${}_n\mathbf{X}_p$, may be written ${}_n\mathbf{Y}_p\mathbf{L}_\rho\mathbf{L}'_p$ where the columns of $\mathbf{L}$ are independent and may be assumed to be orthonormal. (We now drop the conformality notation and write $\mathbf{L}_p$ as the first ρ columns of the $p \times p$ matrix $\mathbf{L}$.) As is obvious geometrically, and is easily verified algebraically, we have

$$\|\mathbf{X} - \mathbf{Y}\mathbf{L}_\rho\mathbf{L}'_\rho\| = \|\mathbf{X} - \mathbf{X}\mathbf{L}_\rho\mathbf{L}'_\rho\| + \|(\mathbf{X} - \mathbf{Y})\mathbf{L}_\rho\mathbf{L}'_\rho\|$$

so that $\|\mathbf{X} - \mathbf{Y}\mathbf{L}_\rho\mathbf{L}'_\rho\|$ is minimum when $\mathbf{X}\mathbf{L}_\rho\mathbf{L}'_\rho = \mathbf{Y}\mathbf{L}_\rho\mathbf{L}'_\rho$. This is valid for *any* orthonormal $\mathbf{L}_\rho$ but PCA has shown that an absolute minimum occurs when $\mathbf{L}_\rho$ is given by certain eigenvectors. Thus, one way of looking at PCA is that it finds the rank ρ matrix $\hat{\mathbf{X}}$ which minimizes $\|\mathbf{X} - \hat{\mathbf{X}}\|$. We have shown that $\hat{\mathbf{X}} = \mathbf{X}\mathbf{V}_\rho\mathbf{V}'_\rho$ where the columns of $\mathbf{V}_\rho$ give the leading eigenvectors of $\mathbf{X}'\mathbf{X}$.

From (A.20) we have

$$\hat{\mathbf{X}} = \mathbf{U\Sigma V'V}_\rho\mathbf{V}'_\rho \tag{A.31}$$

Writing $\mathbf{J}_\rho$ for the first ρ columns of $\mathbf{I}$ we may write (A.31) as

$$\begin{aligned}\hat{\mathbf{X}} &= \mathbf{U\Sigma J}_\rho\mathbf{V}'_\rho \\ &= \mathbf{U\Sigma}_\rho\mathbf{V}'_\rho\end{aligned} \tag{A.32}$$

The result that $\hat{X}$ given by (A.32) minimizes $\|X - \hat{X}\|$ is known as the Eckart–Young theorem and plays a central role in the theory of least-squares approximation of matrices. It was first proved explicitly, using rather complex arguments, by Eckart and Young (1936).

A.4.3 The Eckart–Young theorem in the metric B

When $\mathbf{A} = \mathbf{X'X}$ in the two-sided eigenvalue problem (A.7), then (A.12) shows that $\mathbf{W'X'XW} = \mathbf{\Lambda}$ and is diagonal. Thus, the coordinates $\mathbf{XW}$ are referred to their principal axes and the first ρ eigenvectors are the columns of $\mathbf{J}_\rho$. The PCA of $\mathbf{XW}$ minimizes $\|\mathbf{XW} - \mathbf{XWJ}_\rho\mathbf{J}'_\rho\|$. We have

$$\begin{aligned}&\|\mathbf{XW} - \mathbf{XWJ}_\rho\mathbf{J}'_\rho\| \\ &\quad= \text{trace}(\mathbf{XW} - \mathbf{XWJ}_\rho\mathbf{J}'_\rho)(\mathbf{XW} - \mathbf{XWJ}_\rho\mathbf{J}'_\rho)' \\ &\quad= \text{trace}(\mathbf{X} - \mathbf{XWJ}_\rho\mathbf{J}'_\rho\mathbf{W}^{-1})\mathbf{WW}'(\mathbf{X} - \mathbf{XWJ}_\rho\mathbf{J}'_\rho\mathbf{W}^{-1})' \\ &\quad= \text{trace}(\mathbf{X} - \mathbf{XWJ}_\rho\mathbf{J}'_\rho\mathbf{W}^{-1})\mathbf{B}^{-1}(\mathbf{X} - \mathbf{XWJ}_\rho\mathbf{J}'_\rho\mathbf{W}^{-1})'\end{aligned}$$

Because the left-hand side is minimized over all rank ρ matrices, so is the right-hand side. Thus

$$\hat{\mathbf{X}} = \mathbf{XWJ}_\rho\mathbf{J}'_\rho\mathbf{W}^{-1} \tag{A.33}$$

gives the rank ρ least-squares approximation to $\mathbf{X}$ *with respect to the metric* $\mathbf{B}$. The result (A.33) may be rewritten in several ways, the most simple of which is

$$\hat{\mathbf{X}} = \mathbf{XW}_\rho\mathbf{W}^\rho \tag{A.34}$$

where ρ in $\mathbf{W}^\rho$ denotes the first ρ rows of $\mathbf{W}^{-1}$ rather than an exponent. Substituting the generalized singular value decomposition (A.17) for $\mathbf{X}$ shows that

$$\hat{\mathbf{X}} = \mathbf{U\Sigma W}^{-1}\mathbf{W}_\rho\mathbf{W}^\rho$$

$$= \mathbf{U\Sigma J}_\rho \mathbf{W}^\rho$$

$$= \mathbf{U\Sigma}_\rho \mathbf{W}^\rho \tag{A.35}$$

which may be rewritten as in (A.18), as

$$\hat{\mathbf{X}} = \mathbf{U\Sigma}_\rho \mathbf{W}'_\rho \mathbf{B} \tag{A.36}$$

> *The expressions (A.33), (A.34), (A.35) and (A.36) are different ways of writing the generalized Eckart–Young theorem that $\hat{X}$ minimizes trace $(X - \hat{X})\, B^{-1}(X - \hat{X})$, i.e. with respect to the metric* $\boldsymbol{B}$.

Note that $\mathbf{XW}$ generates inner-products $\mathbf{XB}^{-1}\mathbf{X}'$ and ddistance

$$D^2_{ij} = (\mathbf{x}_i - \mathbf{x}_j)\mathbf{B}^{-1}(\mathbf{x}_i - \mathbf{x}_j)' \tag{A.37}$$

known as Mahalanobis' D^2_{ij}, between the ith and jth rows of $\mathbf{XW}$. Mahalanobis distance is important in applications and hence the interest in the coordinates $\mathbf{XW}$ and their approximation.

A.4.4 Eckart-Young theorem in the metrics P *and* Q

The general decomposition (A.22) gives $\mathbf{S}^{-1}\mathbf{XT} = \mathbf{\Sigma}$ and hence $(\mathbf{S}^{-1}\mathbf{XT})'\ (\mathbf{S}^{-1}\mathbf{XT})$ is diagonal. It follows that a PCA of $\mathbf{S}^{-1}\mathbf{XT}$ minimizes

$$\|\mathbf{S}^{-1}\mathbf{XT} - \mathbf{S}^{-1}\mathbf{XT}\,\mathbf{J}_\rho\mathbf{J}'_\rho\|$$

$$= \text{trace}\ \mathbf{S}^{-1}(\mathbf{XT} - \mathbf{XTJ}_\rho\mathbf{J}'_\rho)(\mathbf{XT} - \mathbf{XTJ}_\rho\mathbf{J}'_\rho)'(\mathbf{S}^{-1})'$$

$$= \text{trace}\ (\mathbf{SS}')^{-1}(\mathbf{X} - \mathbf{XTJ}_\rho\mathbf{J}'_\rho\mathbf{T}^{-1})\mathbf{TT}'(\mathbf{X} - \mathbf{XTJ}_\rho\mathbf{J}'_\rho\mathbf{T}^{-1})'$$

$$= \text{trace}\ \mathbf{P}^{-1}(\mathbf{X} - \mathbf{XTJ}_\rho\mathbf{J}'_\rho\mathbf{T}^{-1})\mathbf{Q}^{-1}(\mathbf{X} - \mathbf{XTJ}_\rho\mathbf{J}'_\rho\mathbf{T}^{-1})' \tag{A.38}$$

Thus, as before, but now relative to the metrics **P** and **Q**

$$\hat{\mathbf{X}} = \mathbf{XTJ}_\rho\mathbf{J}'_\rho\mathbf{T}^{-1} = \mathbf{XT}_\rho\mathbf{T}^\rho \tag{A.39}$$

where **T** is given by (A.24). Replacing **X** by (A.22) gives

$$\hat{\mathbf{X}} = \mathbf{S}_\rho\mathbf{\Sigma}_\rho\mathbf{T}^\rho \tag{A.40}$$

where $\mathbf{\Sigma}_\rho = \text{diag}\ (\sigma_1, \sigma_2, \ldots, \sigma_\rho)$. Recall that (A.24) and (A.26) give direct evaluations of **S** and **T**.

We have the alternative expressions

$$\hat{\mathbf{X}} = \mathbf{S}_\rho \mathbf{S}^\rho \mathbf{X} = \mathbf{X}\mathbf{T}_\rho \mathbf{T}^\rho \tag{A.41}$$

where **S** and **T** remain given by (A.24) and (A.26) as can be verified by substituting (A.22) for **X** and using (A.40). Usually, **P** will be a diagonal matrix of weights for the rows of **X** which will often refer to samples.

Note the (A.38) may be written

$$\|\mathbf{S}^{-1}\mathbf{X}\mathbf{T} - \mathbf{S}^{-1}\mathbf{X}\mathbf{T}\,\mathbf{J}_\rho \mathbf{J}'_\rho\| = \|\mathbf{P}^{-\frac{1}{2}}(\mathbf{X} - \hat{\mathbf{X}})\mathbf{Q}^{-\frac{1}{2}}\| \tag{A.42}$$

and hence that $\mathbf{P}^{-\frac{1}{2}}\hat{\mathbf{X}}\mathbf{Q}^{-\frac{1}{2}}$, and hence $\hat{\mathbf{X}}$ itself, may be obtained from the ρ-dimensional Eckart–Young approximation to $\mathbf{P}^{-\frac{1}{2}}\hat{\mathbf{X}}\mathbf{Q}^{-\frac{1}{2}}$. Any square root matrix may be used but when **P** and/or **Q** is diagonal, the obvious choice will be used.

The expressions (A.40) and (A.41) are different ways of writing the generalized Eckart–Young theorem that $\hat{X}$ minimizes $\| P^{-\frac{1}{2}}(X - \hat{X})Q^{-\frac{1}{2}} \|$ with respect to the metrics P and Q normalized as in (A.23). *(A.43)*

A.4.5 Other least-squares approximations

We consider the minimization over $\mathbf{\Gamma}$ of $\|{}_p\mathbf{Y}_q - {}_p\mathbf{A}_\rho \mathbf{\Gamma}_\rho \mathbf{B}'_q\|$ for given **A** and **B** and diagonal $\mathbf{\Gamma}$. Note that now **A** and **B** are not necessarily square and hence are not symmetric. Thus we must minimize

$$\text{trace}\,[(\mathbf{A\Gamma B'})(\mathbf{B\Gamma A'}) - 2(\mathbf{A'YB})\mathbf{\Gamma}] \tag{A.44}$$

Differentiating the second element of (A.44) with respect to $\mathbf{\Gamma}$ gives 2 [diag(**A′YB**)]**1**. The first element of (A.44) may be written

$$\text{trace}\,[\mathbf{A'A\Gamma B'B\Gamma}]$$

the terms of which that involve γ_1 are

$$(\mathbf{a}'_1\mathbf{a}_1)(\mathbf{b}'_1\mathbf{b}_1)\gamma_1^2 + 2(\mathbf{a}'_1\mathbf{a}_2)(\mathbf{b}'_1\mathbf{b}_2)\gamma_1\gamma_2 + \cdots + 2(\mathbf{a}'_1\mathbf{a}_\rho)(\mathbf{b}'_1\mathbf{b}^\rho)\gamma_1\gamma_\rho.$$

Thus, differentiating this with respect to γ_1, and doing similarly for $\gamma_2, \gamma_3, \ldots, \gamma_\rho$, expressing that (A.44) is a minimum may be written

$$[(\mathbf{A'A})*(\mathbf{B'B})]\hat{\mathbf{\Gamma}}\mathbf{1} = [\mathbf{diag}\,(\mathbf{A'YB})]\mathbf{1}$$

where the symbol * denotes the Hadamard product of two matrices. Thus we have shown the following.

The normal equations for min $\|\boldsymbol{Y}-\boldsymbol{A\Gamma B}'\|$ *are*

$$[(\boldsymbol{A}'\boldsymbol{A})^*(\boldsymbol{B}'\boldsymbol{B})]\hat{\boldsymbol{\Gamma}}\boldsymbol{1} = [\mathit{diag}\,(\boldsymbol{A}'\boldsymbol{YB})]\,\boldsymbol{1} \tag{A.45}$$

We shall also require min $[\Sigma_{i<j}^{L}\|\boldsymbol{Y}_{ij}-\boldsymbol{A}_i\boldsymbol{\Gamma}\boldsymbol{A}_j'\|]$ *which clearly has normal equations*

$$\left[\sum_{i<j}^{L}(\boldsymbol{A}_i'\boldsymbol{A}_i)^*(\boldsymbol{A}_j'\boldsymbol{A}_j)\right]\hat{\boldsymbol{\Gamma}}\boldsymbol{1} = \sum_{i<j}^{L}\mathit{diag}\,(\boldsymbol{A}_i'\boldsymbol{Y}_{ij}\boldsymbol{A}_j)\boldsymbol{1} \tag{A.46}$$

(A.45) and (A.46) immediately given the diagonal matrix $\hat{\boldsymbol{\Gamma}}$ in the form of the vector $\hat{\boldsymbol{\Gamma}}\mathbf{1}$. We note that when $\mathbf{B}'=\mathbf{1}$, then $\mathbf{Y}$ is a column-vector $\mathbf{y}$ and (A.45) becomes

$$(\mathbf{A}'\mathbf{A})\hat{\boldsymbol{\Gamma}}\mathbf{1} = \mathbf{A}'\mathbf{y}$$

the usual normal equations for multiple regression. When $\mathbf{B}=\mathbf{I}$, so that $p=q$, (A.45) becomes

$$\mathrm{diag}\,(\mathbf{A}'\mathbf{A})\hat{\boldsymbol{\Gamma}} = \mathrm{diag}\,(\mathbf{A}'\mathbf{Y})$$

which estimates a set of q independent linear regressions.

A similar least-squares problem requires the minimum of $\|{}_p\mathbf{Y}_q - {}_p\mathbf{A}_s\boldsymbol{\Gamma}_t\mathbf{B}_q'\|$ where $\boldsymbol{\Gamma}$ has rank ρ; however, the similarity is superficial. When $\boldsymbol{\Gamma}$ is unrestricted, the solution is $\boldsymbol{\Gamma}=\boldsymbol{\Gamma}_0$ where

$$\boldsymbol{\Gamma}_0 = (\mathbf{A}'\mathbf{A})^{-1}\mathbf{A}'\mathbf{YB}(\mathbf{B}'\mathbf{B})^{-1} \tag{A.47}$$

and it is easy to see that the residual matrix $\mathbf{R}=\mathbf{Y}-\mathbf{A}\boldsymbol{\Gamma}_0\mathbf{B}'$ satisfies $\mathbf{A}'\mathbf{RB}=0$. It follows that

$$\|\mathbf{Y}-\mathbf{A}\boldsymbol{\Gamma}\mathbf{B}'\| = \|\mathbf{Y}-\mathbf{A}\boldsymbol{\Gamma}_0\mathbf{B}'\| + \|\mathbf{A}\boldsymbol{\Gamma}\mathbf{B}'-\mathbf{A}\boldsymbol{\Gamma}_0\mathbf{B}'\|$$

so that to minimize $\|\mathbf{Y}-\mathbf{A}\boldsymbol{\Gamma}\mathbf{B}'\|$ it is necessary only to minimize $\|\mathbf{A}(\boldsymbol{\Gamma}-\boldsymbol{\Gamma}_0)\mathbf{B}')\|$. Now

$$\|\mathbf{A}(\boldsymbol{\Gamma}-\boldsymbol{\Gamma}_0)\mathbf{B}'\| = \|(\mathbf{A}'\mathbf{A})^{\frac{1}{2}}(\boldsymbol{\Gamma}-\boldsymbol{\Gamma}_0)(\mathbf{B}'\mathbf{B})^{\frac{1}{2}}\|$$

whose minimum is given by the ρ-dimensional approximation to $\boldsymbol{\Gamma}_0$ given by the Eckart–Young theorem in two metrics (A.43) with $\mathbf{P}^{-1}=\mathbf{A}'\mathbf{A}$ and $\mathbf{Q}^{-1}=\mathbf{B}'\mathbf{B}$. When $s=p$ and $t=q$ then $\mathbf{R}=0$ and we return to the unrestricted solution $\|{}_p\mathbf{Y}_q - {}_p\mathbf{U}_\rho\boldsymbol{\Sigma}_\rho\mathbf{V}_q'\|$; this remains valid for $s\geqslant p$ and $t\geqslant q$. When $s<p$ and $t\geqslant q$ then the problem reduces to finding the minimum of $\|{}_p\mathbf{Y}_q - {}_p\mathbf{A}_s\boldsymbol{\Gamma}_q\|$ which is obtained by setting $t=q$ and $\mathbf{B}=\mathbf{I}$ in the above. Thus we have the following result.

Except for pathological cases of overdetermination, briefly discussed above, for all matrices $\boldsymbol{\Gamma}$ or rank ρ, then min $\|\boldsymbol{Y}-\boldsymbol{A\Gamma B}'\|$ is given by

$$\hat{\boldsymbol{\Gamma}}=\boldsymbol{S}_\rho\boldsymbol{\Sigma}_\rho\boldsymbol{T}^\rho$$

the generalized SVD of

$$\boldsymbol{\Gamma}_0=(\boldsymbol{A}'\boldsymbol{A})^{-1}\boldsymbol{A}'\boldsymbol{Y}\boldsymbol{B}(\boldsymbol{B}'\boldsymbol{B})^{-1}$$

in the metrics $\boldsymbol{P}^{-1}=\boldsymbol{A}'\boldsymbol{A}$ and $\boldsymbol{Q}^{-1}=\boldsymbol{B}'\boldsymbol{B}$ with normalizations $\boldsymbol{S}'(\boldsymbol{A}'\boldsymbol{A})\boldsymbol{S}=\boldsymbol{I}$ and $\boldsymbol{T}'(\boldsymbol{B}'\boldsymbol{B})^{-1}\boldsymbol{T}=\boldsymbol{I}$. (A.48)

A.5 Basic properties of distance matrices

A.5.1 Euclidean embedding

We are concerned with symmetric matrices of dissimilarities $\mathbf{D}$, with zero diagonal elements. Consider the **centred form**

$$\mathbf{B}=(\mathbf{I}-\mathbf{1s}')\mathbf{D}(\mathbf{I}-\mathbf{s1}') \tag{A.49}$$

where $\mathbf{s}'\mathbf{1}=1$. When $\mathbf{B}$ is p.s.d then we may write

$$\mathbf{B}=\mathbf{YY}' \tag{A.50}$$

non-uniquely. Regarding the ith row of $\mathbf{Y}$ as the coordinates of a point, then the ddistance between the ith and jth points is given by

$$b_{ii}+b_{jj}-2b_{ij}$$

From (A.49) it follows after some elementary algebraic manipulations that, provided we define $\mathbf{D}=\{-\frac{1}{2}d_{ij}^2\}$:

$$b_{ii}+b_{jj}-2b_{ij}=d_{ij}^2 \tag{A.51}$$

Thus, that (A.49) is p.s.d. guarantees that the coordinates $\mathbf{Y}$ are real and generate the distances d_{ij} (and ddistances d_{ij}^2) thus providing a sufficient condition for the distancs d_{ij} to have a Euclidean representation. That (A.49) is p.s.d. is also a necessary condition for Euclideanarity; an elementary proof is given by Gower (1982). When the distances have such a Euclidean representation, they are said to be **Euclidean embeddable**.

$\boldsymbol{D}$ is embeddable in a Euclidean space if and only if (A.49) is positive semi-definite.

From (A.49), we have that $\mathbf{s}'\mathbf{Bs} = 0$, so the vector $\mathbf{s}'\mathbf{Y} = 0$, giving the centring of the coordinates $\mathbf{Y}$; the different solutions $\mathbf{Y}$ represent different orientations around the centre defined by $\mathbf{s}$. The squares of distances of the n points from the origin are the elements of diag ($\mathbf{B}$), which may be written as a vector $\mathbf{d}$ given by

$$\mathbf{d} = (\mathbf{s}'\mathbf{Ds})\mathbf{1} - 2\mathbf{Ds} \tag{A.52}$$

It follows from (A.52) that the sum-of-squares of the distances from the origin is $\mathbf{1}'\mathbf{d}$ and when the origin is at the centroid of the n points, so that $s = (1/n)\mathbf{1}$, this gives

$$\mathbf{1}'\mathbf{d} = -\frac{1}{n}\mathbf{1}'\mathbf{D1} = \frac{1}{n}\sum_{i<j}^{n} d_{ij}^2$$

A.5.2 Principal coordinates analysis (PCO) (Gower, 1966a)

Of special importance is the choice $\mathbf{s} = \mathbf{1}/n$ which implies that $\mathbf{1}'\mathbf{Y} = 0$ and hence places the origin at the centroid of the generating points. Writing $\mathbf{N} = \mathbf{11}'/n$, then (A.49) becomes

$$\mathbf{B} = (\mathbf{I} - \mathbf{N})\mathbf{D}(\mathbf{I} - \mathbf{N}) \tag{A.53}$$

and, in particular, we require the spectral decomposition

$$\mathbf{B} = \mathbf{YY}' \quad \text{and} \quad \mathbf{Y}'\mathbf{Y} = \mathbf{\Lambda} \tag{A.54}$$

Because $\mathbf{Y}'\mathbf{Y}$ is diagonal, $\mathbf{Y}$ is referred to its principal axes and it follows that $\mathbf{Y}_\rho$, the first ρ-columns of $\mathbf{Y}$, gives a ρ-dimensional PCA of $\mathbf{Y}$.

The rows of settings of $\mathbf{Y}$ *which satisfy (A.53) and (A.54) are known as the principal coordinates of* $\mathbf{D}$.

The **centroid ddistances,** corresponding to (A.52) are

$$\mathbf{d} = \frac{1}{n^2}(\mathbf{1}'\mathbf{D1})\mathbf{1} - \frac{2}{n}\mathbf{D1} \tag{A.55}$$

Note that from (A.55) it follows that the sum-of-squares about the mean (centroid) is

$$\mathbf{1}'\mathbf{d} = -\tfrac{1}{n}\mathbf{1}'\mathbf{D1} = \tfrac{1}{n}\sum_{i<j}^{n} d_{ij}^2 \tag{A.56}$$

We may write (A.54) as

$$\mathbf{B} = \mathbf{YY}' = \mathbf{V\Sigma}^2\mathbf{V}' = \mathbf{V\Lambda V}'$$

where $\mathbf{\Sigma}$ and $\mathbf{\Lambda}$ are, respectively, diagonal matrices of the singular values and eigenvalues of $\mathbf{B}$ and the columns of $\mathbf{V}$ are the eigenvectors. Thus $\mathbf{Y}=\mathbf{V\Sigma}$ and the Moore–Penrose generalized inverse of $\mathbf{B}$ is

$$\mathbf{B}^{-}=\mathbf{V}(\mathbf{\Sigma}^{-})^{2}\mathbf{V}'=\mathbf{V\Sigma}(\mathbf{\Sigma}^{-})^{4}\mathbf{\Sigma V}'=\mathbf{Y}(\mathbf{\Lambda}^{-})^{2}\mathbf{Y}'$$

which is usually written

$$\mathbf{B}^{-}=\mathbf{Y\Lambda}^{-2}\mathbf{Y}' \tag{A.57}$$

We also have

$$(\mathbf{I}-\mathbf{Y\Lambda}^{-1}\mathbf{Y}')\mathbf{B}=(\mathbf{I}-\mathbf{Y\Lambda}^{-1}\mathbf{Y}')\mathbf{YY}'=0 \tag{A.58}$$

where $\mathbf{\Lambda}^{-1}$ remains a Moore–Penrose generalized inverse.

Equation (A.58) shows that the columns of $\mathbf{I}-\mathbf{Y\Lambda}^{-1}\mathbf{Y}'$ are in the null space of $\mathbf{B}$. The matrix $\mathbf{B}_k$, derives from $\mathbf{D}_k$ which is a special case of $\mathbf{D}$ and because $\mathbf{D}$ is assumed to be Euclidean embeddable then so is $\mathbf{D}_k$. It follows that for any vector $\mathbf{m}$ we have $\mathbf{m}'\mathbf{B}_k\mathbf{m}\geqslant 0$ and if $\mathbf{m}$ is a null vector of $\mathbf{B}$ then

$$\mathbf{m}'\mathbf{B}\mathbf{m}=\sum_{k=1}^{p}\mathbf{m}'\mathbf{B}_k\mathbf{m}=0$$

and hence that $\mathbf{m}'\mathbf{B}_k\mathbf{m}=0$ for $k=1,2,\ldots,p$. Now the columns of $\mathbf{I}-\mathbf{Y\Lambda}^{-1}\mathbf{Y}'$ span the null vectors of $\mathbf{B}$ and thus

$$(\mathbf{I}-\mathbf{Y\Lambda}^{-1}\mathbf{Y}')\mathbf{B}_k=0 \quad \text{for } k=1,2,\ldots,p \tag{A.59}$$

A.6 Partitioned distance matrices and distances between centroids

A.6.1 Distance between the centroids of two clouds of points (Digby and Gower, 1981)

Suppose a ddistance matrix is partitioned into two parts of size n_1 and n_2, *respectively, to give*

$$\begin{pmatrix} \boldsymbol{D}_{11} & \boldsymbol{D}'_{12} \\ \boldsymbol{D}_{21} & \boldsymbol{D}_{22} \end{pmatrix}$$

then the ddistance between the centroids of the points that generate $\boldsymbol{D}_{11}$ *and* $\boldsymbol{D}_{22}$, *respectively, is given by*

$$\bar{\boldsymbol{D}}_{11}+\bar{\boldsymbol{D}}_{22}-2\bar{\boldsymbol{D}}_{12} \tag{A.60}$$

where

$$\bar{D}_{ij} = \frac{1}{n_i n_j}(1'_i D_{ij} 1_j) \quad \text{for } i,j = 1,2.$$

The proof is straightforward. Suppose $\mathbf{D}_{11}$ is generated by coordinates $\mathbf{Y}_1$ with centroid $\mathbf{y}_1 = \mathbf{1}'_1\mathbf{Y}_1/n_1$ and $\mathbf{D}_{22}$ is generated by coordinates $\mathbf{Y}_2$ with centroid $\mathbf{y}_2 = \mathbf{1}'_2\mathbf{Y}_2/n_2$. Let $\mathbf{E}_1 = \text{diag}\,(\mathbf{Y}_1\mathbf{Y}'_1)$ and $\mathbf{E}_2 = \text{diag}\,(\mathbf{Y}_2\mathbf{Y}'_2)$; then it follows from (A.51) that

$$-2\mathbf{D}_{12} = \mathbf{E}_1\mathbf{1}_1\mathbf{1}'_2 + \mathbf{1}_1\mathbf{1}'_2\mathbf{E}_2 - 2\mathbf{Y}_1\mathbf{Y}'_2$$

Hence

$$-2\mathbf{1}'_1\mathbf{D}_{12}\mathbf{1}_2 = n_2\mathbf{1}'_1\mathbf{E}_1\mathbf{1}_1 + n_1\mathbf{1}'_2\mathbf{E}_2\mathbf{1}_2 - 2n_1n_2\mathbf{y}_1\mathbf{y}'_2.$$

This implies that

$$(\mathbf{y}_1 - \mathbf{y}_2)(\mathbf{y}_1 - \mathbf{y}_2)' = \bar{\mathbf{D}}_{11} + \bar{\mathbf{D}}_{22} - 2\bar{\mathbf{D}}_{12}.$$

which is the result to be proved.

A.6.2 Distance between the points of one set and the centroid of a second set

Equation (A.60) is a useful result with several applications. We require the special case which gives the ddistances of the centroid of the second set of n_2 points from each of the n_1 points in the first set. Consider P_i, the ith point of the first set, so that $n_1 = 1$. Then the partitioned matrix takes the form

$$\begin{pmatrix} 0 & \mathbf{d}'_{i1} \\ \mathbf{d}_{i1} & \mathbf{D}_{22} \end{pmatrix}$$

where $\mathbf{d}_{i1}$ is the ith column of $\mathbf{D}_{12}$. Then, (A.60) gives the ddistances of P_i from the centroid of the second set as

$$\frac{\mathbf{1}'\mathbf{D}_{22}\mathbf{1}}{n_2^2} - \tfrac{2}{n_2}\mathbf{1}'\mathbf{d}_{i1} \tag{A.61}$$

All n_1 such ddistances are members of the vector

$$\mathbf{g} = \left(\frac{\mathbf{1}'\mathbf{D}_{22}\mathbf{1}}{n_2^2}\right)\mathbf{1} - \tfrac{2}{n_2}\mathbf{D}_{12}\mathbf{1} \tag{A.62}$$

Equation (A.62) may also be written

$$\mathbf{g} = \bar{\mathbf{D}}_{22}\mathbf{1} - \tfrac{2}{n_2}\mathbf{D}_{12}\mathbf{1} \tag{A.63}$$

Figure A.1 shows the vector $\mathbf{g}$ diagramatically.

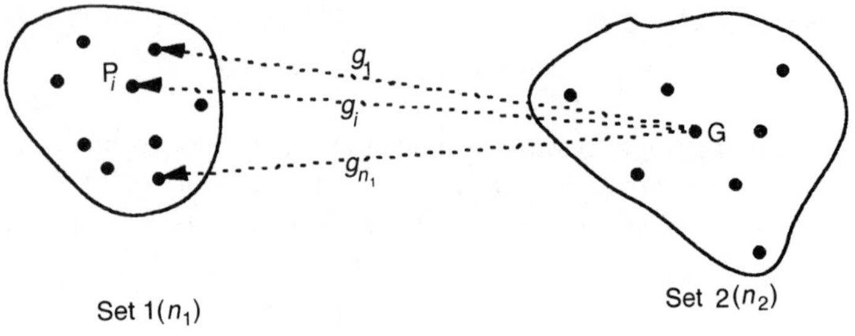

Fig. A.1. *Set 1 is of* n_1 *points and Set 2 of* n_2 *points. G is the centroid of Set 2 and* P_i *is a typical element of Set 1. The diagram shows three of the elements of the vector* **g**.

A.7 Superimposition of a point (interpolation)

A.7.1 Interpolation of one point (Gower, 1968)

We are given the coordinates of n points as the rows of an $n \times m$ *matrix* $\boldsymbol{Y}$*, centred at the origin G, that generates a ddistance matrix* $\boldsymbol{D}$*. A further* $(n+1)$*th point* P_{n+1} *is given only by its ddistances*

$$\boldsymbol{d}_{n+1} = \left\{-\frac{1}{2}d^2_{n+1,i}\right\}$$

from the first set of n points. We show that the coordinates $\boldsymbol{y}$ *of* P_{n+1} *are*

$$\boldsymbol{Y} = (\boldsymbol{Y}'\boldsymbol{Y})^{-1}\boldsymbol{Y}'\left(\boldsymbol{d}_{n+1} - \frac{1}{n}\boldsymbol{D}\boldsymbol{1}\right) \qquad \text{(A.64)}$$

in the first m dimensions plus a coordinate y_{m+1} *in a further dimension given by*

$$y^2_{m+1} = \bar{\boldsymbol{D}} - \frac{2}{n}\boldsymbol{1}'\boldsymbol{d}_{n+1} - \boldsymbol{y}'\boldsymbol{y} \qquad \text{(A.65)}$$

Figure A.2 illustrates the geometry and notation.

Thus $\mathbf{D}$ is bordered to give the ddistance matrix

$$\begin{pmatrix} \mathbf{D} & \mathbf{d}_{n+1} \\ \mathbf{d}'_{n+1} & 0 \end{pmatrix}$$

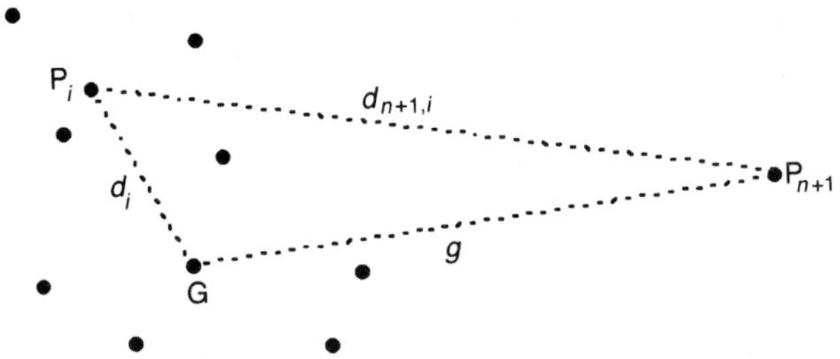

Fig. A.2. *G is the centroid of n points of which P_i is typical. P_{n+1} is the point to be superimposed; g and d_{n+1} are the lengths of the indicated distances, and d_i is the ith centroid distance.*

By definition, the coordinates $(y_1, y_2, \ldots, y_m, y_{m+1}) = (\mathbf{y}, y_{m+1})$ of P_{n+1} satisfy

$$d^2_{n+1,i} = \sum_{k=1}^{m+1} (y_k - y_{ik})^2$$

where $y_{i,m+1} = 0, (i = 1, 2, \ldots, n)$. Now,

$$\sum_{k=1}^{m+1} y_k^2$$

is the ddistance g of P_{n+1} from the centroid G of $\mathbf{Y}$, which is given by (A.61);

$$\sum_{k=1}^{m+1} y_{ik}^2$$

is the sum of the ddistances of the ith point P_i from G and this is given by (A.55). Collecting the elements of $d^2_{n+1,i}$ to form the vector $\mathbf{d}_{n+1}$ and making these substitutions gives

$$\mathbf{d}_{n+1} = -\frac{1}{2}\left[\bar{\mathbf{D}}\mathbf{1} - \tfrac{2}{n}(\mathbf{d}_{n+1}\mathbf{1}) + \frac{1}{n}\bar{\mathbf{D}}\mathbf{1} - \frac{2}{n}\mathbf{D}\mathbf{1} - 2\mathbf{Y}\mathbf{y}\right]$$

$$= -\frac{1}{2}\left[\bar{\mathbf{D}} - \tfrac{2}{n}(\mathbf{d}_{n+1}\mathbf{1}) + \frac{1}{n}\bar{\mathbf{D}}\right]\mathbf{1} + \frac{1}{n}\mathbf{D}\mathbf{1} + \mathbf{Y}\mathbf{y}$$

Because $\mathbf{Y}'\mathbf{1} = 0$ and the quantity in square brackets is a scalar, the previous equation may be rearranged to give

$$\mathbf{y} = (\mathbf{Y}'\mathbf{Y})^{-1}\mathbf{Y}'\left(\mathbf{d}_{n+1} - \frac{1}{n}\mathbf{D}\mathbf{1}\right) \qquad \text{(A.66)}$$

Equation (A.66) is the basic result required, but when **Y** is referred to principal coordinates (A.54), the result simplifies to

$$\mathbf{y} = \mathbf{\Lambda}^{-1}\mathbf{Y}'\left(\mathbf{d}_{n+1} - \frac{1}{n}\mathbf{D1}\right) \tag{A.67}$$

Note that the first ρ dimensions of **y** are given in an obvious notation as

$$\mathbf{y}_\rho = \mathbf{\Lambda}_\rho^{-1}\mathbf{Y}'_\rho\left(\mathbf{d}_{n+1} - \frac{1}{n}\mathbf{D1}\right) \tag{A.68}$$

The remaining dimension is given by the difference between the ddistance g of p_{n+1} from G and the ddistance already fitted by (A.67), which immediately leads to (A.65).

A.7.2 Interpolation of n_2 points

Suppose now that **D** is relabelled $\mathbf{D}_{11}$ of order n_1 and that n_2 points are added to give the ddistance matrix

$$\begin{pmatrix} \mathbf{D}_{11} & \mathbf{D}_{12} \\ \mathbf{D}_{21} & \mathbf{D}_{22} \end{pmatrix}$$

$\mathbf{D}_{11}$ and **Y** are given, as before but now n_2 points are superimposed with known distances from each of the basic n_1 points given in $\mathbf{D}_{12}$. Then the coordinates of the added points are given by replacing $\mathbf{d}_{n+1}$ in (A.67) by successive rows of $\mathbf{D}_{12}$ and averaging to give

$$\mathbf{y} = \frac{1}{n_2}\mathbf{\Lambda}^{-1}\mathbf{Y}'\left(\mathbf{D}_{12}\mathbf{1} - \frac{1}{n_1}\mathbf{D}_{11}\mathbf{1}\right) \tag{A.69}$$

The ddistance y_{m+1} between the two centroids is given by (A.60) and so the extra dimension is now

$$y_{m+1}^2 = \bar{\mathbf{D}}_{11} + \bar{\mathbf{D}}_{22} - 2\bar{\mathbf{D}}_{12} - \mathbf{y}'\mathbf{y} \tag{A.70}$$

A.8 Orthogonal projection

We consider a point **x** in $\mathscr{R}_n$ with its orthogonal projection **y** onto a ρ-dimensional subspace spanned by vectors given as independent columns of a matrix **L**. Because **y** lies in $\mathscr{L}$ there exist constants, not all zero, which may be assembled into a vector **c** such that

$$\mathbf{y} = \mathbf{Lc}$$

Because of the orthogonality

$$\mathbf{L}'(\mathbf{y} - \mathbf{x}) = 0$$

It follows that

$$\mathbf{L}'\mathbf{Lc} = \mathbf{L}'\mathbf{x}$$

and hence

$$\mathbf{y} = \mathbf{L}(\mathbf{L}'\mathbf{L})^{-1}\mathbf{L}'\mathbf{x} \tag{A.71}$$

Equation (A.71) is the classical result for orthogonal projection and gives the ρ-dimensional coordinates of the projection in $\mathscr{L}$ relative to the coordinate system of $\mathscr{R}_n$. The matrix $\mathbf{H} = \mathbf{L}(\mathbf{L}'\mathbf{L})^{-1}\mathbf{L}'$ is idempotent ($\mathbf{H}^2 = \mathbf{H}$), showing that repeated projections have no further effect, expressing the obvious geometric fact that projected points stay where they are. When the columns of **L** represent an orthogonal basis in $\mathscr{L}$, then $\mathbf{L}'\mathbf{L} = \mathbf{I}$ and (A.7) simplifies to

$$\mathbf{y} = \mathbf{LL}'\mathbf{x} \tag{A.72}$$

In (A.71) it is assumed that the inverse exists. When $\mathbf{L}'\mathbf{L}$ is a singular matrix, then (A.71) may be replaced by

$$\mathbf{y} = \mathbf{L}(\mathbf{L}'\mathbf{L})^{-}\mathbf{L}'\mathbf{x} \tag{A.73}$$

in which $(\mathbf{L}'\mathbf{L})^{-}$ is any generalized inverse (g-inverse) of $\mathbf{L}'\mathbf{L}$. The transformation (A.73) remains idempotent and continues to represent a unique projection irrespective of the particular choice of g-inverse. Uniqueness is obvious geometrically and may be shown algebraically by writing **L** in its SVD form $\mathbf{L} = {}_n\mathbf{U}_o\boldsymbol{\Sigma}_o\mathbf{V}'_\rho$ where $\sigma < \rho$. All the g-inverses $\mathbf{A}^{=}$ of a matrix **A** may be expressed in terms of a particular g-inverse $\mathbf{A}^{-}$ as

$$\mathbf{A}^{=} = \mathbf{A}^{-} + (\mathbf{I} - \mathbf{AA}^{-})\mathbf{C} + \mathbf{D}(\mathbf{I} - \mathbf{AA}^{-})$$

where **C** and **D** are arbitrary. Taking the Moore–Penrose inverse

$$(\mathbf{L}'\mathbf{L})^{-} = {}_\rho\mathbf{V}_\sigma\boldsymbol{\Sigma}_\sigma^{-2}\mathbf{V}'_\rho$$

shows that $\mathbf{L}(\mathbf{L}'\mathbf{L})^{=}\mathbf{L}' = \mathbf{UU}'$ and hence that (A.7) is invariant to the particular choice of g-inverse.

We are also interested in the projection expressed in terms of the coordinate system of $\mathscr{L}$ itself. The matrix **L** defines the directions of coordinate axes in $\mathscr{L}$ which is embedded in $\mathscr{R}$. Writing

$$\mathbf{y}_{\mathscr{L}} = (\mathbf{L}'\mathbf{L})^{-1}\mathbf{L}'\mathbf{x} \tag{A.74}$$

then (A.74) becomes $\mathbf{y} = \mathbf{L}\mathbf{y}_{\mathscr{L}}$ showing that $\mathbf{y}_{\mathscr{L}}$ defines the linear combination of the columns of $\mathbf{L}$ that gives $\mathbf{y}$. It follows that $\mathbf{y}_{\mathscr{L}}$ gives the coordinates of the projection relative to the directions $\mathbf{L}$. Normally, $\mathbf{L}$ defines non-orthogonal axes but when $\mathbf{L}$ is orthonormal, corresponding to (A.72), (A.74) simplifies to

$$\mathbf{y}_{\mathscr{L}} = \mathbf{L}'\mathbf{x} \tag{A.75}$$

As well As (A.71), (A.72) and (A.73), we also refer to (A.74) and (A.75) as defining an orthogonal projection even though the corresponding transformations are not idempotent; indeed, they are not even represented by a square matrix. Nevertheless, they define the same geometric point; only the reference axes differ.

A.8.1 Orthogonal projection with an offset

In the above, $\mathscr{L}$ represents a vector space and therefore contains the origin. We are also interested in orthogonal projections where $\mathscr{L}$ is offset from the origin by some vector $\mathbf{p}$, say. Thus, any point in $\mathscr{L}$ has the form $\mathbf{y} = \mathbf{p} + \mathbf{L}\boldsymbol{\lambda}$ where $\boldsymbol{\lambda}$ is arbitrary. The nearest point $\mathbf{y}$ in $\mathscr{L}$ to $\mathbf{x}$ minimizes $\|\mathbf{x} - \mathbf{p} - \mathbf{L}\boldsymbol{\lambda}\|$ which (A.71) gives as $\mathbf{L}\boldsymbol{\lambda} = \mathbf{L}(\mathbf{L}'\mathbf{L})^{-1}\mathbf{L}'(\mathbf{x} - \mathbf{p})$. Thus, $\mathbf{y}_{\mathscr{L}} = \mathbf{p} + \mathbf{L}(\mathbf{L}'\mathbf{L})^{-1}\mathbf{L}'(\mathbf{x} - \mathbf{p})$ which on rearrangement gives

$$\mathbf{y}_{\mathscr{L}} = \mathbf{L}(\mathbf{L}'\mathbf{L})^{-1}\mathbf{L}'\mathbf{x} + [\mathbf{I} - \mathbf{L}(\mathbf{L}'\mathbf{L})^{-1}\mathbf{L}']\mathbf{p} \tag{A.76}$$

and when $\mathbf{L}$ is orthonormal

$$\mathbf{y}_{\mathscr{L}} = \mathbf{L}\mathbf{L}'\mathbf{x} + [\mathbf{I} - \mathbf{L}\mathbf{L}']\mathbf{p} \tag{A.77}$$

Geometrically, (A.76) and (A.77) represent a projection of $\mathbf{x}$ onto a space parallel to $\mathscr{L}$ containing the origin, plus the projection of $\mathbf{p}$ onto the orthogonal complement of $\mathscr{L}$. Note that when $\mathbf{p}$ is chosen to be orthogonal to $\mathscr{L}$ then (A.76) and (A.77) simplify to (A.71) and (A.73) modified by the offset $\mathbf{p}$.

A.8.2 Orthogonal projection onto the intersection of two spaces

Suppose $\mathscr{L}_1$ and $\mathscr{L}_2$ are two spaces. We require the projection of $\mathbf{x}$ onto $\mathscr{L}_1 \cap \mathscr{L}_2$. We shall assume that these spaces are spanned by the columns of $\mathbf{L}_1$ and $\mathbf{L}_2$ and that $\mathbf{N}_1$ and $\mathbf{N}_2$ are their orthogonal complements (i.e. $\mathbf{L}_1\mathbf{L}_1' + \mathbf{N}_1\mathbf{N}_1' = \mathbf{L}_2\mathbf{L}_2' + \mathbf{N}_2\mathbf{N}_2' = \mathbf{I}$ where, without loss of generality, all these matrices are now assumed orthonormal). $\mathscr{L}_1 \cap \mathscr{L}_2$ is spanned by vectors orthogonal to $\mathbf{N}_1$ and $\mathbf{N}_2$. Independent vectors in the spanning space are the columns of $\mathbf{N}_1$ and that part of $\mathbf{N}_2$ that is independent of $\mathbf{N}_1$, i.e $(\mathbf{I} - \mathbf{N}_1\mathbf{N}_1')\mathbf{N}_2 = \mathbf{L}_1\mathbf{L}_1'\mathbf{N}_2$. Thus projections onto $\mathscr{L}_1 \cap \mathscr{L}_2$ are given by

$$\mathbf{I} - \mathbf{N}_1\mathbf{N}_1' - (\mathbf{L}_1\mathbf{L}_1')\mathbf{N}_2(\mathbf{N}_2'\mathbf{L}_1\mathbf{L}_1'\mathbf{N}_2)^{-}\mathbf{N}_2'(\mathbf{L}_1\mathbf{L}_1')$$

which becomes

$$\left.\begin{array}{l} \mathbf{L}_1\mathbf{L}_1' - (\mathbf{L}_1\mathbf{L}_1')\mathbf{N}_2(\mathbf{N}_2'\mathbf{L}_1\mathbf{L}_1'\mathbf{N}_2)^{-}\mathbf{N}_2'(\mathbf{L}_1\mathbf{L}_1') \\ \text{or} \\ \mathbf{L}_2\mathbf{L}_2' - (\mathbf{L}_2\mathbf{L}_2')\mathbf{N}_1(\mathbf{N}_1'\mathbf{L}_2\mathbf{L}_2'\mathbf{N}_1)^{-}\mathbf{N}_1'(\mathbf{L}_2\mathbf{L}_2') \end{array}\right\} \quad (A.78)$$

where the alternative form, with the suffix numbers interchanged, is obtained by reversing the roles of $\mathbf{N}_1$ and $\mathbf{N}_2$ when determining the independent spanning vectors.

In (A.78), we have used a generalized inverse to allow for the possibility of common dimensions in the two sets of spanning vectors; in many applications of the formula, a unique ordinary inverse will exist.

A.9 Back projection

Similarly to Section A.8, we consider a point $\boldsymbol{x}$ *in* $\mathscr{R}_n$. $\boldsymbol{x} \in \mathscr{M}$ and $\mathscr{N}$ is *normal to* $\mathscr{M}$ *at* $\boldsymbol{x}$. *The basic result that we require is the orthogonal projection* $\mathbf{y}$ *of* $\boldsymbol{x}$ *onto the intersection of an r-dimensional space* $\mathscr{L}$ *and an* s*-dimensional space* $\mathscr{N}$. *Thus,* $\mathbf{y}$ *is the point in* $\mathscr{L} \cap \mathscr{N}$ *that is nearest* $\boldsymbol{x}$. *We assume that the origin is in* $\mathscr{L}$ *and that* $\mathscr{M}$ *has an orthogonal displacement* $\boldsymbol{q}$ *from the origin. Figure A.3. shows the geometry. The basic results are given by (A.83) or its alternative (A.85), when* $\mathscr{M}$ *is one-dimensional and/or contains the origin, these simplify to (A.86), (A.88) and (A.89).*

We assume that $\mathscr{L}$ is spanned by orthonormal vectors that are columns of $\mathbf{L}$ and that $\mathscr{N}$ is spanned by orthonormal vectors $\mathbf{N}$. We shall also require the orthogonal complements $\mathbf{K}$ and $\mathbf{M}$

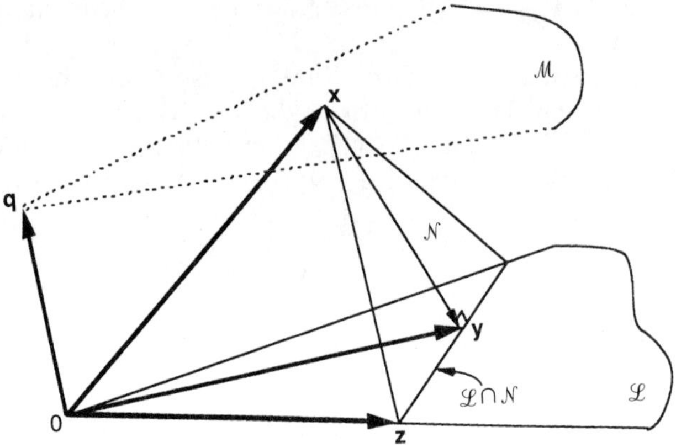

Fig. A.3. *The geometry of back-projection.* **y** *is the back-projection of* **x** *in* $\mathcal{L}$; **q** *is the offset of* $\mathcal{M}$.

defined such that

$$\mathbf{KK}' + \mathbf{LL}' = \mathbf{I} \quad \text{and} \quad \mathbf{MM}' + \mathbf{NN}' = \mathbf{I} \tag{A.79}$$

where $\mathbf{K}'\mathbf{K}$, $\mathbf{L}'\mathbf{L}$, $\mathbf{M}'\mathbf{M}$ and $\mathbf{N}'\mathbf{N}$ are all unit matrices of appropriate sizes and $\mathbf{K}'\mathbf{L} = \mathbf{M}'\mathbf{N} = 0$.

Thus, that $\mathbf{y} \in \mathcal{L}$ and $\mathbf{y} \in \mathcal{N}$ give, respectively

$$\mathbf{yKK}' = \mathbf{0} \quad \text{and} \quad \mathbf{yMM}' = \mathbf{xMM}' \tag{A.80}$$

Thus the problem is to find **y** that is the projection of **x** onto $\mathcal{L} \cap \mathcal{N}$ and this may be obtained by combining (A.77) and (A.78). We may set $\mathbf{L}_1 = \mathbf{L}$, $\mathbf{L}_2 = \mathbf{N}$, $\mathbf{N}_1 = \mathbf{K}$ and $\mathbf{N}_2 = \mathbf{M}$ and use the first form of (A.78) to give the matrix **P** for projections onto $\mathcal{L} \cap \mathcal{N}$ as

$$\mathbf{P} = \mathbf{LL}' - \mathbf{LL}'\mathbf{M}(\mathbf{M}'\mathbf{LL}'\mathbf{M})^{-}\mathbf{M}'\mathbf{LL}' \tag{A.81}$$

To apply (A.78) requires an offset **p** for $\mathcal{L} \cap \mathcal{N}$. Any point **p** satisfies

$$\mathbf{M}'\mathbf{p} = \mathbf{M}'\mathbf{x} \quad \text{and} \quad \mathbf{K}'\mathbf{p} = 0 \tag{A.82}$$

Inserting (A.81) and (A.83) into (A.76) gives the back projection **y** as

$$\mathbf{y} = \mathbf{Px} + (\mathbf{I} - \mathbf{P})\mathbf{p}$$

Now using the orthogonality relationships (A.79), $\mathbf{K'L} = 0$ and (A.82) we have

$$\begin{aligned}
\mathbf{y} &= [\mathbf{LL'} - \mathbf{LL'M}(\mathbf{M'LL'M})^{-}\mathbf{M'LL'}]\mathbf{x} \\
&\quad + [\mathbf{KK'} + \mathbf{LL'M}(\mathbf{M'LL'M})^{-}\mathbf{M'LL'}]\mathbf{p} \\
&= [\mathbf{LL'} - \mathbf{LL'M}(\mathbf{M'LL'M})^{-}\mathbf{M'LL'}]\mathbf{x} \\
&\quad + [\mathbf{LL'M}(\mathbf{M'LL'M})^{-}\mathbf{M'}]\mathbf{x} \\
&= [\mathbf{LL'} + \mathbf{LL'M}(\mathbf{M'LL'M})^{-}\mathbf{M'KK'}]\mathbf{x} \\
&= \mathbf{LL'}[\mathbf{I} + \mathbf{M}(\mathbf{M'LL'M})^{-}\mathbf{M'KK'}]\mathbf{x} \qquad \text{(A.83)}
\end{aligned}$$

This is the basic result, but an alternative expression derives from the second form of (A.78). This gives the projection matrix onto the intersection as

$$\mathbf{y} = \mathbf{Qx} + (\mathbf{I} - \mathbf{Q})\mathbf{p}$$

where

$$\mathbf{Q} = \mathbf{NN'} - \mathbf{NN'K}(\mathbf{K'NN'K})^{-}\mathbf{K'NN'} \qquad \text{(A.84)}$$

Making the substitutions gives

$$\begin{aligned}
\mathbf{y} &= [\mathbf{NN'} - \mathbf{NN'K}(\mathbf{K'NN'K})^{-}\mathbf{K'NN'}]\mathbf{x} \\
&\quad + [\mathbf{MM'} + \mathbf{NN'K}(\mathbf{K'NN'K})^{-}\mathbf{K'NN'}]\mathbf{p} \\
&= [\mathbf{NN'} - \mathbf{NN'K}(\mathbf{K'NN'K})^{-}\mathbf{K'NN'}]\mathbf{x} \\
&\quad + [\mathbf{MM'} - \mathbf{NN'K}(\mathbf{K'NN'K})^{-}\mathbf{K'MM'}]\mathbf{x} \\
&= [\mathbf{I} - \mathbf{NN'K}(\mathbf{K'NN'K})^{-}\mathbf{K'}]\mathbf{x} \qquad \text{(A.85)}
\end{aligned}$$

When $\mathscr{M}$ is one-dimensional, as with nonlinear biplots, (A.83) simplifies to

$$\mathbf{y} = \left[\mathbf{LL'} + \frac{(\mathbf{LL'm})(\mathbf{KK'm})'}{\mathbf{m'LL'm}}\right]\mathbf{x} \qquad \text{(A.86)}$$

Of special interest is when $\mathbf{M}$ contains the origin. Now $\mathbf{MM'x} = \mathbf{x}$ and (A.83) may be written

$$\begin{aligned}
\mathbf{y} &= \mathbf{LL'}[\mathbf{MM'} + \mathbf{M}(\mathbf{M'LL'M})^{-}\mathbf{M'}(\mathbf{I} - \mathbf{LL'})'\mathbf{MM'}]\mathbf{x} \\
&= \mathbf{LL'MR^{-}M'x} + \mathbf{r} \qquad \text{(A.87)}
\end{aligned}$$

where

$$\mathbf{R} = \mathbf{M'LL'M} \quad \text{and} \quad \mathbf{r} = \mathbf{LL'M}[\mathbf{I} - \mathbf{R^{-}R}]\mathbf{M'x}$$

Then

$$\mathbf{r}'\mathbf{r} = \mathbf{x}'\mathbf{M}[\mathbf{I} - \mathbf{R}^{-}\mathbf{R}]'\mathbf{M}'\mathbf{L}\mathbf{L}'\mathbf{L}\mathbf{L}'\mathbf{M}[\mathbf{I} - \mathbf{R}^{-}\mathbf{R}]\mathbf{M}'\mathbf{x}$$
$$= \mathbf{x}'\mathbf{M}[\mathbf{I} - \mathbf{R}^{-}\mathbf{R}]'\mathbf{R}[\mathbf{I} - \mathbf{R}^{-}\mathbf{R}]\mathbf{M}'\mathbf{x} = 0$$

Hence $\mathbf{r} = 0$ and (A.87) becomes

$$\mathbf{y} = \mathbf{L}\mathbf{L}'\mathbf{M}(\mathbf{M}'\mathbf{L}\mathbf{L}'\mathbf{M})^{-}\mathbf{M}'\mathbf{x} \tag{A.88}$$

The special case of (A.88) when $\mathscr{M}$ is one-dimensional corresponds to a Cartesian coordinate axis. Then, replacing $\mathbf{M}$ by a vector $\mathbf{m}$ and recalling that $\mathbf{m}\mathbf{m}'\mathbf{x} = \mathbf{x}$, gives

$$\mathbf{y} = \frac{\mathbf{L}\mathbf{L}'\mathbf{x}}{\mathbf{m}'\mathbf{L}\mathbf{L}'\mathbf{m}} \tag{A.89}$$

which is the most simple example of back-projection and represents a scaled orthogonal projection onto $\mathscr{L}$.

A.9.1 Orthogonality relationships

There are several orthogonalities between the vectors of Fig. A.3. We shall show that

$$\begin{aligned} \mathbf{x} - \mathbf{z} &\perp \mathbf{x} - \mathbf{q} && \text{(i)} \\ \mathbf{x} - \mathbf{y} &\perp \mathbf{x} - \mathbf{q} && \text{(ii)} \\ \mathbf{y} - \mathbf{z} &\perp \mathbf{x} - \mathbf{q} && \text{(iii)} \\ \mathbf{y} - \mathbf{z} &\perp \mathbf{y} - \mathbf{q} && \text{(iv)} \\ \mathbf{y} - \mathbf{z} &\perp \mathbf{y} - \mathbf{x} && \text{(v)} \end{aligned} \tag{A.90}$$

where now $\mathbf{q}$ is taken to be the projection from the origin onto $\mathscr{M}$. Thus $\mathbf{x} - \mathbf{q} = \mathbf{M}\mathbf{M}'\mathbf{x}$. Also $\mathbf{z}$ is an arbitrary point in $\mathscr{L} \cap \mathscr{N}$, and hence $\mathbf{M}\mathbf{M}'\mathbf{z} = \mathbf{M}\mathbf{M}'\mathbf{x}$. Most of these are obvious on geometrical grounds. For example, in (i), (ii) and (iii) all the left-hand vectors lie in $\mathscr{N}$ and $\mathbf{x} - \mathbf{q}$ is in $\mathscr{M}$. Here we give algebraic proofs, more as a verification of the previous less obvious algebraic results rather than for their intrinsic interest.

To establish (i) we note that

$$(\mathbf{x} - \mathbf{z})'(\mathbf{x} - \mathbf{q}) = (\mathbf{x} - \mathbf{z})'\mathbf{M}\mathbf{M}'\mathbf{x}$$
$$= (\mathbf{M}\mathbf{M}'\mathbf{x} - \mathbf{M}\mathbf{M}'\mathbf{z})'\mathbf{x} = 0$$

To establish (ii), we note that from (A.85)

$$(\mathbf{x} - \mathbf{y})'(\mathbf{x} - \mathbf{q}) = \mathbf{x}'[\mathbf{K}(\mathbf{K}'\mathbf{N}\mathbf{N}'\mathbf{K})^{-}\mathbf{K}'\mathbf{N}\mathbf{N}']\mathbf{M}\mathbf{M}'\mathbf{x} = 0$$

Result (iii) is established by subtraction of (i) and (ii). Thus

$$(\mathbf{y}-\mathbf{z})'(\mathbf{x}-\mathbf{q})=(\mathbf{x}-\mathbf{z})'(\mathbf{x}-\mathbf{q})-(\mathbf{x}-\mathbf{y})'(\mathbf{x}-\mathbf{q})=0$$

To establish (iv), we start by noting that the setting of $\mathbf{p}$ is arbitrary, provided it lies in $\mathscr{L}\cap\mathscr{N}$; hence we may set $\mathbf{p}=\mathbf{z}$ and $\mathbf{y}=\mathbf{P}\mathbf{x}+(\mathbf{I}-\mathbf{P})\mathbf{z}$. Then $\mathbf{y}-\mathbf{z}=\mathbf{P}(\mathbf{x}-\mathbf{z})$ and

$$\begin{aligned}(\mathbf{y}-\mathbf{z})'(\mathbf{y}-\mathbf{q})&=(\mathbf{x}-\mathbf{z})'\mathbf{P}(\mathbf{P}\mathbf{x}+(\mathbf{I}-\mathbf{P})\mathbf{z}-\mathbf{N}\mathbf{N}'\mathbf{x})\\&=(\mathbf{x}-\mathbf{z})'(\mathbf{P}-\mathbf{P}\mathbf{N}\mathbf{N}')\mathbf{x}\\&=(\mathbf{x}-\mathbf{z})'\mathbf{P}\mathbf{M}\mathbf{M}'\mathbf{x}\end{aligned}$$

It is easily verified that $\mathbf{PM}=0$, thus establishing the result.

Result (v) is established by subtracting (iii) from (iv). Thus

$$(\mathbf{y}-\mathbf{z})'(\mathbf{y}-\mathbf{x})=(\mathbf{y}-\mathbf{z})'(\mathbf{y}-\mathbf{q})-(\mathbf{y}-\mathbf{z})'(\mathbf{x}-\mathbf{q})=0$$

Note that if this result is regarded as obvious on geometric grounds, then (iv) may be derived by adding (iii) to (v).

A.10 Procrustes problems

Procrustes problems seek to minimize over $\mathbf{T}$ an L_2-norm criterion $\|{}_n\mathbf{Z}_q-{}_n\mathbf{X}_p\mathbf{T}_q\|$. The form of the transformation matrix $\mathbf{T}$ varies for different Procrustes problem. We are concerned with only three in this book; see Gower (1984) for a brief review of Procrustes problems. Most Procrustes problems do not have solutions in algebraic closed form and two of our problems fall into this class, so we present algorithmic solutions.

A.10.1 Orthogonal Procrustes analysis

It is required to minimize $\|\mathbf{Z}-\mathbf{XQ}\|$ over orthogonal matrices $\mathbf{Q}$; hence $p=q$. We have

$$\begin{aligned}\|\mathbf{Z}-\mathbf{XQ}\|&=\operatorname{trace}(\mathbf{Z}-\mathbf{XQ})(\mathbf{Z}-\mathbf{XQ})'\\&=\operatorname{trace}(\mathbf{ZZ}'+\mathbf{XX}'-2\mathbf{ZQ}'\mathbf{X}')\end{aligned}$$

Thus we must maximize $\operatorname{trace}(\mathbf{ZQ}'\mathbf{X}')=\operatorname{trace}(\mathbf{Z}'\mathbf{XQ})$. Write $\mathbf{Z}'\mathbf{X}=\mathbf{U}\sum\mathbf{V}'$ in terms of its SVD, so we have to maximize

$$\begin{aligned}\operatorname{trace}(\mathbf{U}\Sigma\mathbf{V}'\mathbf{Q})&=\operatorname{trace}(\Sigma\mathbf{V}'\mathbf{Q}\mathbf{U})\\&=\operatorname{trace}(\Sigma\mathbf{H})\end{aligned}$$

where $\mathbf{H} = \mathbf{V}'\mathbf{Q}\mathbf{U}$ is an orthogonal matrix. Because $\mathbf{\Sigma}$ is non-negative and diagonal, trace$(\mathbf{\Sigma H})$ is maximized when $\mathbf{H} = \mathbf{I}$, i.e.

$$\mathbf{Q} = \mathbf{V}\mathbf{U}' \tag{A.91}$$

A.10.2 An algorithm for minimal projection error Procrustes analysis

It is required to find the projection matrix $\mathbf{P}$ which minimizes $\|\mathbf{Z} - \mathbf{XP}\|$, where now we assume $p > q$. Green and Gower (1979) proposed the following algorithm; see also Gower (1994).

(i) Add $p - q$ zero columns to $\mathbf{Z}$.
(ii) Find the orthogonal Procrustes solution $\mathbf{XQ}$ for $\mathbf{X}$ and the augmented version of $\mathbf{Z}$.
(iii) Replace the final $p - q$ columns of $\mathbf{Z}$ by the final $p - q$ columns of $\mathbf{XQ}$.
(iv) Stop if the residual sum-of-squares has stabilized, else return to step (ii) with the current versions of $\mathbf{Z}$ and $\mathbf{XQ}$ (which now replaces $\mathbf{X}$). (A.92)

At each step, the algorithm reduces the residual sum-of-squares, which is bounded below by zero. Hence convergence is guaranteed, though not necessarily to the global minimum. The first columns of $\mathbf{Z}$ remain unaltered throughout and the first q columns of $\mathbf{XQ}$ converge to the desired projection $\mathbf{XP}$; $\mathbf{P}$ is given by the first q columns of the orthogonal matrix $\mathbf{Q} = (\mathbf{P}, \bar{\mathbf{P}})$. The rationale of the algorithm can be summed up in the sequence

$$\begin{aligned}\|\mathbf{Z} - \mathbf{XP}\| &= (\mathbf{Z}, \mathbf{X}\bar{\mathbf{P}}) - \mathbf{X}(\mathbf{P}, \bar{\mathbf{P}})\| \\ &= \|(\mathbf{Z}, \mathbf{X}\bar{\mathbf{P}}) - \mathbf{XQ}\|\end{aligned}$$

where the right-hand term is an ordinary orthogonal Procrustes problem. Thus if the right-hand criterion is minimized over all orthogonal matrices $\mathbf{Q}$, then so is the left-hand criterion over all projection matrices $\mathbf{P}$.

A.10.3 Maximal group-average Procrustes analysis

It is required to find the projection matrix $\mathbf{P}$ which maximizes $\|\mathbf{Z} + \mathbf{XP}\|$, where we assume $p > q$. Note that this variant is a maximization problem. In orthogonal Procrustes analysis, the minimization of $||\mathbf{Z} - \mathbf{XQ}||$ and the maximization of $||\mathbf{Z} + \mathbf{XQ}||$ give the same result, both requiring the maximization of trace $(\mathbf{ZQ}'\mathbf{X}')$. This is not so when the orthogonal matrix $\mathbf{Q}$ is replaced by a projection matrix $\mathbf{P}$ so that minimal projection error Procrustes analysis and maximal group-average Procrustes analysis have different solutions. The following provides an algorithmic solution.

(i) Add $p - q$ zero columns to $\mathbf{Z}$.
(ii) Find the orthogonal Procrustes solution $\mathbf{XQ}$ for $\mathbf{X}$ and the augmented version of $\mathbf{Z}$.
(iii) Do a PCA of the current $(\mathbf{Z}, \mathbf{X})$ jointly. That is treat $\mathbf{Z}, \mathbf{X}$ as if it were a sample of size $2n$. Replace the first q columns of $\mathbf{Z}, \mathbf{X}$ by their corresponding PCA values.
(iv) Stop if the residual sum-of-squares has stabilised, else return to step (iii) with the current versions of $\mathbf{Z}$ and $\mathbf{X}$. (A.93)

A.11 Perron–Frobenius theorem

If $\mathbf{A}$ is a square matrix of non-negative elements then:

(i) The largest eigenvalue λ of $\mathbf{A}$ is positive, and
(ii) The left and right eigenvectors associated with λ have all positive elements.

It follows from (i) that when $\mathbf{A}$ is p.s.d. and symmetric then all its eigenvalues are non-negative and do not exceed λ. Because of the orthogonality of vectors all other eigenvectors must contain negative elements. Thus any eigenvector of $\mathbf{A}$ with all positive elements must correspond to the maximum eigenvalue of $\mathbf{A}$.

There is more to the Perron–Frobenius theorem than this but the above results are all that will be needed here. Proofs are difficult (Seneta, 1973). A direct consequence of the Perron–Frobenius theorem is as follows.

If $\boldsymbol{X} = \boldsymbol{U\Sigma V}'$ is the SVD (A.20) of a rectangular matrix and $\boldsymbol{XX}'$ and $\boldsymbol{X}'\boldsymbol{X}$ are both non-negative matrices, then the vectors $\boldsymbol{u}_1$ and $\boldsymbol{v}_1$ corresponding to the biggest singular value σ_1, are non-negative. Conversely, if $\boldsymbol{u}$ and $\boldsymbol{v}$ are a non-negative vector pair in the decomposition of $\boldsymbol{X}$, then the corresponding singular value is the maximum.

This result is useful when **X** is itself a non-negative matrix such as a contingency table.

A.12 Properties of the space $\mathscr{R}^+$

The space $\mathscr{R}^+$ is an ordinary m-dimensional Euclidean space $\mathscr{R}$, supplemented by a continuum of extra dimensions which are different for every point of the space but which may be subsumed into a single dimension. When $P_1(\mathbf{x}_1, x_1)$ and $P_2(\mathbf{x}_2, x_2)$ are two points in $\mathscr{R}^+$ then the coordinate values x_1 and x_2 refer to different extra dimensions which have zero values for all other points. Thus a more conventional coordinate representation would be $P_1(\mathbf{x}_1, x_1, 0, 0, 0, \ldots)$ and $P_2(\mathbf{x}_2, 0, x_2, 0, 0, \ldots)$ and the ddistance between P_1 and P_2 is given by

$$d_{12}^2 = (\mathbf{x}_1 - \mathbf{x}_2)'(\mathbf{x}_1 - \mathbf{x}_2) + x_1^2 + x_2^2 \qquad \text{(A.94)}$$

We are not normally interested in distance between two points in $\mathscr{R}^+$ but only between points in $\mathscr{R}$ and points in $\mathscr{R}^+$. If $P(\mathbf{x}, 0)$ is in $\mathscr{R}$ then its ddistance from P_1 is $(\mathbf{x} - \mathbf{x}_1)'(\mathbf{x} - \mathbf{x}_1) + x_1^2$.

In Chapter 7, we are interested in the intersection with $\mathscr{R}$ of the bisecting plane of P_1 and P_2. To investigate whether the coordinate representation affects the result, we express the condition that P is equidistant from P_1 and P_2 in both representations. In both cases we have

$$(\mathbf{x} - \mathbf{x}_1)'(\mathbf{x} - \mathbf{x}_1) + x_1^2 = (\mathbf{x} - \mathbf{x}_2)'(\mathbf{x} - \mathbf{x}_2) + x_2^2$$

representing the plane in $\mathscr{R}$, which simplifies to

$$2\mathbf{x}'(\mathbf{x}_1 - \mathbf{x}_2) = \mathbf{x}_1'\mathbf{x}_1 - \mathbf{x}_2'\mathbf{x}_2 + x_1^2 - x_2^2 \qquad \text{(A.95)}$$

In $\mathscr{R}^+$, the bisecting plane of P_1 and P_2 is normal to the direction $(\mathbf{x}_1 - \mathbf{x}_2, x_1, -x_2)$ and passes through their mid-point $\frac{1}{2}(\mathbf{x}_1 + \mathbf{x}_2, x_1, x_2)$ and hence has equation

$$2\mathbf{x}'(\mathbf{x}_1 - \mathbf{x}_2) + 2ux_1 - 2vx_2 = \mathbf{x}_1'\mathbf{x}_1 - \mathbf{x}_2'\mathbf{x}_2 + x_1^2 - x_2^2 \qquad \text{(A.96)}$$

where u and v refer to the extra dimensions. When $u = v = 0$, we have the intersection with $\mathscr{R}$ and then (A.96) coincides with (A.95).

For most practical purposes, these results justify treating the contracted coordinate representation of $\mathscr{R}^+$ as conventional Cartesian coordinates. This greatly simplifies things, justifying conventional algebraic presentation in methodology and algorithms.

A.13 Algorithms for constructing prediction regions

Given a set of c points $C_1, C_2, \ldots, C_c$ (CLPs) in $\mathscr{R}$, or $\mathscr{R}^+$ (section A.12), the convex nearest-neighbour regions $\mathscr{F}_1, \mathscr{F}_2, \ldots, \mathscr{F}_c$ to these points are the prediction regions. We are interested in constructing the intersections of $\mathscr{F}_1, \mathscr{F}_2, \ldots, \mathscr{F}_c$ with $\mathscr{L}$. We shall be especially concerned with the most important case which is when the dimensionality of $\mathscr{L}$ is $\rho = 2$. The boundaries are linear, and in two dimensions the intersections define polygonal prediction regions (Gower, 1993). In Chapter 7, we gave an algorithm couched in terms of pixels in $\mathscr{L}$, 'colouring' each pixel according to its nearest CLP. Although simple, this algorithm is inefficient and better ones are needed. Below, we sketch one way of proceeding.

The bisecting plane of C_i, C_j meets $\mathscr{L}$ in a line which we label (i,j). Line (i,j) separates the points of $\mathscr{L}$ which are nearest C_i and C_j. Let (i,j) meet (i,k) at a point (i,j,k) then this point is equally near C_i, C_j and C_k and is potentially one of the corners where three of the polygonal prediction regions meet in $\mathscr{L}$. Whether or not it *is* such a corner depends on whether there is another CLP $C_h (h \neq i,j,k)$ that is nearer to (i,j,k); if there is, then the triple (i,j,k) is termed a **virtual point**, otherwise (i,j,k) is a **real point**.

We now define the meaning of **join** of two points (i,j,h) and (i,j,k) as follows (Fig. A.4):

(i) if both points are real then join has its usual meaning on the line (i,j);
(ii) if both points are virtual then the join is null; and
(iii) if one point is real and the other null, then join is along (i,j) starting from the real point and away from the virtual point.

To compute the triple (i,j,k), it is best first to express all coordinates in terms of axes orthogonal to and contained in $\mathscr{L}$; principal axes in and orthogonal to $\mathscr{L}$ will usually be chosen. Then the bisecting planes of all pairs C_i and C_j can be found as described in section A.12 and then the terms in the first two coordinates define the line (i,j) –note that it is immaterial whether these operations occur in $\mathscr{R}$ or in $\mathscr{R}^+$. The coordinates of (i,j,k) can then be found as the intersection of (i,j) and (i,k); this point must also lie on (j,k).

Thus the algorithm consists of:

(i) compute the coordinates of all $\binom{n}{3}$ triplets (i,j,k);

(ii) decide whether each triplet is real or virtual;
(iii) join all $\binom{n}{2}$ pairs sharing two suffices, interpreting 'join' as described above; and
(iv) label the prediction regions.

A good implementation of the algorithm would have to cope with parallel or coincident pairs of lines. Gower (1993) discusses some further points that need consideration.

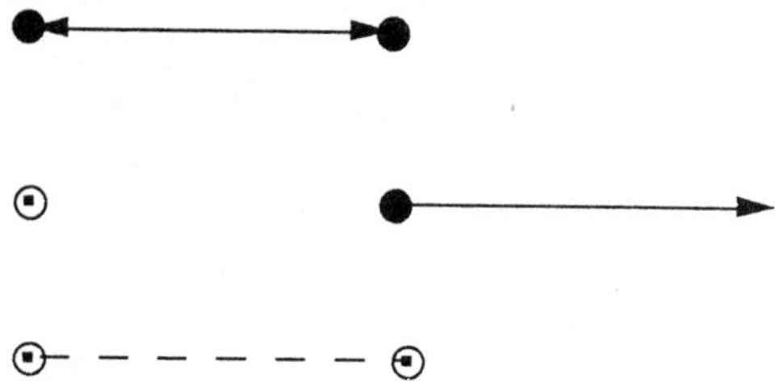

Fig. A.4. *The definition of join in the algorithm for constructing prediction regions. Key: • point; ⊙ virtual point.*

References

Bailey, R.A. and Gower, J.C. (1990) Approximating a symmetric matrix. *Psychometrika*, **55**, 665–75.

Beale, E.M.L. and Little, R.J.A. (1975) Missing values in multivariate analysis. *Journal of the Royal Statistical Society, B*. **37**, 129–45.

Benzécri, J.-P. (1973) *L'Analyse des Données*. Vol. 1: *La Taxinomie*. Vol. 2 *L'Analyse des Correspondances*. Dunod, Paris.

Bradu, D. and Gabriel, K.R. (1978) The biplot as a diagnostic tool for models of two-way tables. *Technometrics*, **20**, 47–68

Carlier, A. and Kroonenberg, P.M. (1995) Decompositions and biplots in three-way correspondence analysis. *Psychometrika*, (in press).

Carroll, J.D. (1968) Generalization of canonical correlation analysis to three or more sets of variables. *Proceedings of the 76th Annual Convention of the American Psychological Association*, **3**, 227–8.

Carroll, J.D. and Chang, J-J. (1970) Analysis of individual differences in multidimensional scaling via an N-way decomposition of the 'Eckart-Young' decomposition. *Psychometrika*, **32**, 283–319.

Carroll, J.D., Green, P.E. and Schaffer, C.M. (1986) Interpoint distance comparisons in correspondence analysis. *Journal of Marketing Research*, **23**, 271–80.

Carroll, J.D., Green, P.E. and Schaffer, C.M. (1987) Comparing interpoint distances in correspondence analysis: a clarification. *Journal of Marketing Research*, **24**, 445–50.

Constantine, A.G. and Gower, J.C. (1978) Graphical representation of asymmetry. *Applied Statistics*, **27**, 297–304.

Cook, R.D. and Weisberg, S. (1982) *Residuals and influence in regression*, Chapman & Hall, London.

Coppi, R. and Bolasco, S. (eds) (1991) *Multiway Data Analysis*, North Holland, Amsterdam.

Cox, T.F. and Cox, M.A.A. (1994) *Multidimensional Scaling*, Chapman & Hall, London.

de Falguerolles, A. and Francis, B. (1994) An algorithmic approach to bilinear models for two-way contingency tables, In *New Approaches in Classification and Data Analysis*, (eds E. Diday, Y. Lechevallier, M. Schader, P. Bertrand and B. Burtschy), Springer-Verlag, Berlin, pp. 518–24.

de Leeuw, J. (1982) Generalized eigenvalue problems with positive semi-definite matrics. *Psychometrika*, **47**, 87–93.

de Leeuw, J. and van der Heijden, P.G.M. (1988) Correspondence analysis of incomplete contingency tables. *Psychometrika*, **53**, 223–33.

Denis, J-B. and Gower, J.C. (1994) Asymptotic covariances for the parameters of biadditive models. *Utilitas Mathematica*, **46**, 193–205.

Denis, J-B. and Gower, J.C. (1995) Asymptotic confidence regions for biadditive models: interpreting genotype-environment interactions. *Applied Statistics*, (submitted).

Digby, P.G.N and Gower, J.C. (1981) Ordination between and within groups applied to soil classification, in *Down to Earth Statistics: Solutions Looking for Geological Problems.* (ed. D.F. Merriam), Syracuse University Geology Contributions, 53–75.

Dijksterhuis, G.B. and Gower, J.C. (1991/2) The interpretation of generalized Procrustes analysis and allied methods. *Food Quality and Preference*, **3**, 67–87.

Eckart, C. and Young, G. (1936) The approximation of one matrix by another of lower rank. *Psychometrika*, **1**, 211–18.

Fisher, R.A. (1938) *Statistical Methods for Research Workers* (7th Édn) Oliver and Boyd, Edinburgh.

Fisher, R.A. and Mackenzie, W.A. (1923) Studies in crop variation: 2. The manurial response of different potato varieties. *Journal of Agricultural Science*, **13**, 311–20.

Flury, B.N. (1984) Common principal components in k groups. *Journal of the American Statistical Association*, **79**, 892–8.

Flury, B.N. (1987) Two generalizations of the common principal component model. *Biometrika*, **74**, 59–69.

Gabriel, K.R. (1971) The biplot graphical display of matrices with application to principal component analysis. *Biometrika*, **58**, 453–67.

Gabriel, K.R. (1972) Analysis of meteorological data by means of canonical decomposition and biplots. *Journal of Applied Meteorology*, **11**, 1071–7.

Gabriel, K.R. and Zamir, S. (1979) Lower rank approximation of matrices by least squares with any choice of weights. *Technometrics*, **21**, 489–98.

Gauch, H.D. (1992) *Statistical Analysis of Yield Trials: AMMI Analysis of Factorial Designs*, Elsevier, Amsterdam.

Gifi, A. (1990) *Nonlinear Multivariate Analysis*, John Wiley & Sons, Chichester.

Gower, J.C. (1966a) Some distance properties of latent root and vector methods used in multivariate analysis. *Biometrika*, **53**, 325–38.

Gower, J.C. (1966b) A Q-technique for the calculation of canonical variates. *Biometrika*, **53**, 588–9.

Gower, J.C. (1968) Adding a point to vector diagrams in multivariate analysis. *Biometrika*, **55**, 582–5.

Gower, J.C. (1971) A general coefficient of similarity and some of its properties. *Biometrics*, **27**, 857–71.

Gower, J.C. (1975) Generalized Procrustes analysis. *Psychometrika*, **40**, 33–51.

Gower, J.C. (1977) The analysis of asymmetry and orthogonality, in *Recent Developments in Statistics*, (eds J. Barra *et al.*) North Holland Press, Amsterdam, pp. 109–23.

Gower, J. C. (1982) Euclidean distance geometry. *The Mathematical Scientist*, **7**, 1–14.

Gower, J C. (1984) Multivariate analysis: ordination, multidimensional scaling and allied topics, in *Handbook of Applicable Mathematics: Vol. VI Statistics* (ed E. H. Lloyd), J. Wiley and Sons, Chichester, 727–81.

Gower, J. C. (1989) Generalized canonical analysis, in *Multiway Data Analysis*, (eds R. Coppi and S. Bolasco), Elsevier (North Holland), Amsterdam, 221–32.

Gower, J. C. (1990a) Three dimensional biplots. *Biometrika*, **77**, 773–85.

Gower, J. C. (1990b) Fisher's optimal scores and multiple correspondence analysis. *Biometrics*, **46**, 947–61.

Gower, J. C. (1992) Generalized biplots. *Biometrika*, **79**, 475–93.

Gower, J. C. (1993) The construction of neighbour-regions in two dimensions for prediction with multi-level categorical variables, in *Information and Classification: Concepts–Methods–Applications*

Proceedings 16th Annual Conference of the Gesellschaft fur Klassifikation, Dortmund, April 1992 (eds O. Opitz, B. Lausen and R. Klar), Springer-Verlag. Heidelberg–Berlin, 174–89.

Gower, J. C. (1994) Orthogonal and projection Procrustes analysis, in *Recent Advances in Descriptive Multivariate Analysis,* (ed. W. J. Krzanowski), Clarendon Press, Oxford.

Gower, J.C. and Dijksterhuis G. (1994) Coffee images: A study in the simultaneous display of multivariate quantitative and categorical variables for several assessors. *Quality and Quantity,* **28**, 165–84.

Gower, J. C. and Harding, S. (1988) Nonlinear biplots. *Biometrika,* **75**, 445–55.

Gower, J. C. and Legendre, P. (1986) Metric and Euclidean properties of dissimilarity coefficients. *J. Classification,* **3**, 5–48.

Green, B. F. and Gower, J. C. (1979) A problem with congruence. Paper presented at the annual meeting of the Psychometric Society, Monterey, California.

Greenacre, M. J. (1984) *Theory and Applications of Correspondence Analysis,* Academic press, London.

Greenacre, M. J. (1988) Correspondence analysis of multivariate categorical data by weighted least squares. *Biometrika,* **75**, 457–67.

Greenacre, M. J. (1989) The Carroll-Green-Schaffer Scaling in correspondence analysis: a theoretical and empirical appraisal. *Journal of Marketing Research,* **26**, 358–65.

Greenacre, M. J. (1990) Some limitations of multiple correspondence analysis. *Computational Statistics Quarterly,* **3**, 249–56.

Greenacre, M. J. (1991) Interpreting multiple correspondence analysis. *Applied Stochastic Models and Data Analysis,* **7**, 195–210.

Guttman, L. (1994) The quantification of a class of attributes: a theory and method of scale construction, in *The Prediction of Personal Adjustment,* (eds P. Horst *et al.*) 319–48, Bulletin No. 48. New York: The Social Science Research Council.

Hand, D. J., Daly, F., Lunn, A. D., McConway, K. J. and Ostrowski, E. (1994) *A Handbook of Small Data Sets,* Chapman & Hall, London.

Healy, M. J. R. and Goldstein, H. (1976) An approach to the scaling of categorised attributes. *Biometrika,* **63**, 219–29.

Hills, M. (1969) On looking at large correlation matrices. *Biometrika,* **56**, 249–53.

Hirschfeld, H. O. (1935) A connection between correlation and contingency. *Proc. Camb. Phil. Soc.*, **31** 520–24.

Hotelling, H. (1933) Analysis of a complex of statistical variables into principal components. *Journal of Educational Psychology*, **24**, 417–41; 498–520.

Iloni, K. (1991) Biplot graphical display techniques. MSc. dissertation, University of Cape Town.

Jolliffe, I. T. (1986) *Principal Component Analysis*, Springer-Verlag, New York.

Jongman, R. H. G., ter Braak, C. J. F. and van Tongeren, O. F. R. (1987) *Data Analysis in Community and Landscape Ecology*, PUDOC (Centre for Agricultural Publishing and Documentation), Wageningen.

Kempton, R. A. and Talbot, M. (1988) The development of new crop varieties. *Journal of the Royal Statistical Soceity, A*. **151**, 327–41.

Kruskal, J. B. and Wish, M. W. (1978) *Multidimensional Scaling*, Sage University Paper series on Quantitative Applications in the Social Sciences, series no. 07–011. Sage Publications, Beverley Hill.

Krzanowski, W. J. (1979) Between groups comparisons of principal components. *Journal of the American Statistical Association*. **74**, 703–7. (Correction in **76**, 1022.)

Krzanowski, W. J. (1988) *Principles of multivariate analysis*, Oxford University Press, Oxford.

Lawley, D. N. and Maxwell, A. E. (1971) *Factor Analysis as a Statistical Method* (Second edn), Butterworths, London.

Lebart, L., Morineau, A. and Warwick, K. M. (1984) *Multivariate Descriptive Statistical Analysis*, New York: John Wiley & Sons.

Lyons, R. (1980) A Review of Multidimensional Scaling. MSc. dissertation, University of Reading.

McCullagh, P. and Nelder, J. A. (1989) *Generalized Linear Models* (2nd edn), Chapman & Hall, London.

Meulman, J. J. (1986) *A distance approach to non-linear multivariate analysis*, DSWO Press, Leiden.

Meulman, J. J. (1992) The integration of multidimensional scaling and multivariate analysis with optimal transformations. *Psychometrika*, **57**, 539–65.

Meulman, J. J. and Heiser, W. J. (1993) Nonlinear biplots for nonlinear mappings, in *Information and Classication: Concepts–*

methods–Applications Proceedings 16th Annual Conference of the Gesellschaft fur Klassifikaiton, Dortmund, April 1992 (eds O. Opitz, B. Lausen and R. Klar) Springer-Verlag: Heidelberg–Berlin, pp. 201–213.

Nishisato, S. (1980) *Analysis of Categorical Data: Dual Scaling and its Applications*, University of Toronto Press, Toronto.

Pearson, K. (1901) On lines and planes of closest fit to a system of points in space. *Philosophical Magazine*, **2**, (Series 6), 559–72.

Seneta, E. (1973) *Non-negative Matrices*, George Allen and Unwin Ltd, London.

Shepard, R. N. (1962a) The analysis of proximities: multidimensional scaling with an unknown distance function I. *Psychometrika*, **27**, 125–40.

Shepard R. N. (1962b) The analysis of proximities: multidimensional scaling with an unknown distance function II. *Psychometrika*, **27**, 219–46.

Sneath, P. M. A. and Sokal, R. R. (1973) *Numerical Taxonomy*, W. H. Freeman, San Francisco.

Summers, R. W., Underhill, L. G., Pearson, D.J. and Scott, D. A. (1987) Wader migration systems in south and eastern Africa and western Asia. *Wader Study Group Bulletin*, **49**, Supplement, 15–34.

Tenenhaus, M. and Young, F. W. (1985) An analysis and synthesis of multiple correspondence analysis, optimal scaling, dual scaling, homogeneity analysis and other methods for quantifying categorical multivariate data. *Psychometrika*, **50**, 91–119.

ter Braak, C. J. F. (1983) Principal components biplots and alpha and beta diversity. *Ecology*, **64**, 454–62.

ter Braak, C. J. F. (1986) Canonical correspondence analysis: a new eigenvector technique for multivariate direct gradient analysis. *Ecology*, **67**, 1167–69.

ter Braak, C. J. F. (1990a) CANOCO–a Fortran program for canonical community ordination by [partial] [detrended] [canonical] correspondence analysis, principal components analysis and redundancy analysis (version 3.10). Ithaca, Microcomputer Power, New York.

ter Braak, C. J. F. (1990b) Interpreting canonical correlation analysis through biplots of structure and weights. *Psychometrika*, **55**, 519–32.

ter Braak, C. J. F. (1994a) Biplots in reduced-rank regression. *Biometrical Journal*, **8**, 983–1003.

ter Braak, C. J. F. (1994b) Canonical community ordination. Part 1: Basic theory and linear methods. *Ecoscience*, **1**, 127–40.

Titterington, D. M. (1984) Recursive parameter estimation using incomplete data. *Journal of the Royal Statistical Society, B*. **46**, 257–67.

Titterington, D. M. and Jiang, J.-M. (1983) Recursive estimation procedures for missing-data problems. *Biometrika*, **70**, 613–24.

Torgerson, W. S. (1958) *Theory and Methods of Scaling*. John Wiley & Sons, New York.

Underhill, L. G. (1990) The coefficient of variation biplot. *Journal of Classification*, **7**, 41–56.

van Eeuwijk, F. A. (1995) Multiplicative interaction in generalized linear models. *Biometrics*, (in press).

van Eeuwijk, F. A. and Keizer, L. C. P. (1995) On the use of diagnostic biplots in model screening for genotype by environment tables. *Statistics and Computing*, **5**, 141–153.

van der Heijden, P. G. M. and de Leeuw, J. (1985) Correspondence analysis used complementary to log linear analysis. *Psychometrika*, **50**, 429–47.

van der Heijden, P. G. M., de Falguerolles, A. and de Leeuw, J. (1989) A combined approach to contingency table analysis using correspondence analysis and log linear analysis (with discussion). *Applied Statistics*, **38**, 249–92.

Wilkinson, J. (1965) *The Algebraic Eigenvalue Problem*, Clarendon Press, Oxford.

Index

Page numbers appearing in **bold** refer to figures and page numbers appearing in *italic* refer to tables.